Management of Disease in Wild Mammals

Richard J. Delahay • Graham C. Smith
Michael R. Hutchings
Editors

Management of Disease in Wild Mammals

Springer

Editors
Richard J. Delahay
Central Science Laboratory
Wildlife Disease Ecology Team
Sand Hutton, York
YO41 1LZ, United Kingdom

Graham C. Smith
Central Science Laboratory
Wildlife Disease Ecology Team
Sand Hutton, York
YO41 1LZ, United Kingdom

Michael R. Hutchings
Scottish Agricultural College (SAC)
Disease Systems Team
West Mains Road, Edinburgh
EH9 3JG, United Kingdom

ISBN: 978-4-431-99844-0
e-ISBN: 978-4-431-77134-0
DOI: 10.1007/978-4-431-77134-0

Printed in Japan

Printed on acid-free paper

Springer is a part of Springer Science+Business Media
springer.com

Preface

In recent years nobody could have failed to notice the frequent and often sensationalist media headlines warning of the latest global disease threat to humankind. But behind all the hyperbole lie real challenges related to dealing with the increasing incidence of emerging zoonotic disease events, the majority of which are thought to originate in wildlife (Jones et al. 2008). There are also many important diseases of domestic livestock which also occur in wildlife (e.g. foot and mouth disease and classical swine fever in wild boar, bovine tuberculosis in deer, badgers or possums), some of which can have a devastating impact on the farming industry, the wider rural economy and ultimately the public purse. But we should also not forget that wildlife diseases may have serious implications for the conservation of biodiversity. For some of the rarest, most endangered species (such as the Ethiopian wolf) disease may pose the greatest threat to their survival. If we are to avoid or reduce these impacts then we must improve our ability to detect and manage the risks associated with disease in wildlife populations. This is a challenge that will require expertise from many different disciplines: veterinary, ecological, medical, economic, political and zoological. In such an interdisciplinary field it is difficult to stay up to date with contemporary ideas and with techniques that may be rapidly evolving. We hope that in some small way this book contributes to informing people from a range of disciplines on our current state of knowledge and potential future directions in the management of disease in wildlife.

Largely because of our personal interests and expertise we have focused in this book on disease in wild mammals, although much that is discussed will be relevant to other wild fauna. Our aim has been to present and discuss the main issues related to disease management in wild mammals, and in doing so we have inevitably drawn upon the opinions of experts in a range of fields. We have attempted to be as inclusive as possible, in the knowledge that this is a topic at the interface between several scientific disciplines. We also acknowledge the important role that scientific knowledge plays in underpinning policy, and have therefore produced a text that is hopefully also accessible to those without a scientific training, but who are nevertheless important players in the development and implementation of disease management plans.

The editors have worked in the field of wildlife diseases for many years and whilst we maintain interests in other fields we continue to have close links with each other, particularly in the area of bovine tuberculosis in wildlife. We have seen at first hand how opinions change over time (albeit slowly in some cases), and how this process depends on the views and foresight of a wide diversity of experts. We have thus sought to include the opinions of many additional experts in different fields and would formally like to acknowledge their invaluable contributions. The co-authors not only gave generously of their time and expertise in helping to write individual chapters, but in many cases also improved the book by commenting on and correcting errors throughout the text. In addition we would like to thank Fred Landeg, Hamish McCallum, Menna Jones, Pete Robertson and Robbie McDonald for reviewing parts of the text and giving us additional perspectives. GCS and RJD would also like to thank Chris Cheeseman for his support and enthusiasm over the years. Many of the authors are involved in the Wildlife Disease Association and in particular with the European Section and we wish to collectively acknowledge the important contributions this organisation has made to promoting scientific endeavour in this field.

Contents

Contributors

Marc Artois
Université J. Fourier, Laboratoire TIMC-IMAG, Unité Environnement et Prévision de la Santé des Populations F-38000 Grenoble; Ecole Nationale Vétérinaire de Lyon, France

Roy Bengis
Veterinary Investigation Centre, P.O. Box 12, Skukuza 1350, South Africa

Richard Bennett
School of Agriculture, Policy and Development, University of Reading, Reading, RG6 6AR, UK

Jean Blancou
11, rue Descombes 75 017 Paris, France

Andrew C. Breed
School of Veterinary Science, Australian Biosecurity Cooperative Research Centre, University of Queensland, Brisbane, Australia

Peter Caley
National Centre for Epidemiology & Population Heath, Australian National University, ACT 0200, Australia

Stephen P. Carter
Central Science Laboratory, Sand Hutton, York, YO41 1LZ, UK

Mark A. Chambers
TB Research Group, Department of Statutory and Exotic Bacterial Diseases, Veterinary Laboratories Agency, Weybridge, New Haw, Addlestone, Surrey, KT15 3NB, UK

Sarah Cleaveland
The Roslin Institute/Royal (Dick) School of Veterinary Studies, University of Edinburgh, Easter Bush, Roslin, Midlothian, EH25 9RG, UK

Dave P. Cowan
Central Science Laboratory, Sand Hutton, York, YO41 1LZ, UK

Paul C. Cross
U.S. Geological Survey, Northern Rocky Mountain Science Center, 229 AJM Johnson Hall, Bozeman MT 59717, USA, and Montana State University, Department of Ecology, Bozeman MT, USA

Andrew A. Cunningham
Institute of Zoology, Zoological Society of London, Regent's Park, London, NW1 4RY, UK

Richard J. Delahay
Central Science Laboratory, Sand Hutton, York, YO41 1LZ, UK

Julian Drewe
Wildlife Health and Conservation Medicine Group, Department of Veterinary Medicine, University of Cambridge, Madingley Road, Cambridge, CB3 0ES, UK

Marie-José Duchêne
AFSSA LERRPAS, Technopôle Agricole et Vétérinaire, B.P. 40009, 54220 Malzéville, France

J. Paul Duff
Veterinary Laboratories Agency Diseases of Wildlife Scheme (VLADoWs), VLA Penrith, Penrith, Cumbria, CA11 9RR, UK

Gareth Edwards-Jones
School of the Environment and Natural Resources, University of Wales, Bangor, Gwynedd, LL57 2UW, UK

Dirk Eisinger
Helmholtz Centre for Environmental Research – UFZ, Permoserstraße 15, D-04318 Leipzig, Germany

Ezio Ferroglio
Dipartimento Produzioni Animali, Epidemiologia ed Ecologia Via Leonardo da Vinci, 44-10095 Grugliasco (TO), Italy

David Gardner-Roberts
The Mountain Gorilla Veterinary Project, BP 105 Musanze District, North Province, Rwanda

Emmanuelle Gilot-Fromont
Université de Lyon; Université Lyon 1; CNRS;UMR5558, Laboratoire de Biométrie et Biologie Evolutive, 43 Boulevard du 11 Novembre 1918, Villeurbanne F-69622, France

Christian Gortazar
Instituto de Investigación en Recursos Cinegéticos IREC (CSIC-UCLM-JCCM), Ronda de Toledo s.n. E-13071, Ciudad Real, Spain

Dan T. Haydon
Department of Environmental and Evolutionary Biology, Graham Kerr Bldg, University of Glasgow, Glasgow, Scotland, G12 8QQ, UK

David T. S. Hayman
Cambridge Infectious Diseases Consortium, University of Cambridge, Cambridge, CB3 0ES, UK

Steve Houghton
Veterinary Vaccines Consultancy, Coles Cottage, Plum Park Lane, Paulerspury, Northants, NN12 7NN, UK

Selene Huntley
Central Science Laboratory, Sand Hutton, York, YO41 1LZ, UK

Michael R. Hutchings
SAC, West Mains Road, Edinburgh, EH9 3JG, UK

Vicky S. Jackson
Central Science Laboratory, Sand Hutton, York, YO41 1LZ, UK

Weihong Ji
Massey University, Albany, Private Bag 102 904, North Shore Mail Centre, Auckland, New Zealand

Volker Kaden
Friedrich-Loeffler-Institut, Bundesforschungsinstitut für Tiergesundheit, Institut für Infektionsmedizin, 17493 Greifswald-Insel Riems, Germany

Darryn L. Knobel
The Roslin Institute/Royal (Dick) School of Veterinary Studies, University of Edinburgh, Easter Bush, Roslin, Midlothian, EH25 9RG, UK

Richard A. Kock
Conservation Programmes, Zoological Society of London, Regent's Park, London, NW1 4RY, UK

Mark S. Lambert
Central Science Laboratory, Sand Hutton, York, YO41 1LZ, UK

Frederick A. Leighton
Canadian Cooperative Wildlife Health Centre, Department of Veterinary Pathology, University of Saskatchewan, Saskatoon, Saskatchewan S7N 5B4, Canada

Glenn Marion
Biomathematics Statistics Scotland, The Kings Buildings, Edinburgh, EH9 3JG, UK

Giovanna Massei
Central Science Laboratory, Sand Hutton, York, YO41 1LZ, UK

Fieke M. Molenaar
Institute of Zoology, Zoological Society of London, Regent's Park, London, NW1 4RY, UK

Torsten Mörner
National Veterinary Institute, 751 89 Uppsala, Sweden

Victoria Patrek
Montana State University, Department of Ecology, Bozeman MT, USA

Gareth Pearce
Wildlife Health and Conservation Medicine Group, Department of Veterinary Medicine, University of Cambridge, Madingley Road, Cambridge, CB3 0ES, UK

Dirk Pfeiffer
The Royal Veterinary College, London, Hawkshead Lane, North Mymms, Hatfield, Hertfordshire, AL9 7TA, UK

Raina K. Plowright
Consortium for Conservation Medicine, 460 West 34th Street, New York, NY 10001, USA

Sophie Rossi
Unité Sanitaire de la Faune, Micropolis, la Bérardie, Belle Aureille, 05000 GAP, France

Sugoto S. Roy
Central Science Laboratory, Sand Hutton, York, YO41 1LZ, UK

Steve Rushton
University of Newcastle upon Tyne, Newcastle upon Tyne, NE1 7RU, UK

Anthony W. Sainsbury
Institute of Zoology, Zoological Society of London, Regent's Park, London, NW1 4RY, UK

Michael D. Samuel
U.S. Geological Survey, Wisconsin Cooperative Wildlife Research Unit, University of Wisconsin, Madison, WI 53706, USA

Graham C. Smith
Central Science Laboratory, Sand Hutton, York, YO41 1LZ, UK

Mike A. Taylor
Central Science Laboratory, Sand Hutton, York, YO41 1LZ, UK

Hans H. Thulke
Helmholtz Centre for Environmental Research – UFZ, Permoserstraße 15, D-04318 Leipzig, Germany

Alex Tomlinson
Central Science Laboratory, Sand Hutton, York, YO41 1LZ, UK

Kurt C. VerCauteren
USDA, APHIS, Wildlife Services, National Wildlife Research Center,
Fort Collins, CO, USA

W. David Walter
USDA, APHIS, Wildlife Services, National Wildlife Research Center,
Fort Collins, CO, USA

Alastair I. Ward
Central Science Laboratory, Sand Hutton, York, YO41 1LZ, UK

Ken Willis
School of Architecture, Planning and Landscape, University of Newcastle,
NE1 7RU, UK

Gavin J. Wilson
Central Science Laboratory, Sand Hutton, York, YO41 1LZ, UK

Rosie Woodroffe
Institute of Zoology, Zoological Society of London, Regent's Park, London,
NW1 4RY, UK

Chapter 1
The Science of Wildlife Disease Management

Richard J. Delahay, Graham C. Smith, and Michael R. Hutchings

1.1 What is Disease?

In its widest sense disease can be regarded as any impairment of normal functions. However, for the purposes of this book we will mostly restrict our discussion to infectious diseases, the agents of which are often described as parasites or pathogens. For convenience, these organisms are often split into two categories that reflect their broad characteristics, and their relative size. The macroparasites are multi-cellular organisms that live in or on the host, such as helminths and arthropods, while microparasites include viruses, bacteria, fungi and protozoa. The main functional differences between the two relate to their generation times, with microparasites exhibiting relatively higher within-host reproductive rates and shorter generation times than macroparasites. As a result microparasites are frequently associated with acute disease, although they can induce long-lived immunity to re-infection in recovered hosts. Macroparasites by contrast are more likely to produce chronic infections often characterised by short-lived immunity in heavily infected hosts, and re-infection. Macroparasites may also have distinct life stages that can survive outside the host (e.g. eggs or larvae) and sometimes require other host species to complete their life cycle. Two important groups of pathogens fall outside this classification: rogue proteins (prions) implicated in transmissible spongiform encephalopathies (TSEs) and infectious cancers, of which Tasmanian devil facial tumour disease is a well known example. However, in broad respects these are most usefully considered as microparasites, often producing acute clinical signs without host immunity.

Disease can affect individual hosts by reducing growth rates or fecundity, increasing metabolic requirements, changing patterns of behaviour and ultimately may cause death. Sub-lethal effects of pathogens may also enhance mortality rates by for example, increasing the susceptibility of the infected host to predation. However,

R.J. Delahay and G.C. Smith
Central Science Laboratory, Sand Hutton, York, YO41 1LZ, UK

M.R. Hutchings
SAC, West Mains Road, Edinburgh, EH9 3JG, UK

R.J. Delahay et al. (eds.), *Management of Disease in Wild Mammals*,
DOI:10.1007/978-4-431-77134-0_1, © Springer 2009

the intimate relationships between hosts and parasites have in many instances evolved over time into subtle and potentially complex interactions, such that infection does not in itself necessarily lead to disease. Many parasites have little detrimental effect on their hosts for most of the time, only causing pathological damage if this delicate balance is upset, for example when the parasites become too numerous or when the immunological capability of the host is impaired. This balance could be influenced by many factors including nutrition, concomitant infections and a variety of physiological stressors.

Parasites are natural components of ecosystems. They influence the structure of ecological communities (Wood et al. 2007) and are important agents of evolutionary change (Clayton and Moore 1997; Little 2002). Hosts and their parasites are locked in an evolutionary arms race, an endless game of 'hide and seek', which finds its ultimate expression in the complex immune systems of mammals. So fundamental is the role of parasitism in the development of biological systems that the imperative to avoid disease may have been an important driver for the evolution of sexual reproduction, which provides a means for recombination of genetic material and the inheritance of protective genes.

Disease is a ubiquitous characteristic of ecosystems. In humans (the most comprehensively studied mammal) over 1,400 diseases have been identified, in our livestock we know of over 600 and in domestic carnivores nearly 400 have been recorded (Cleaveland et al. 2001). Over 60% of human diseases are zoonotic, and for those considered to be of emerging importance, the figure rises to 75% (Taylor et al. 2001). By inference alone there are likely to be many thousands of diseases affecting the 5,400 or so mammal species in the world. Nevertheless, despite the clear implication that they are likely to play an important role in the epidemiology of some diseases of importance to human health and livestock, information on the pathogens of wild mammals is relatively poor.

1.2 The Significance of Wildlife Diseases

There is no doubt that recent years have seen a growing recognition of the potential importance of wild mammals in the epidemiology of diseases that impact on global human health, agriculture and biodiversity. In terms of public health, this has been manifest in high profile reports of hanta virus, Lyme disease and SARS-associated coronavirus in humans, and their links to wild mammals. In some countries, wild mammals are implicated in the persistence of bovine tuberculosis and brucellosis infection in cattle, which have impacted severely on the welfare and productivity of domestic animals and imposed high costs on stakeholders. Some such diseases are the subject of eradication programmes as their potential impact on human activities is so acute. But wildlife populations themselves may also be threatened by disease, particularly if they are already fragmented, and vulnerable to extinction from stochastic events. This is illustrated by examples such as the impact of rabies on populations of the endangered African wild dog (*Lycaon pictus*) and Ethiopian wolf (*Canis simensis*), and of facial tumour disease in Tasmanian devils (*Sarcophilus harrisii).*

In this book we focus on the management of disease in wild mammals, although many of the issues and approaches discussed here will apply to other wildlife. Wild mammals are of particular interest because they share so many common pathogens with domestic livestock and humans, and consequently play a prominent role in the dynamics of diseases of public health and agricultural concern. Most known zoonotic diseases infect carnivores, livestock and commensal rodents, probably as a result of the historical and evolutionary associations with humans. Mammals are also of particular value as sensitive barometers of ecosystem health, sitting as they do at, or near the top of, trophic food chains. For this reason they have often served as key species for conservation initiatives, under the, often unstated, assumption that their protection will safeguard the habitats that they and many other species inhabit.

The growing importance of diseases in wild mammals to a range of human activities has occurred against the background of a rapidly changing world, in which the interface between human and wildlife populations has been profoundly modified by urbanisation, agricultural intensification, climate change and habitat degradation. Some wild mammals have proven extremely adaptable in the face of anthropogenic changes to the environment. The most adaptive species tend to be those with generalist diets and opportunistic habits. Some have increased in abundance and distribution, as they have become habituated to agricultural and urban environments. Examples include red foxes (*Vulpes vulpes*) and Eurasian badgers (*Meles meles*) in the UK, both of which have successfully adapted to life in highly urbanised environments. Furthermore, the high densities of badgers observed in some rural areas of the UK are in no small part due to the abundance of food afforded them by the modern pastoral farming landscape. In several instances the direct management of wild mammals for hunting or game farming has resulted in localised concentrations of unsustainably high density. Wild boar (*Sus scrofa*) and red deer (*Cervus elaphus*) in parts of Central and Western Europe, and white-tailed deer (*Odocoileus virginianus*) in some regions of the North-Eastern USA are notable examples. But the wild mammals with which we have the longest standing and most intimate relationships are undoubtedly the commensal rodents with whom we share our homes and farmland across the globe. Within modified environments, these adaptive species may frequently live in close proximity to humans and our domesticated animals, thus enhancing opportunities for inter-specific transmission of pathogens. For most wild mammals however, human activities have had a devastating impact, largely through the destruction and degradation of their habitat, but also through direct exploitation and pollution. The result is that many species of wild mammal survive in diminished and fragmented populations that are vulnerable to the effects of disease (Chapter 11). Out of 5,416 species of wild mammal, 1,094 were regarded as 'threatened' (i.e. vulnerable, endangered or critically endangered) with extinction by the International Union for Conservation of Nature (IUCN 2007).

In recent decades there has been an unprecedented increase in the global transport of people, animals and animal-derived products. International air travel now provides the opportunities for a disease that would once have taken many months or years to traverse a single continent, to be carried to the far corners of the globe within a matter of hours. Live wild animals are translocated in the interests of the pet trade, game management and conservation, and their products are distributed in the form of, often illegal, bushmeat, 'medicines', trophies and other merchandise. The associated risks of introducing new diseases to previously isolated and naïve populations can have potentially catastrophic consequences. Nearly 38 million live wild vertebrates were legally imported into the USA between 2000 and 2004 (Marano et al. 2007), including 23,000 mammals and at least 263 non-native species. One widely reported consequence of these imports was the 2003 outbreak of the zoonotic monkeypox virus which was initiated by infection in exotic African rodents imported for the pet trade (Guarner et al. 2004). Such events emphasise the need to develop contingency plans to ensure some level of preparedness to deal with disease introductions that could establish in endemic wildlife populations (Chapter 9).

The perpetual movement of people, animals and products around the world is not the only anthropogenic process that creates opportunities for enhanced disease transmission. Environmental degradation in a wide variety of guises may also be a driving factor in the emergence of wildlife diseases. Airborne pollution, habitat fragmentation and the eutrophication of aquatic ecosystems have for example all been linked to disease outbreaks in wildlife (Dobson and Foufopoulos 2001). But, the most pervasive and potentially damaging environmental impact to arise from human activity is undoubtedly global climate change. The consequences for global ecosystems will clearly have significant implications for the ecology of wild mammals and their pathogens (Epstein 2001), as well as presenting major challenges to human activities. Changes in global weather patterns are likely to be accompanied by an increasing tendency for the emergence (and re-emergence) of pathogens and their vectors in new geographic areas and in novel hosts. The development of methods to predict such events and of co-ordinated systems to provide appropriate responses, are major challenges for the international community.

1.3 Managing Disease in Wild Mammals

It is important to consider the question of when disease in a wildlife population requires management intervention. After all, diseases are natural components of ecosystems, although it is often a moot point as to whether a particular pathogen would have existed in a wild population in the absence of its purported introduction by humans or livestock. Human modification of the environment has been so substantial and widespread that the question often arises as to what constitutes a natural ecosystem and, perhaps more importantly, what we can consider to be a natural disease event. The question of when and when not to manage, essentially rests on the extent to which the disease endangers human health, wealth, welfare or conservation aspirations, and the likelihood that intervention will have a beneficial effect. Opinions on the point at which a line is crossed and management becomes necessary, may vary widely between stakeholders of differing perspectives, and the search for 'common ground' is a continuing challenge for policy makers and politicians. However, even when a problem is identified as sufficient to warrant management, this may not necessarily mean that intervention is best directed at the wildlife population or the pathogen. In many cases changes to other components of the system (e.g. human behaviour) may be more effective. This may be particularly true when such approaches are targeted at the more tractable elements of the system (e.g. livestock husbandry), which can be managed using the existing socio-economic and legislative framework.

Once the decision to intervene has been reached then the objective of management will need to be determined. This may be prevention or control of disease, or even local or global eradication of the pathogen. The appropriate approach will depend on the characteristics of the problem and in particular on the correct identification of reservoirs of infection (see below). Inevitably, prevention and control are generally more easily achieved than eradication, not least because the latter requires the accurate identification of all reservoirs of infection. The appropriate target of disease management may be the pathogen itself (Chapter 6), one or more host populations (Chapter 7), or some element of the environment that influences transmission (Chapter 8). In this book we will discuss each in turn, although in practice a combination of approaches may be most successful.

Despite the clear requirement to develop effective means of dealing with wildlife disease issues, advances in practical management have lagged far behind the development of disease ecology theory. In particular, managers have been slow to respond to the need to understand and accommodate the ecological complexities of wild mammal populations in intervention plans. And yet, understanding wildlife disease problems is invariably as much an ecological as it is a veterinary challenge. This is elegantly illustrated by an example from the UK where in 1997 the Government convened an Independent Scientific Group (ISG) of experts in veterinary science, ecology, epidemiology, statistics and economics, to investigate the effects of badger culling on bovine tuberculosis in cattle. The results of the large scale field experiment and related research they initiated, showed that attempts to reduce disease in cattle by culling badgers caused changes in the behaviour of the wild host that under certain circumstances were counter-productive for disease control (Independent Scientific Group 2007). Their findings illustrate the fundamental importance of understanding host ecology and social behaviour (Chapter 2) for the development of disease control strategies, and the clear need to identify, characterise and quantify the key ecological processes that drive disease transmission and persistence (Chapter 3) in wildlife populations. Hence we need to look critically at existing assumptions of disease control and management, particularly where they are underpinned by experience in dealing with disease in domestic animals. The development of successful approaches to the management of disease in wild populations will require careful consideration of the entire host community, of the economic dimensions, and of the practical challenges of successfully implementing any intervention. Where management of disease involving wildlife was once the almost exclusive domain of veterinarians, it is now increasingly recognised that it requires a multi-disciplinary approach involving ecologists, epidemiologists, experts in public health, mathematical modellers, geographic information specialists, statisticians and economists. Such an approach is essential if we are to further our understanding of the dynamics of disease in wildlife and to develop sustainable strategies for their management.

The key to developing effective tools for the management of disease involving wildlife is a sufficient understanding of the conditions required for the persistence of pathogens. Many important diseases infect multiple hosts, some of which will constitute persistent sources of infection for other species, whilst others will not. Unfortunately, many past attempts to manage disease in wildlife populations have failed to recognise this distinction and have instead been rooted in a poor or even misguided understanding of the host community and the likely impact of intervention on disease dynamics. Central to our understanding of any disease system is the concept of the reservoir host. An over-abundance of definitions of disease reservoirs can be found in the literature, each emphasising different aspects, and together leading to no small amount of confusion. A clearer conceptual framework may be achieved by taking an ecological community-based approach which defines a reservoir as "*one or more epidemiologically connected populations or environments in which the pathogen can be permanently maintained, and from which infection is transmitted to the defined target population*" (Haydon et al. 2002b). Past attempts to manage disease involving wildlife have all too often been aimed at 'suspected' reservoirs with little hard evidence that they represented the most important source of infection. That said, it can be difficult to unequivocally identify a reservoir host population. Although correlative and risk-based associations can provide strong circumstantial evidence, only interventions that can isolate target populations can produce experimental evidence, and these are rarely possible.

Effective management of wildlife diseases needs to be based on sound science and developed on the basis of the objective review of previous evidence. This evidence-based approach has led to a radical change in the way human medicine is influenced by previous experience. Systematic review of the effectiveness of previous practices is now widely accepted as standard practice in public health and has been advocated for conservation management (Sutherland et al. 2004). There is a clear need to develop and maintain systems to support evidence-based practice in wildlife disease management. This implies a fundamental change from what has been common practice in the past, such that in the future the outcomes of disease management interventions should be systematically monitored, collated and made available to others. Inevitably however, even with unfettered access to evidence from past experiences of dealing with disease in wildlife, many unanswered questions regarding the potential impact of management interventions will remain. Some important areas of data shortfall may be addressed through systematic scientific investigations and experimentation, although in some cases this may be practically difficult, prohibitively expensive, or there may be insufficient time given the magnitude of the problem. As a consequence, the reality is that we will often be required to make decisions in the face of substantial uncertainty. In such circumstances mathematical modelling can provide a powerful tool, both for increasing our understanding and for generating predictions of the likely outcome of interventions (Chapter 4). Mathematical simulations provide the opportunity to play out various scenarios under different conditions and to incorporate the known uncertainties of the system under investigation. If the modelled outcome of management decisions is robust to different underlying assumptions, then we can be more confident of its utility. If management decisions rely heavily on assumptions, then we have to make a decision based on the relative risk, and cost of each potential outcome. With sufficient understanding of the underlying assumptions, the limitations and levels of uncertainty associated with outputs, then the results of mathematical models of disease dynamics and management interventions can make valuable contributions to the decision-making process.

Modelling can therefore be used to help define interventions that are likely to give a positive benefit, in terms of reducing disease prevalence. However, the most effective techniques to reduce the burden of disease will likely require the most effort, and so be more costly. As resources are always limited, a balance needs to be struck between desired outcomes and their financial costs. This is where the application of economic analyses can help (Chapter 5). The costs and benefits of each potential strategy can be compared in terms of cost-effectiveness or the cost–benefit ratio, and so help to identify an 'optimum' strategy.

In the world of commerce it is widely recognised that you cannot manage what you do not measure. This is equally relevant to disease management. Unless we are able to identify changes in disease occurrence in wildlife populations through monitoring and surveillance (Chapter 10), we will not be able to identify situations that require action, and if we cannot monitor the impact of interventions, then we will not know whether they are working. This seems obvious enough, but in practice surveillance for diseases of wildlife is poorly developed in most countries. Also, past endeavours to control disease in wildlife have often been characterised by a failure to adequately monitor progress, describe the baseline pre-intervention situation against which to measure progress, or indeed to clearly state the objectives of the intervention. An appropriate programme of monitoring should therefore always accompany any wildlife disease management intervention, and should be designed so as to assess its effectiveness in achieving the stated objectives. Further development of methods for the surveillance and monitoring of pathogens and hosts is intrinsic to the future successful management of diseases in wildlife.

1.4 Conclusions

Management of disease in wild mammals should be sustainable, based on sound epidemiological and ecological knowledge, and must balance the requirements for preserving biodiversity, and protecting human health and economic well-being. Striking the appropriate balance between these interests will be a major challenge for the development of future national and international policies. The magnitude of this task grows as the unrelenting processes of globalisation gradually move us in the direction of a free mixing population in which the opportunities for disease transmission and persistence are profoundly enhanced. At the same time, environmental degradation and habitat loss continue to reduce global biodiversity, and themselves contribute to the emergence of pathogens in wildlife. In the face of this growing threat to the health of humans, domestic animals and wildlife, there is an increasing awareness amongst many researchers, managers and stakeholders of the need to change the way we deal with these problems. All too often the management of wildlife diseases has in the past been characterised by reactive, unsustainable and ill-informed interventions that have ignored the fundamental importance of the ecology of hosts, pathogens and vectors, and have been out of step with the global imperative to conserve biodiversity. The conservation of species and preservation of healthy ecosystems are inextricably linked to sustained human well-being. Consequently the retention of biodiversity and the potential for adverse ecological impacts must become material considerations when choosing how we manage disease in wildlife. We need to start treating wildlife diseases as wildlife management issues, and to develop a greater capacity to predict and prepare for potential problems. To these ends we must ensure that we employ the appropriate contemporary tools such as mathematical modelling, risk assessment, economic analysis and GIS. And perhaps most importantly, we need to recognise the role that human activities play in perpetuating disease in wildlife, and the potential for changes in human attitudes and behaviour to reduce opportunities for disease emergence. The world has changed immeasurably in recent decades and so our approaches to managing disease in wildlife must change too.

Chapter 2
Wildlife Population Structure and Parasite Transmission: Implications for Disease Management

Paul C. Cross, Julian Drewe, Victoria Patrek, Gareth Pearce, Michael D. Samuel, and Richard J. Delahay

2.1 Introduction

Emerging infectious diseases have become an important challenge for wildlife ecologists and managers. Management actions to control these diseases are usually directed at the parasite, the host population, or a key component of the environment, with the goal of reducing disease exposure and transmission. Control methods directed at the host population, however, remain limited in approach (e.g. vaccination, population reduction, test-and-remove) and scope, by financial, logistical, ethical and political constraints. Furthermore, these control methods have often been implemented without due consideration of how host ecology and behaviour may influence disease dynamics. This chapter highlights how host population structure and social organisation affect parasite transmission and prevalence.

Traditionally, variation in disease prevalence among species, genders, and ages may have been explained by immunological differences in susceptibility. However,

P.C. Cross
U.S. Geological Survey, Northern Rocky Mountain Science Center, 229 AJM Johnson Hall, Bozeman MT 59717, USA, and Montana State University, Department of Ecology, Bozeman MT, USA

J. Drewe
Wildlife Health and Conservation Medicine Group, Department of Veterinary Medicine, University of Cambridge, Madingley Road, Cambridge, CB3 0ES, UK

V. Patrek
Montana State University, Department of Ecology, Bozeman MT, USA

G. Pearce,
Wildlife Health and Conservation Medicine Group, Department of Veterinary Medicine, University of Cambridge, Madingley Road, Cambridge, CB3 0ES, UK

M.D. Samuel
U.S. Geological Survey, Wisconsin Cooperative Wildlife Research Unit, University of Wisconsin, Madison, WI 53706, USA

R.J. Delahay
Central Science Laboratory, Sand Hutton, York, YO41 1LZ, UK

R.J. Delahay et al. (eds.), *Management of Disease in Wild Mammals*,
DOI:10.1007/978-4-431-77134-0_2, © Springer 2009

ecological and behavioural factors can also affect the rates and routes of parasite transmission and potential control options. Using this information, future control efforts may be improved by focusing on subsets of individuals, areas, environmental factors, or times of year that are most important in the propagation and persistence of a pathogen.

The social systems of mammal populations exhibit structure at several levels. Individuals vary by age, sex, reproductive status, genetic relatedness, position in a dominance hierarchy, social interactions and patterns of space-use. Group sizes can vary within and among species, from solitary individuals that only interact during mating, to socially complex groups or aggregations of over a million individuals. Within a group, the sex, age and social status of an individual, as well as the season, will often affect the number and type of intra-specific contacts experienced, thus affecting exposure and transmission of parasites. Meanwhile, the transmission of a parasite among groups may depend on group size, composition, territoriality and levels of inter-group movement or contact. This chapter explores how the characteristics of host social systems may interact with parasite life-history characteristics to affect parasite transmission, prevalence and dynamics, and hence the effectiveness of disease management strategies.

2.2 Intra-Group Factors

The gender, age, dominance and reproductive status of hosts are some of the characteristics that affect parasite prevalence and transmission within a group of individuals. Most studies of these host characteristics have focused on differences in prevalence, while only a few have compared incidence rates (Begon et al. 1999; Caley and Hone 2002; Heisey et al. 2006). Disease prevalence depends on the transmission rate, disease-induced mortality, duration of infection (or duration of antibodies for seroprevalence), and the length of time a disease has been present in the population. On the other hand, incidence measures the rate of infection per unit time. The distinction between prevalence and incidence is important because differences in prevalence are often assumed to correspond to differences in incidence. In some cases, however, differences in prevalence may instead be driven by disease-induced mortality or infectious periods that vary by sex, age and dominance.

2.2.1 Sex

Several studies suggest male-biased infection for bovine tuberculosis (*Mycobacterium bovis* infection; bTB) and chronic wasting disease (CWD) in deer (see Box 2.1) (Shang et al. 2002; Miller and Corner 2005; Grear et al. 2006), cowpox in rodents (Burthe et al. 2006), and nematode infections in chamois (*Rupicapra r. rupicapra*) (Citterio et al. 2006). Analyses using data collated from studies on a range of mammal species also report male-biased prevalence and intensity of parasitism (Poulin

Box 2.1 Chronic wasting disease (CWD) in deer

CWD belongs to a family of diseases known as transmissible spongiform encephalopathies (TSEs) which affect a wide range of mammals including humans (Williams et al. 2002b). The causative agent of TSEs is most likely an abnormal prion protein that is consistently associated with the disease (Prusiner 1991). CWD is the only TSE that affects free-ranging cervids (Miller et al. 2000). The origins of the disease are unknown, but in North America it was first recognised in the 1960s in captive cervids, and since 1981 in free-ranging deer. Clinical signs of illness develop about 1.5 years after infection, and no captive or wild cervid has subsequently recovered (Williams et al. 2002b).

Studies of CWD in captive deer indicate that direct contact (Miller and Williams 2003), and contact with prion contamination of the environment (Miller MW et al. 2004) are potential routes of transmission, although their relative importance in wild populations is poorly understood. The route of transmission, the role of social groups, and the spatial scale over which transmission occurs are factors that will affect whether CWD behaves like a frequency or density-dependent disease (Gross and Miller 2001; Schauber and Woolf 2003). For example, groups of female deer may overlap spatially, but have limited direct contact with other groups (Schauber et al. 2007) and risks of infection with bTB and CWD increase with the level of genetic relatedness (Blanchong et al. 2007; Grear 2006). These social boundaries may limit the rate of direct transmission between groups because contact is reduced. Thus, indirect transmission of CWD may be an important route of between group infections, but direct contact and indirect transmission may be important routes within social groups.

Adult male mule deer (*Odocoileus hemionus*) and white-tailed deer (*Odocoileus virginianus*) tend to have a higher prevalence of CWD than adult females and this increases with age (Miller and Conner 2005; Grear et al. 2006). Because there are no indications that adult males are more susceptible or harbour the disease for longer, this suggests that differences in social structure and behaviour of males and females may influence disease transmission. Several hypotheses have been suggested to explain the increased risk of CWD infection in males compared to females. First, males are typically more social than females, especially outside the breeding season when they form single sex groups, within which unrelated males readily groom one another. In contrast, female grooming usually takes place between mother-daughter pairs or among individuals from the same matrilineal group. Second, transmission to susceptible males may increase during the breeding season when they contact infected females or visit scent stations used by infected males. These behaviours may expose breeding males to prions, which are shed through the alimentary tract. In addition, males may be at greater risk of contact with prions in the environment than females owing to their larger home range size and breeding season movements.

(continued)

Box 2.1 (continued)

Differences in movement and dispersal between male and female white-tailed deer may also be a significant component of CWD distribution across the landscape, especially in areas where animals do not show seasonal migration. Between 50% to 80% of yearling males disperse distances of 10 to 30 km, depending on habitat characteristics (Long et al. 2005), whereas less than 20% of females disperse (Rosenberry et al. 1999). Infected yearling males are therefore more likely to spread CWD into new areas. Prevalence of CWD in yearling males and females is similar and considerably lower than in adult males. If environmental transmission is an important route of infection in free-ranging deer, then adult males have the potential to contaminate a wider area than females because of their larger home ranges and increased movements during breeding.

To a limited extent, movement and dispersal information have been used to establish CWD surveillance zones and assess local disease prevalence. In addition, movement distances and spatial scales for disease transmission have been used to identify areas for intensive culling or disease detection around new CWD positive deer or in areas of high infection risk (e.g., infected game farms). However, culling strategies to reduce numbers of adult males (which have higher rates of infection) or yearling males (which have higher rates of dispersal) may deserve further consideration. Whether strategies that focus on these higher risk components of the deer population could reduce transmission or spread of CWD is currently unknown, as is the geographical scale over which control should be implemented. In many cases, implementation of such male-biased culling strategies to control CWD will conflict with goals for trophy deer management and make public support for this approach challenging. Because of the long-term chronic nature and slow transmission of CWD in deer, epizootics are likely to last for decades making control a long-term problem, and emphasising the need for prevention or early detection and eradication.

1996; Schalk and Forbes 1997; Moore and Wilson 2002). Several studies have identified positive correlations between host body weight and the intensity of parasite infection (Poulin 1995; Arneberg et al. 1998; Ezenwa 2004; Burthe et al. 2006). These findings have produced a variety of hypotheses to explain male-biased parasitism. Larger hosts may provide more space or a greater diversity of niches for parasites. They may also present a larger target for vectors, and the greater nutritional requirements of larger hosts could increase their exposure to parasites that can be transmitted by ingestion. In many species, males have larger home ranges, which may also lead to increased exposure. Sex-related differences in physiology and behaviour may also produce differences in exposure and susceptibility to pathogens. In male mammals, increased stress levels during the breeding season and the physiological effects of testosterone may be linked to immunosuppression and increased susceptibility to disease (Zuk and McKean 1996).

Mating behaviour is also likely to have important implications for parasite exposure, particularly when considering sexually transmitted diseases (STDs). In this case transmission rates are likely to depend more on the prevalence, or frequency, of the infectious individuals rather than their overall density, because the number of sexual contacts experienced by each individual is likely to be constant across a wide range of population densities. Mammal mating systems range from monogamy (one male mates with one female) to polygynandry (both sexes mate with multiple partners). Polygamy (one male mates with several females) is the most common mating system among mammals. This strategy tends to increase the variance in mating success amongst males; such that some males mate with many females whilst others fail to mate with any. Theoretical investigations suggest that this reproductive variation may increase the prevalence of disease amongst females and reduce prevalence in males, as the few reproductive males are more likely to acquire and transmit infection to their partners, while non-reproductive males remain uninfected (Thrall et al. 2000). Although few empirical studies have been conducted, the prevalence of STDs was significantly higher amongst adult females in studies of STDs in primates (Nunn and Altizer 2004) and koalas (*Phascolarctos cinereus*) (Jackson et al. 1999).

2.2.2 *Age*

The relationship between age and prevalence is related to host characteristics and disease or parasite life histories. Assuming that hosts do not recover from infection and disease-induced mortality is low, prevalence often increases with age because older individuals have been exposed for longer (Fig. 2.1, Heisey et al. 2006). This has been demonstrated for bTB in bison (*Bison bison*) (Joly and Messier 2004) and African buffalo (*Syncerus caffer*) (Jolles et al. 2005), and for CWD in mule deer (*Odocoileus hemionus*) (Miller and Corner 2005; Grear et al. 2006; see Box 2.1). When antibody titres persist, seroprevalence (i.e. prevalence based on serological test results) is also likely to increase with age. In these cases, seroprevalence reflects past exposure rather than current infection. The form of the relationship between age and prevalence is also influenced by changes in immunity, age-dependent exposure, and both host and parasite mortality (Heisey et al. 2006). For example, if parasite-induced mortality increases with time since infection then prevalence may be lower in older age categories than in juveniles because older individuals are likely to have had the disease for longer and as a result older infected individuals die at a faster rate (Fig. 2.1).

When hosts can recover from infection and become immune, juveniles may have a higher prevalence than adults because many adults may have already been exposed and recovered (Cattadori et al. 2005). Age-dependent changes in immunity may also influence host susceptibility to disease. Infants may initially be protected by maternal antibodies, but once passive immunity wanes they may become susceptible, as recorded for rabbit haemorrhagic disease (Cooke 2002) and tapeworm infestation in mice (Theis and Schwab 1992). Furthermore, senescent individuals

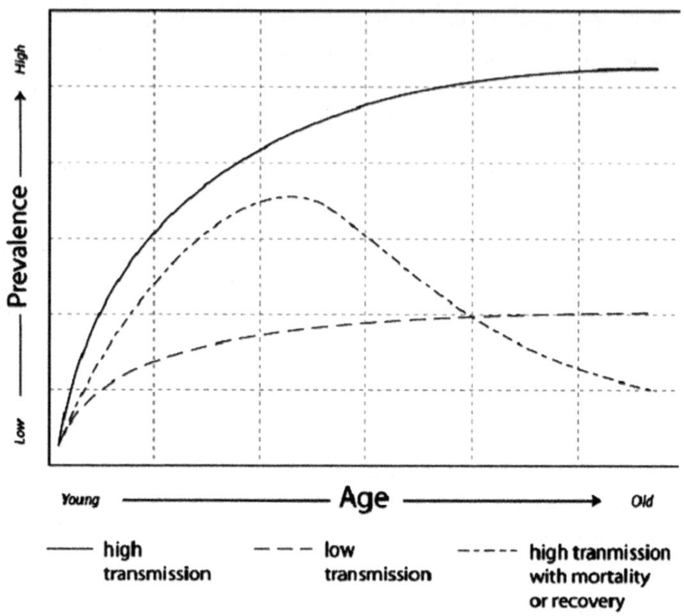

Fig. 2.1 Prevalence generally increases with age for many pathogens when individuals are born susceptible and do not recover. Higher transmission rates correspond to higher prevalence (compare solid and dashed lines). Disease recovery or disease-induced mortality may reduce prevalence in older ages (dot-dashed line)

may be more susceptible to disease due to declining immune function. Parasite-induced immunity may also affect age-prevalence patterns by either suppressing the immune system or priming the host for a stronger response to subsequent exposure. The latter seems to be the case for *Nematodirus gazellae* infections in saiga antelope (*Saiga tatarica tatarica*), in which parasite intensity peaked in 2–3 year olds but declined thereafter (Morgan et al. 2005).

2.2.3 Dominance

The influence of social dominance on parasitism is complicated by breeding behaviour, rank stability, and coping mechanisms for subordinates. Dominance is likely to affect exposure rates as well as stress. In general, mild and transient stressors enhance immunity, particularly innate immunity. Chronic stress, however, can suppress the immune system, but it remains unclear whether these changes are sufficient to increase the risk of infection (Dhabhar and McEwen 1999; Sapolsky 2005). Furthermore, those individuals that experience the most stress may be at either the top or the bottom of the dominance hierarchy, depending on the stability of the hierarchy and potential coping mechanisms (Sapolsky 2005). A study of captive cynomolgus monkeys (*Macaca fascicularis*) showed that low ranking individuals had higher rates of adenovirus infection (Cohen et al. 1997), whereas subordinate males in a koala population had lower levels of STDs than dominant individuals (Jackson et al. 1999). At this point it is difficult to determine whether these differences are driven by contact patterns, routes of transmission, stress, susceptibility or combinations of these factors. Further research is necessary before information on dominance hierarchies can be used by managers to help control disease.

2.2.4 Superspreaders

Researchers, managers and disease modellers have in the past often assumed that all hosts are equally susceptible and infectious for microparasites. However, studies of some human diseases have shown that the distribution of the number of infections caused by an individual is also strongly skewed, whereby most individuals do not infect anyone, whilst a few infect many. As a result, it has been estimated that focusing half of all control effort on the most infectious 20% of cases may be up to threefold more effective than random control (Lloyd-Smith et al. 2005a). Such heterogeneities are also likely to apply to wildlife populations (Cross et al. 2007b) and so offer the potential for more effective management strategies if these so-called 'superspreaders' can be targeted. It is not yet clear to what extent this heterogeneity is due to differences in the immunological status of hosts, or to variations in contact rates arising from behaviour, or both. Therefore, there are significant logistical and diagnostic difficulties in identifying superspreaders in wildlife populations, which will require the development of new theoretical and diagnostic tools. In addition, it is not clear whether managers could focus control efforts on 'superspreader groups' and achieve similar improvements in effectiveness of control. Intuitively however, it seems reasonable to focus attention on individuals (or classes of animals) that have long infectious periods or high rates of contact with susceptibles, as they are likely to be significant in the spread of disease.

2.3 Inter-Group Factors

2.3.1 Territoriality

Territorial defence often involves aggressive encounters that may increase exposure to parasites. Defensive behaviours are energetically costly and may increase stress and testosterone levels, which can then suppress immune function (Zuk and McKean 1996). Acts of aggression may also enhance transmission by biting or scratching. Territorial species may also encounter a high rate of contact with infectious pathogens within their territory through environmental contamination with parasite-laden faeces. A study of strongyle nematodes in African bovids found higher levels of infection in territorial than in non-territorial species, most likely as a result of environmental contamination with faeces (Ezenwa 2004). On the other hand, territoriality may also serve to reduce parasite transmission by reducing the overall level of direct contact between individuals or groups. This may be particularly pronounced amongst species that use indirect communication (e.g. scent marking and vocalisations) to minimise the need for direct contact. Individuals that occupy territories may also have access to more desirable resources making them less susceptible to parasitism. However, not all individuals within a single population will necessarily display territorial behaviour. Those that are unable to control a territory may "float" from one occupied territory to another, increasing their own exposure rates and facilitating the spread of disease across territories.

2.3.2 Group Size and Population Density

Hosts living in large aggregations are likely to have more direct contacts than those in small groups. When parasite transmission is a function of direct contacts, then prevalence is likely to increase with group size or population density (McCallum et al. 2001). The relationship between transmission rate and host density has profound implications for disease management. If transmission rates increase with density then reducing population size or density may be an effective management option. The distinction between population size and density is important (De Jong et al. 1995). In many cases, host population size may be strongly correlated with the extent of area occupied, such that as population size increases so too does the area occupied, resulting in minimal changes to density and contact rates (Begon et al. 2002). Although it is logical to assume that contact and transmission rates increase with density, the relationship may be confounded by host behaviour (e.g. territoriality or hosts seeking contacts at low densities). Also, it is seldom clear how to estimate the area occupied (i.e. the denominator), as even in the simple case of a fenced park, not all habitats may be accessible or usable by a given host species. For group-living species, contact rates are more likely to be related to local group size than overall population density.

The aggregation of animals at experimental feeding sites has been associated with significant increases in the prevalence of endoparasites in raccoons (*Procyon lotor*) (Wright and Gompper 2005), *M. bovis* in white-tailed deer (*Odocoileus virginianus*) (Chaddock 1998), and brucellosis (*Brucella abortus*) in elk (*Cervus elaphus*) (Cross et al. 2007c). A population size of 200 susceptible animals in an area of 220 km^2 has been suggested as the threshold density necessary for the maintenance of classical swine fever virus in populations of free-living wild boar (*Sus scrofa*) (Artois et al. 2002). However, population size rather than density was important in determining whether cowpox would invade and persist in a field study of wood mice (*Apodemus sylvaticus*) and bank voles (*Clethrionomys glareolus*) (Begon et al. 2003). Meta-analyses have shown nematode parasite richness, abundance and prevalence to be positively associated with population density in mammals (Arneberg 2002). Group size has also been implicated in promoting parasitism. A meta-analysis covering diverse taxa showed a positive association between group size, prevalence and intensity of contagious parasites (Côté and Poulin

1995). The relationship between parasite species richness and group size however appears highly variable, with studies showing positive, negative and absence of association between the two factors.

For directly-transmitted parasites in a single-host system, the relationship between population density and parasite transmission may be complicated by several factors. One theoretical study showed that the probability of a pandemic occurring depended on rates of host movement among groups, group size and the duration of infectiousness (Cross et al. 2005). Chronic infections with long infectious periods (e.g. bTB) required less movement among groups to create a pandemic (i.e. an epidemic that propagates across a large region and hence many groups) than those causing acute conditions, because they were able to persist for longer within the local group. Longer persistence within a group increases the likelihood that an infectious individual moves to another group. Larger group sizes and higher movement rates amongst groups facilitated the invasion of acute infections (e.g. rabies and rinderpest). This suggests that group sizes and movement rates are likely to affect the spread of acute diseases, such as rabies, more than chronic infections, such as tuberculosis. However, parasites causing acute disease often persist in other ways, such as in the environment or alternative hosts, or by causing latent infections in some individuals.

Transmission rates that vary seasonally or annually are also likely to affect the relationship between host population size and parasite prevalence. Seasonal variation in host social behaviour, such as breeding or wintering aggregations of deer and migrations of wildebeest in East Africa, may introduce temporal patterns in disease transmission. For example, brucellosis induces abortions in elk and bison prior to and during the calving season (Cheville et al. 1998). Other individuals become infected by licking or consuming the contaminated foetus. In northwestern Wyoming, USA, brucellosis seroprevalence was higher at sites where elk were provided with supplementary feed later into spring, because the timing and duration of host aggregation coincided with peak transmission (Cross et al. 2007c). This sort of complexity in the relationship between host population size or density and parasite transmission may be common to many wild mammal disease systems.

The effects of group size and population density appear to vary widely for indirectly transmitted parasites. Studies of malaria in primates have shown a higher prevalence of infection in larger groups, possibly because more hosts increase the strength of olfactory cue to mosquito vectors (Davies et al. 1991; Nunn and Heymann 2005). In contrast, other studies have provided evidence that the prevalence of parasitised individuals can be negatively associated with host group size when the parasite has a mobile vector. In feral horses (*Equus cabalus*) this phenomenon probably arises as a result of their tendency to aggregate when biting flies are most abundant (Côté and Poulin 1995).

Many parasites are neither specific to one host species nor directly transmitted amongst individuals. In primates, 68% of recorded parasites infected more than one host species and 43% were transmitted indirectly (e.g. via fomites, contaminated soil or water), 32% by arthropod vectors, 15% by intermediate hosts and 34% could be transmitted by multiple routes (Pedersen et al. 2005). Parasites that are transmitted by vectors or have intermediate or alternate hosts add further complexity to the relationships between host social structure and parasite dynamics. Consequently, in the many cases where multiple hosts share a parasite, the relationship between group size and prevalence in each host may be weak.

2.4 Mathematical Modelling of Host Population Structure

2.4.1 Contacts, Transmission and Host Density

Chapters 3 and 4 provide details on the construction and analyses of epidemiological models, while this chapter broadly addresses the importance of population structure in a mechanistic understanding of disease dynamics. Relationships between population or group size and parasite transmission and prevalence, play a critical role in efforts to mathematically model host-parasite systems and to develop effective disease management strategies (Lloyd-Smith et al. 2005b). If contact rates among hosts increase with population size, then the transmission and prevalence of directly transmitted parasites are also likely to increase. This density-dependent relationship implies a threshold host population size below which the disease is unable to persist (Kermack and McKendrick 1927; Bartlett 1957). This is the logic that underpins management strategies aimed at reducing the density of susceptible individuals below some threshold by culling, sterilisation or vaccination. However, few studies have evaluated the functional relationship between contact rates and density and the evidence for host population thresholds in wildlife disease systems remains limited (Lloyd-Smith et al. 2005b).

The paucity of evidence supporting density-dependent transmission and population thresholds is however not surprising considering the difficulties in collecting contact and transmission data at a range of densities, and over the seasonal fluctuations common in many mammal populations. Furthermore, for many parasites, it is not clear what constitutes an infectious contact (with the possible exception of STDs) nor is it simple to determine the probability of a contact resulting in infection of a susceptible host. Many species have home ranges that limit contact between infected animals and the remainder of the population. As a result, factors that drive parasite transmission such as contact, density and environmental sources of infection are likely to operate only at the local scale that affects the rate of infection across the population. For directly transmitted pathogens, contact rates are probably related to local group sizes, the spatial scale of transmission (i.e. aerosol transmission vs direct contact) and the amount of movement among groups.

The relationship between group size and total population size can also suggest how contact rates are likely to change in the face of management actions that reduce the number of hosts. For many species, the distribution of group sizes is strongly right-skewed. For example, ungulate populations in the Kruger National Park (KNP) in South Africa, often contain many small groups with a few much larger groups (Fig. 2.2). Aerial surveys are likely to miss small groups more often than large ones, which

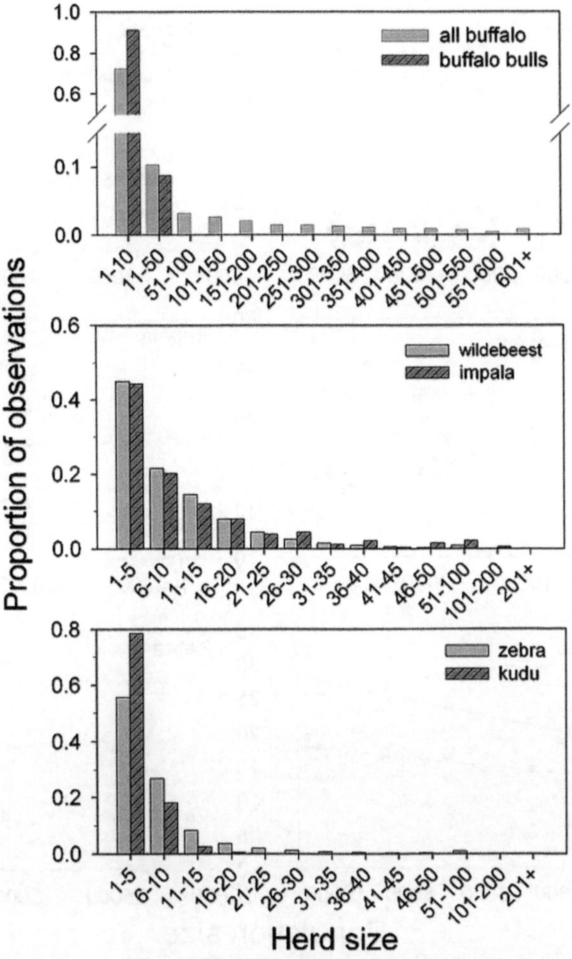

Fig. 2.2 Histograms of group sizes for ungulates in Kruger National Park, South Africa. Buffalo data originate from annual helicopter surveys from 1985 to 2003, and other data is from annual fixed-wing surveys from 1980 to 1993

can further contribute to the right-skew of group size distributions. Arithmetic means are often used in studies relating parasitism to group size but the expected group size of a randomly chosen individual (Lloyd 1967; Krause and Ruxton 2002) may be a more relevant measure for disease studies. This parameter,

$$\left(\sum_i n_i^2 \Big/ \sum_i n_i\right) \quad (2.1)$$

(where n_i is the ith group size) is essentially a weighted group size, which represents the average group size experienced by each individual and more closely relates to the average per-capita risk of infection.

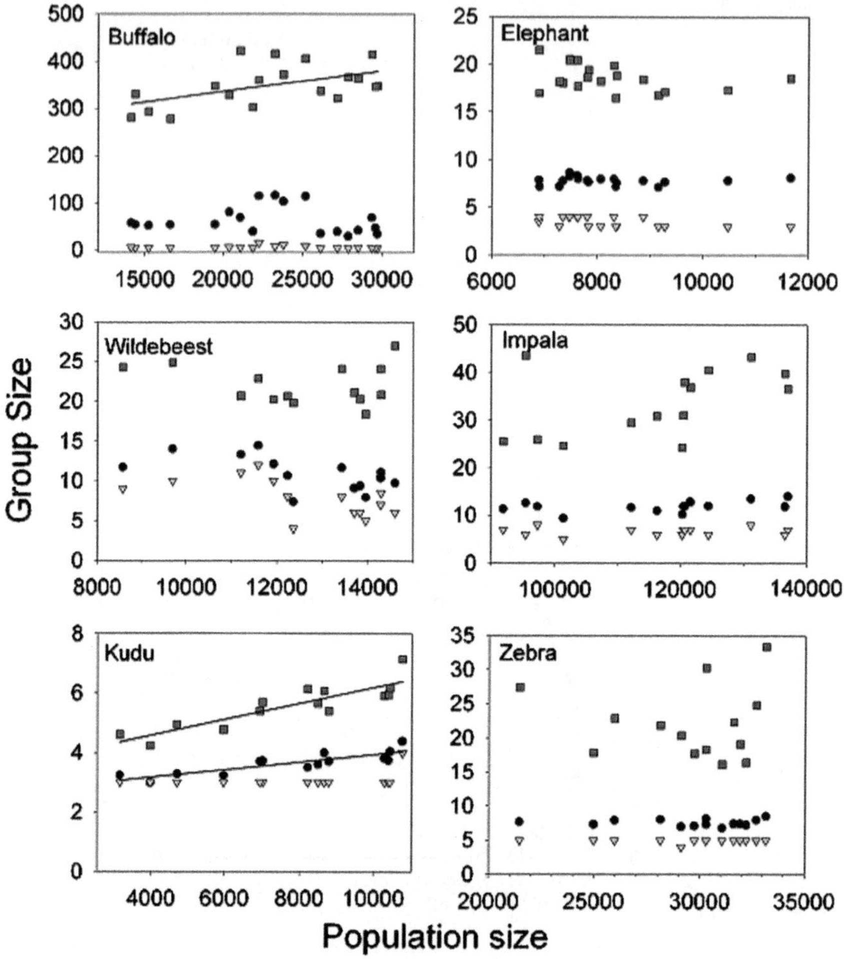

Fig. 2.3 Mean (circles), weighted mean (squares), and median (triangles) group size as a function of total population size for ungulates in the Kruger National Park, South Africa. Weighted means (i.e. $\sum_i n_{ijk}^2 / \sum_i n_{ijk}$, where n_{ij} is the ith group size observation of species j in year k) represent the expected group size experienced by a randomly chosen individual. Buffalo and elephant data were derived from annual helicopter surveys from 1985 to 2003, while data for other species are from annual fixed-wing surveys from 1980 to 1993 (solid lines depict statistically significant correlations, i.e. $p < 0.05$)

For species with right-skewed distributions, the weighted mean will be much larger than either the median or mean group size (Fig. 2.3), indicating that, although average group sizes may be small, most individuals experience groups of intermediate size. For many of the ungulates studied in the KNP little or no association was found between total population size and any measure of group size (Fig. 2.3). Group sizes for African buffalo and kudu (*Tragelaphus strepsiceros*) were weakly correlated with total population size, such that a doubling of the population was only associated with an increase in the weighted group size of about 25% (Fig. 2.3). Because the perimeter of KNP was

Box 2.2 Bovine tuberculosis and the social structure of badger populations

The Eurasian badger (*Meles meles*) is implicated in the transmission of bTB to cattle in the UK and Ireland. However, the extent to which badgers contribute to the persistent reservoir of infection in cattle herds is still uncertain. Badgers are social animals, and in the UK they live in groups of typically three to ten individuals (Neal and Cheeseman 1996), although this varies widely with population density. Each social group defends a territory, within which will be several burrow systems (setts), one of which is likely to be their principal residence (the main sett). In medium to high density badger populations social group territories may be largely contiguous, and boundaries are characterised by latrine sites where faeces and other scent marks are deposited. This structured system of social organisation determines patterns of movement, contact rates and hence the distribution of infection in the population.

The dynamics of bTB infection in badgers has been the subject of a long-term study at Woodchester Park, Gloucestershire in southwest England. In this 11 km^2 study area of lowland pastoral farmland and mixed woodland, the resident badgers have been regularly captured, marked, examined and released. In addition, bait marking (see Section 2.4.3) was carried out each year to determine the territorial configuration of the resident social groups. In this high density population, fluctuations in badger numbers were driven largely by changes in social group size, whilst the number of groups and their territorial configuration has remained relatively stable (Cheeseman et al. 1987; Rogers et al. 1997). Initially during the study, the badger population increased in size, but subsequently stabilised. These changes in host density did not however correlate with the incidence of infection detected in the population (Rogers et al. 1999). Furthermore, as group size was also not related to the incidence of infection (Delahay et al. 2000a) it appeared that host density did not drive TB dynamics either at the scale of the population or the social group.

The territorial behaviour of badger social groups inhibits the free movement of individuals, encouraging them to remain within clearly defined ranges and limiting levels of inter-group contact. Such a highly structured system of social organisation is likely to have a profound influence on the dynamics of disease distribution. As a consequence, in the Woodchester Park badger population, infection remained spatially restricted for many years, with only limited spread beyond a cluster of persistently affected groups (Fig. 2.4) (Delahay et al. 2000a). Nevertheless, there was some movement of individuals between social groups (Rogers et al. 1998), probably largely stimulated by the pursuit of breeding opportunities (Carpenter et al. 2005). Interestingly, these movements (detected from trapping records) were highly correlated with the incidence of infection, such that years with high rates of movement between social groups were followed by years with an increase in the number of new cases of disease detected (Rogers et al. 1998). Although the presence of other infected individuals

(continued)

Box 2.2 (continued)

in a group was the most important predictor of further infections, this relationship became less important as levels of immigration and emigration in a group increased (Vicente et al. 2007a). Hence, even individuals in groups that were diminishing in size experienced an enhanced risk of infection. This may be related to the observation that movement of an individual badger may be a protracted affair, during which it may split its time between two groups for several days or weeks before settling (Roper et al. 2003). Clearly, the associated repeated movement back and forth represents a window of enhanced association and increased transmission risk between the groups.

The limited movement of badgers among social groups in relatively high density populations probably limits disease spread. This is an important consideration for developing strategies for managing bTB in badgers, as some interventions such as culling, may have counter-productive effects if they disrupt this social system and enhance movement rates (see Box 7.2).

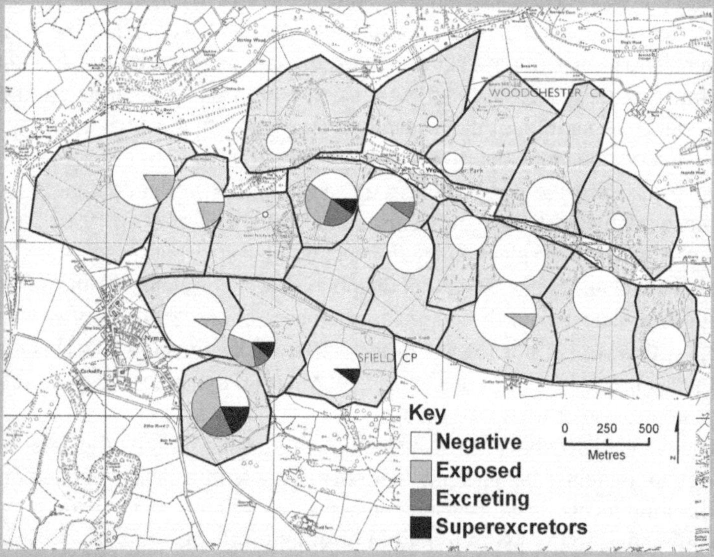

Fig. 2.4 The spatial distribution of bovine TB infection in the Woodchester Park badger population in 1996. Polygons represent social group territories. Pie charts are scaled relative to group size and show the proportion of residents falling into different disease status categories (exposed = seropositive, excretors and superexcretors = infectious). This spatially clustered pattern of infection persisted during a 15 year period of study

entirely fenced during this study, recorded population size was correlated with density. Hence, in this system, increases in population size were accompanied by an increase in the number of groups, while group size generally remained constant. Thus, in this case one would not expect per capita contact rates to increase with population size, and so disease management efforts focused at reducing the population size as a whole might not be effective. However, contact among groups may increase with the number of groups, facilitating the invasion of a disease even if group size remains constant (Cross et al. 2005).

Similar patterns may also occur for other species where social behaviour limits the frequency or intensity of contact. For example, adult female white-tailed deer are more likely to contact other females within their matrilineal social group (Schauber et al. 2007), therefore, increases in population density may not substantially alter the number of female-to-female contacts and direct pathogen transmission may be limited. Eurasian badgers (*Meles meles*) are group-living across most of the UK, and population density may be driven by changes in group size, whilst the number of groups remains relatively constant (Cheeseman et al. 1987; Rogers et al. 1999). Interestingly, bTB infection in badgers does not appear to be a simple function of population density, suggesting that other factors may drive transmission (see Box 2.2).

2.4.2 Group Structure

Early disease models often assumed that the host population was homogeneously mixed (Anderson and May 1991) so that each individual was equally likely to contact every other individual per unit time. Because these conditions do not hold for many human or wildlife populations, alternative methods have been developed to account for the effects of spatial heterogeneity or social structure on contact rates (Hess 1996b; Swinton et al. 1998). One approach is to combine individuals within categories, which may be based on sex, age, dominance or core risk groups (e.g. drug addicts sharing needles), and then incorporate data on contact rates within and among classes of individuals using a mixing matrix (Blower and McLean 1991) to scale transmission rates within and among categories or subpopulations of individuals.

Researchers have also used network models as a flexible method of capturing the socio-spatial structure of populations (Keeling 1999; Watts 1999; Ferrari et al. 2006). While traditional transmission models assume that the risk of infection depends on the prevalence or density of infectious individuals in the local (or global) population, network models explicitly incorporate information about relationships among individuals, and calculate infection risk as a function of known contacts with infectious cases (see Section 4.2.4). These models have been used primarily to describe the dynamics of sexually transmitted infections where contacts among individuals may be limited and variable. One strength of network modelling is its inherent flexibility to represent a wide range of social or spatial structures. In fact, metapopulation or patch models of disease can be thought of as a subset of network models where everyone within a group is connected and between group connections are infrequent. To date, most network models have been static due to the lack of empirical data on temporal changes in network structure. However, these models can be used to illustrate how the contact network evolves over time as individuals become infected and die. Individuals with the most connections are likely to be

Box 2.3 Bovine tuberculosis and social networks in meerkats

Meerkats (*Suricata suricatta*) are desert-adapted social mongooses living in groups of up to 40 animals in southern Africa. In the South African Kalahari, 300 individually-identifiable wild meerkats living in 14 social groups were habituated to researchers investigating social interactions between individuals and groups in a long-term study (Clutton-Brock et al. 2004). In the late 1990s, a pathogenic mycobacterial infection was diagnosed in the study population (Alexander et al. 2002) which was subsequently confirmed to be *Mycobacterium bovis,* the causal agent of bTB. Infection in meerkats is invariably fatal following the onset of clinical signs, and appears to have been responsible for the extinction of four social groups in the study population between 1995 and 2005.

Investigation of infectious disease transmission in wild animals is often constrained by lack of empirical data on social interactions. As the meerkat study had a long history of behavioural research, it offered a unique opportunity to examine the role of social interaction in the transmission of bTB in a wild mammal. As part of a longitudinal study, meerkats were routinely caught and tested for bTB using serology and culture techniques (Drewe et al. 2009). Social Network Analysis (SNA) was used to determine the role of host population structure in the transmission of bTB. This approach is used to describe the social position of each individual using both direct and indirect interactions, and so offers a way of determining which individuals within an at-risk population are more likely to be involved in disease transmission (Corner et al. 2003). A set of precise formal definitions for SNA measures has been produced (Wasserman and Faust 1994) and the techniques can be applied to quantify interactions both between and within groups of animals.

Adult male meerkats frequently visit other groups in search of mates before returning to their original group (Young et al. 2005) and this behaviour may be important for inter-group transmission of bTB. The temporary inter-group movements of male meerkats and corresponding bTB transmission dynamics can be illustrated using network diagrams (Fig. 2.5). Meerkats are often aggressive towards other members of their resident social group and the outcome of these interactions determines host population structure. Common causes of aggression include competition to become the dominant male or female, and dominance assertion between any two group-members. Subordinate females are often forcibly evicted by the dominant female when she is about to give birth (Stephens et al. 2005). Data from a group of meerkats in the Kalahari study population showed that the incidence of aggressive encounters temporarily increased following the death of a dominant meerkat, as others competed for the vacancy (Fig. 2.6). Disease appears to disproportionately affect dominant individuals, although this does not appear to be simply age-related, as younger and similar aged subordinates are also susceptible. Stress-induced immunosuppression of dominant meerkats is one possible explanation. In one group, a single female (F43) was involved in the greatest

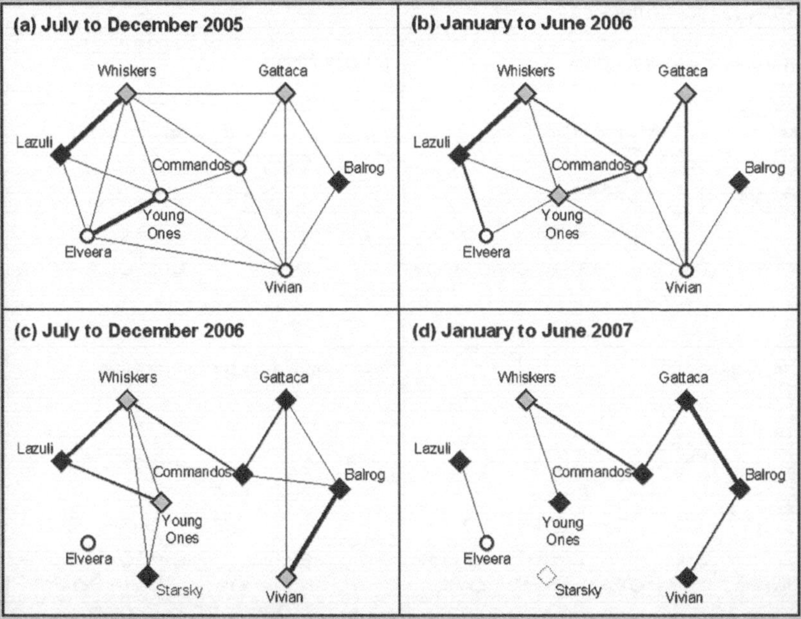

Fig. 2.5 Intergroup movements of meerkats and the spread of bovine tuberculosis (bTB) between eight social groups over a two-year period (open circle = uninfected group, grey diamond = seropositive (i.e. exposed) group, black diamond = group with clinical bTB). Line thickness is proportional to the number of intergroup movements. For visual clarity, only eight of the fourteen meerkat groups studied are shown. During the study period individuals transferred from uninfected status to seropositive (indicating exposure to bTB), to clinically positive and ultimately death. Harsh environmental conditions during 2007 markedly reduced the frequency of intergroup movements and this is reflected in network diagram (d) by fewer lines connecting meerkat groups compared with the beginning of the study

number of aggressive encounters (Fig. 2.5), shown as the highest number of lines connecting her to other group members. If intra-group aggression were responsible for bTB transmission between meerkats, incidence within this group should have fallen in the months following the death of F43. However, subsequently high levels of infection suggested that transmission may have already occurred, but was not detected before this female died, or that an interplay of several social interactions (e.g. aggression, grooming, feeding) determined bTB transmission in wild meerkats.

SNA has rarely been applied to the study of wildlife diseases although it has potential to significantly improve our current understanding and contribute to the development of effective management strategies. The examples shown here illustrate how SNA may be used to elucidate the role of specific behaviours in generating spatial and temporal variation in bTB transmission within and between meerkat social groups. Differences in transmission patterns within

(continued)

Box 2.3 (continued)

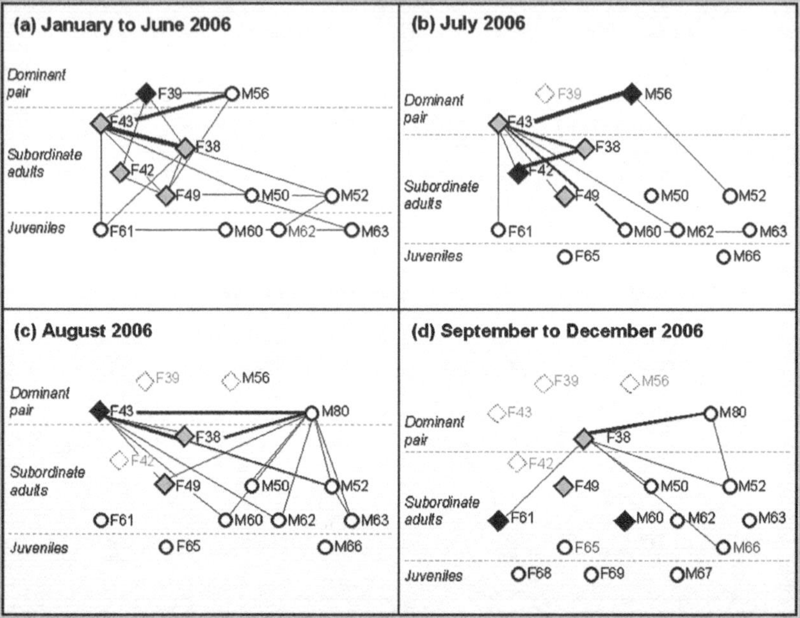

Fig. 2.6 Intragroup aggression networks and transmission of bTB in meerkats (M = male, F = female, open circles = uninfected individuals, grey diamonds = seropositive (i.e. exposed) individuals, black diamonds = individuals with clinical bTB, dashed diamonds = deceased individuals). Line thickness is proportional to the number of aggressive interactions between individuals. Male 80 (M80) immigrated from another group in August 2006

meerkat groups are beginning to be quantified by epidemiological modelling of social behaviour data. These data are being used in the development of a predictive model for quantifying the risk of transmission, which is likely to be useful in informing policy for the management of bTB in other social mammals.

infected first, leaving a more sparsely connected network of susceptibles and hence making disease persistence more difficult (Ferrari et al. 2006).

Network models are often under-pinned by a matrix of pairwise contact probabilities, where the element in row i and column j of the matrix describes the connection (or lack thereof) between individuals i and j. These connections are often assumed to be binary in that contact either does or does not occur. Alternatively, values of the matrix may reflect the relative strength of the connections between individuals or populations. In a study of African buffalo, the proportion of time that pairs of individuals spent in the same herd was estimated from radio-tracking data (Cross et al. 2004). These contact indices were multiplied by infection rates or probabilities, to simulate disease transmission dynamics. Properties of the contact network may be particularly important for acute infections where the disease can become extinct within a local group prior to any connections forming between groups. For chronic diseases the network structure connecting different groups may be less important because disease persistence is long relative to the rate of new connections between groups. However, the importance of network structure for intra-group transmission of bTB in meerkats (*Suricata suricatta*) (see Box 2.3) suggests important species-specific differences in this relationship.

Network and metapopulation models can also be used to investigate the roles of connectivity and group size in disease dynamics. Metapopulation models assume that populations are distributed over a number of patches, or areas, which are connected by dispersal. This approach has been adapted to diseases where host groups or individuals represent suitable habitat patches (Hess 1996a). These models can be used to ask questions about the spread of disease between populations and the likely effectiveness of implementing different management strategies, such as quarantine in some subpopulations and not others.

Early work using metapopulation models showed that host movement may facilitate recolonisation of unoccupied habitat. However, host movement may also facilitate parasite invasion (Hess 1996a). Metapopulation models show that the probability of a disease pandemic (i.e. parasite spread among many groups) may not be a simple function of host or parasite characteristics, but a more complex interaction between the two (Cross et al. 2005; Cross et al. 2007b). Consequently, acute diseases may require more frequent host movement compared to chronic diseases, in order to create a pandemic. Assuming that all individuals in a group become infected, then the movement rate, recovery rate and group size determine the expected number of infectious dispersers, which must be greater than one for a pandemic to occur.

Despite the flexibility of network models to accurately represent complex host social structures, their utility in investigations of wildlife disease systems is currently limited. A particular problem is that it is not clear how to scale up the network from a sub-sample of the population such that it represents the entire population of interest. Rare linkages that allow a parasite to move from one group to another may be absent from the sampled population. Raccoons hitchhiking on refuse trucks is an interesting example of potentially rare but important long-distance movements that may have a significant impact on disease spread (Real et al. 2005). Theoretical work has shown that just a few such connections can radically alter the structure of a network and may be critical to understanding disease dynamics (Watts 1999). It is however empirically challenging to document these potentially rare but influential connections at spatial and temporal scales that are relevant to many management problems (but see Box 2.3).

2.4.3 Describing Host Social Structure

Understanding disease transmission in most wild populations is difficult because it usually involves three steps: exit from the host, passage across an external environment to a new host and infection of the new host. Determining when and where these events occur in cryptic wild mammals and with parasites that are difficult to detect can be demanding. As a result, understanding the additional complexity of host social structure on wildlife disease dynamics has proved challenging. Logistic and financial limitations in research studies often require a trade-off between the collection of detailed data on local movements, rates of contact and infection on a limited spatial and temporal scale, and coarse data on dispersal and migration at a broader scale.

Where the target species is conspicuous and diurnally active (e.g. meerkats), direct observation may be possible. However, this is seldom the case, and more often researchers have relied on traditional ecological methods such as mark-recapture trapping programmes and radio-tracking to provide information on movements and population structure. Live trapping has been widely employed to generate demographic data using capture-mark-recapture models (see Thompson et al. 1998), although this can be labour intensive. For some species (e.g. small mammals, badgers) live trapping can provide useful information on individual movements with relatively large sample sizes, albeit at a lower resolution to that obtained from radio-tracking. Mark-recapture studies that incorporate disease infection status of captured animals can also be used to estimate transmission rates and evaluate the impact of parasites on host demographics (Samuel et al. 1999; Faustino et al. 2004).

Researchers have traditionally monitored mammal movements using very high frequency (VHF) radiotelemetry or more recently global positioning systems (GPS). VHF transmitters are cheaper but more labour-intensive and may result in data that is spatially coarse and temporally sparse. GPS collars, on the other hand, yield very fine-resolution spatio-temporal data but the costs can be prohibitive. The recent development of proximity collars or dataloggers that record when tagged individuals are within a certain range presents new opportunities to investigate how wild mammals associate and contact one another (Ji et al. 2005).

Another approach to monitoring mammal movements is to use bait laced with a persistent physical or chemical tag to mark the excretory products. This approach has been usefully employed in small mammals (e.g. Randolph 1973) and is commonly used to delineate the social group territories of the Eurasian badger (Delahay et al. 2000b). This bait-marking technique has potential applications for monitoring the spatial organisation of other mammals, particularly where the faeces are used to mark territories.

Whatever method is used, a sufficient number of individuals are required to maximise the chances of recording relatively rare long-range dispersal events and transient short-distance movements. Once the data are collected, a further significant challenge remains in terms of interpreting how contact data relate to transmission risks.

In many cases, unambiguous determination of group membership may be difficult, particularly as levels of inter-group movement increase. In this case, cluster analysis can provide an alternative approach to describing social structure (Cross et al. 2004). Association indices based on the proportion of time or observations where pairs of individuals are together can be used to construct an association matrix where each row and column represents an individual. Cluster analyses can then be used to objectively group individuals according to their levels of association. Otherwise, the association matrix can be used directly to create a network model. The time interval used to construct the association values is critical to the resulting structure and should be similar to the infectious period of the parasite (Cross et al. 2005).

The application of population genetics provides an alternative means of estimating potential connectivity between subpopulations across a larger spatial and temporal scale. For example, gene flow in white-tailed deer populations was used to evaluate potential barriers to their dispersal and has been correlated with the spatial spread of CWD from a focus of infection (Blanchong et al. 2008). As it may be possible to collect genetic samples (either from trapped animals, carcasses or faeces) over a wider area than is often practical for radio-tracking, this approach may allow researchers to investigate patterns of host connectivity and to predict direction of disease spread on a much larger scale. However, patterns of host gene flow may often reflect translocations, and historical rather than contemporary movements, which may limit the value of this approach for investigating disease dynamics. Tracking the evolution of pathogen populations may be a useful approach in studies of host population structure. For example RNA viruses evolve rapidly compared to the host, so their phylogeography may provide insights into relatively recent demographic changes and movement patterns in mammal populations. Hence, feline immunodeficiency virus (FIV) was used as a host marker to investigate the genetic structure of mountain lion (*Puma concolor*) populations across the northern Rocky Mountains, USA (Biek et al. 2006a).

2.5 Conclusions

The integration of wildlife ecology, behaviour and disease dynamics is a relatively new area of research. Key scientific information on host-pathogen-environment interactions, disease impacts and appropriate management strategies are frequently unknown. As a result, although this chapter presents many patterns, the observations often apply to only a limited number of situations and there are few general principles that relate to a wide range of hosts or parasites. In many cases, it is the interaction of host and parasite life-histories that will drive disease dynamics and hence determine management options. In this chapter, we have highlighted several factors that are likely to be important with respect to host behaviour and social organisation (e.g. sex, age, group structure, and dispersal). However, their importance will depend on the parasite in question, and whether it has an intermediate host or is directly-transmitted, and whether the host recovers from infection. For example, small group sizes may help to exclude some directly-transmitted diseases, but have little effect on the persistence of a parasite that has a vector and multiple alternative hosts. Hence it is crucially important that the ecological community comprising the pathogen, hosts and their interactions, is considered as a whole, during the formulation of strategies to manage disease in wild mammals.

Chapter 3
Assessment of Transmission Rates and Routes, and the Implications for Management

Peter Caley, Glenn Marion, and Michael R. Hutchings

3.1 Introduction

3.1.1 Preamble

Being able to estimate disease transmission rates and determine the underlying mechanisms of transmission is fundamental to the effective management of wildlife disease – transmission rates drive disease dynamics and persistence, and thus determine the level of control or vaccination necessary to achieve disease eradication, or predict the likely impact of a biocontrol agent. The mechanisms of transmission determine where management efforts can be targeted. Not knowing and not being able to estimate transmission rates when trying to manage disease in wildlife is analogous to managing overpopulated wildlife without knowing the intrinsic rate of population increase. Being able to estimate transmission rates allows us to determine whether management actions are achieving their aims. This chapter looks at the measures of disease transmission and how they can be calculated. We recommend that the non-mathematical readers skim through Section 3.2 without trying to follow the mathematics, and refer back to it when needed.

3.1.2 Measures of Transmission

The term disease transmission means many things and can be quantified in different ways. Exactly what measure is required will depend on the aims of the investigator/manager. The following terms are all measures that result from disease transmission:

P. Caley
National Centre for Epidemiology & Population Heath, Australian National University, ACT 0200, Australia

G. Marion
Biomathematics Statistics Scotland, The Kings Buildings, Edinburgh, EH9 3JG, UK

M.R. Hutchings
SAC, West Mains Road, Edinburgh, EH9 3JG, UK

R.J. Delahay et al. (eds.), *Management of Disease in Wild Mammals*,
DOI:10.1007/978-4-431-77134-0_3, © Springer 2009

Force of Infection (λ) – the instantaneous *per capita* rate at which susceptible individuals acquire infection. Also called the instantaneous incidence.
Basic Reproduction Number (R_0) – the expected number of secondary infections produced by a typical infected individual over the course of their infectious period when among a population where every individual is equally susceptible. Also called the basic reproduction ratio and basic reproduction rate.
Effective Reproduction Number (R) – the actual number of secondary infections produced by an infectious individual.
Disease Prevalence (p) – the proportion of the population that is infected at a given time.
Attack Rate (α) – the proportion of the population infected over the course of an epizootic.
Transmission Coefficient (β) – the model-dependent constant that as part of the transmission function, determines the rate at which susceptible individuals become infected in the population.

Note that despite being related, knowing the value of one measure of transmission does not necessarily mean any other is also known. Also, in general, measures such as p and λ are dependent on the prevailing conditions (e.g. numbers of infectives and susceptibles) – they are not fixed parameters. Conversely, R_0, whilst essentially being a fixed parameter that underlies the number of secondary infections an infected individual produces (which is a random variable), is often specific to the population from which it is estimated, and usually changes with host density or numbers. Hence the usefulness of R_0, for all its laudings, becomes tempered when applied to wildlife populations inhabiting different environments and/or locations from those used in its estimation. An analogous problem occurs with the intrinsic (maximum) rate of increase (r_m) of a wildlife population, which is specific to the particular environment in which it is measured (Caughley and Birch 1971). To get any measure of transmission that can be generalised to changed conditions (e.g. post-intervention or to a different population) requires that we relate these measures to an underlying model of transmission (described by the transmission function) that can account for changed conditions. In its most basic form, the transmission function describes how transmission scales with population size and/or density and is where the transmission coefficient β is found. As such, β is typically the only intrinsic measure of transmission. It is also the most difficult to estimate; estimation necessarily being achieved via a model. Commonly considered forms of the transmission function are shown in Table 3.1. Clearly these functions do not accommodate variation in transmission relating to factors such as environmental conditions influencing pathogen survival, strain-specific differences in transmission, population immunity/susceptibility or local influences on the spatial arrangement (and hence mixing) of hosts. Most of these influences are all subsumed within β, which is typically assumed a constant (or if using Bayesian statistics, a distribution expressing belief in it's likely value). Violation of these assumptions may go partway to explaining why transmission rates often differ between sites for unknown reasons; underlining the simplistic nature of our models in many cases. It should also be stressed that where spatial information is available it is possible to infer contact rates within spatial

Table 3.1 Proposed forms for the transmission function. Adapted from McCallum et al. (2001) and references therein (reproduced with permission)

Number	Function[a]	Comments
1	βsi	Density-dependent transmission (also termed mass action)
2	$\beta si/n$	Frequency-dependent transmission
3	$\beta s^p i^q$	Power relationship; constants: $0 < p < 1$, $0 < q < 1$. Phenomenological. Sometimes considered to account for spatial effects such as local depletion of susceptibles
4	$\beta i(n - 1/q)$	Constant: $0 < q < 1$. Embodies a refuge effect (q = proportion of population potentially susceptible)
5	$ks \ln\left(1+\frac{\beta i}{k}\right)$	Negative binomial. Small k corresponds to highly aggregated infection. As $k \rightarrow \infty$, expression reduces to Function 1
6	$\frac{n}{1-\varepsilon+\varepsilon n}\frac{F(s,i)}{n}$	Asymptotic contact function separated from the mixing term $F(s,i)$ which may be any of Functions 1–5 above. If the constant $\varepsilon = 0$, the contact rate is proportional to n. If $\varepsilon = 1$, contacts are independent of n
7	$\frac{\beta si}{c+s+i}$	Asymptotic transmission where c is a constant

[a] i is the density of infected hosts, s is the density of susceptible hosts, and n is the total host density. β is the transmission coefficient. Other parameters are described where necessary under comments.

models. Additionally, if a management intervention aims to change the behaviour of animals (e.g. increasing mating frequency as reported by Caley and Ramsey (2001)), then clearly β will change and can no longer be considered fixed.

3.1.3 Practical Difficulties in Field Measurement

Disease transmission is typically an unobservable event – even if we observed a known infected "contacting" a known susceptible, we would be none the wiser as to whether transmission occurred. Thus we have to infer transmission from observable data such as evidence of prior or current infection (e.g. diagnostic testing) or surrogate markers for infection such as the onset of clinical signs or death. Such data usually requires that animals can be captured and samples collected, or easily observed. Obtaining such data for free-ranging mammals is often problematic, making large-scale replicated field experiments difficult and smaller pen studies the only feasible type of experimentation. Considerable difficulties, however, are often experienced when extrapolating transmission rates estimated from experimental conditions to field conditions (McCallum 2000).

Estimating epidemic threshold parameters (e.g. critical host population size N_T or critical host density K_T) from whether an introduced pathogen establishes and gives rise to a major epizootic as opposed to a minor epizootic is difficult to achieve experimentally (Lloyd-Smith et al. 2005b). The reasons being the epizootics are by nature dichotomous (either the epizootic is major with many cases or is minor with a trivial number of cases) and stochastic (an epidemic may not take off despite $R_0 > 1$). Many experiments may therefore be required to estimate where the threshold may lie with any kind of precision. The result is that many researchers are forced to undertake observational experiments of host/pathogen systems as a means of quantifying disease transmission.

Exactly what measure of disease transmission is estimated will depend on the aims of the investigation and logistical constraints. If the aim is simply to determine whether a management intervention is reducing transmission or whether a particular pathway of transmission occurs (a hypothesis testing question), then bias may not be problem and precision more important. Fitting more parsimonious models is a way of achieving this (though increasing bias). For example, ignoring the effect of disease-induced mortality when modelling age-prevalence data biases estimates of the force of infection downwards, though it facilitates straightforward model fitting (Caley and Hone 2002). If the bias of an estimator is consistent across experimental treatments, then such an estimator may suffice for estimating relative changes in underlying transmission. If the purpose of the investigation is to identify risk factors contributing to disease transmission (as typically measured by either the prevalence or time to infection), then robust statistical frameworks such as logistic regression (e.g. Joly and Messier 2004) or Cox's proportional hazards model (e.g. Calvete et al. 2004b) will suffice. Such models typically do not explicitly include a transmission component and hence cannot be used to estimate rates of transmission. Conversely, if the aim is to investigate predicted changes to the host(s)/pathogen system of a mechanistic nature (e.g. introducing vaccination), then unbiased estimates of transmission coefficients will be required along with knowledge of the correct underlying transmission function, and models will need to be specified accordingly.

3.2 Estimating Transmission Rates for Directly Transmitted Pathogens

Quantifying disease transmission is simplest for directly transmitted pathogens, particularly if only one or two hosts are involved, and this is the focus of this section.

3.2.1 Estimating the Force of Infection (λ)

The force of infection experienced by a susceptible individual will depend on the infection status of other individuals that the susceptible mixes with (as quantified by prevalence of infection or density of infectives), and the form of the transmission function. For this reason, estimates of λ in isolation are of little use for quantifying underlying transmission rates. However, relating λ via a model to host density and the relative abundance of susceptibles and infecteds, in combination with other demographic parameters, is a practical approach for estimating parameters (e.g. β) that determine transmission rates (McCallum et al. 2001).

Using Age-Prevalence Data In general, methods to estimate disease transmission rates from age-prevalence data assume steady-state (c.f. epizootic) conditions.

This is a strong assumption that needs to be applied with care, as it is difficult to distinguish between age-dependent and time-dependent variation in disease incidence. Most models developed for analysing age-specific prevalence data were developed for diseases of humans, and assume that mortality due to infection can be ignored (e.g. Farrington et al. 2001). This is less often the case with wildlife diseases, and accounting for disease-induced mortality introduces additional complications. Disease-induced mortality tends to flatten age-prevalence curves (Heisey et al. 2006) as does loss of evidence of prior infection (or recovery from infection for chronic diseases), resulting in the force of infection being underestimated if ignored. This may not be a problem in a hypothesis testing application (e.g. answering "does the intervention significantly reduce transmission?"), but will be an issue if estimation is the main aim of the investigation (Caley and Hone 2002).

If disease-induced mortality can be ignored, and the system is in equilibrium, then the modelled probability of an individual being infected (or showing signs of past infection) at age a when subjected to age-dependent force of infection $\lambda(a)$ is

$$p(a) = 1 - \exp\left(- \int_{t=0}^{t=a} \lambda(t)dt \right) \tag{3.1}$$

The form of $\lambda(a)$ may be as simple or complex (data willing) as the scientific investigation requires, and may change as a function of age, time and other covariates. The underlying form chosen for $\lambda(a)$ may be flexible (e.g. Grenfell and Anderson 1985; Heisey et al. 2006) or consistent with how transmission is known or hypothesised to occur (e.g. Caley and Hone 2002 and see Box 3.1). For simple forms of $\lambda(a)$ it is often possible to express Eq. (3.1) as a generalised linear model and obtain estimates of λ and factors influencing it directly (see Box 3.2). For more complex forms of $\lambda(a)$ and if additional demographic parameters are included, analytical solutions for the prevalence usually do not exist and numerical methods are used, although the parameters may still be estimated via standard maximum likelihood techniques. For n samples of individuals of ages a_j ($j = 1, \ldots, n$) where each sample contains N_j individuals of which I_j are infected (or shows signs of previous infection), the likelihood assuming that the probability of infection for a given age is binomially distributed is

$$L = \prod_{j=1}^{n} p\left(a_j\right)^{I_j} \left(1 - p\left(a_j\right)\right)^{N_j - I_j} \tag{3.2}$$

Maximum likelihood estimates of the parameters are obtained by minimising the negative of the log-likelihood function with respect to the parameters that determine $p(a)$:

$$-\ln(L) = -\sum_{j=1}^{n} \left[I_j \ln\left(p\left(a_j\right)\right) + \left(N_j - I_j\right) \ln\left(1 - p\left(a_j\right)\right) \right]. \tag{3.3}$$

This is usually achieved numerically by a standard numerical algorithm. Likelihood theory also enables estimation of the precision of these estimates, and comparison of models via likelihood ratio tests or information-theoretic methods (e.g. Akaike's Information Criterion). Alternatively, the likelihood function may be used within a Bayesian estimation framework (e.g. Markov Chain Monte Carlo) to obtain posterior

Box 3.1 Estimating the rates of rabbit to rabbit transmission of *Mycobacterium avium* subspecies *paratuberculosis (Map)*

European rabbits (*Oryctolagus cunniculus*) have been increasingly linked to the persistence of *Mycobacterium avium* subspecies *paratuberculosis* (*Map*) (Johne's disease) in domestic ruminants in the UK. Quantifying the routes of rabbit to rabbit transmission of *Map* is a key step to establishing whether rabbits are a persistent source of infection (i.e. a reservoir). Judge et al. (2006) fitted an SI (Susceptible-Infected) epidemiological model to field data to estimate the probabilities of vertical (vertical + pseudo-vertical) and horizontal transmission. *Map* was isolated from various tissues and excreta from a study site in Scotland suggesting the potential for vertical, pseudo-vertical and horizontal rabbit-to-rabbit transmission routes. The overall prevalence of *Map* in rabbits was high at both sites studied, with an average of 39.7%.

Estimating rates of transmission: A maximum likelihood fitting procedure was used to fit the SI model to the data on the proportion of infected rabbits per age group (2 month blocks) from the random sample to derive probabilities of vertical/pseudo-vertical and horizontal transmission (Fig. 3.1).

In order to model the variation of the mean infection prevalence with age, Judge et al. (2006) assumed that both the number of individuals at any given age and the number of infected individuals at any given age remain constant at least on the time scale of an individual's lifetime. This was consistent with the finding that the overall prevalence of infection in rabbits did not increase across the years of sampling (Judge et al. 2005a). Given this assumption it was then possible to pool the prevalence data taken on each visit and treat the inferred prevalence

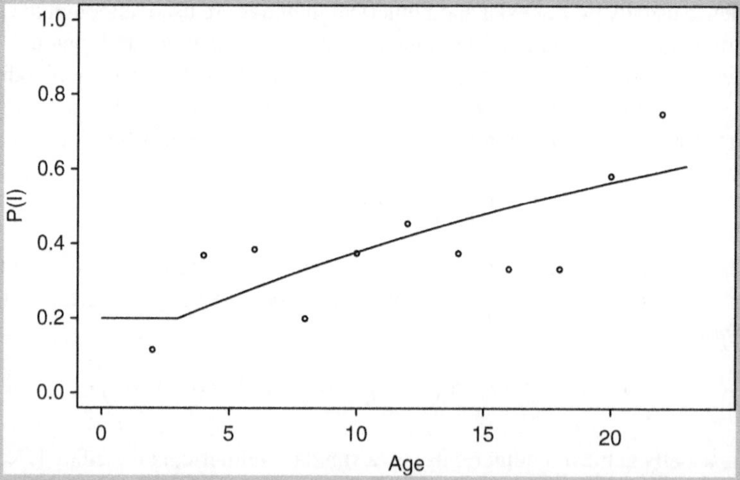

Fig. 3.1 Fitted prevalence of *Map* infection in rabbits as a function of age. Data are categorised in age ranges of two months (from Judge et al. 2006)

for each age as being equal to the prevalence that would be measured if it were possible to track a cohort of individuals from birth to death, measuring the prevalence in that cohort. A model for the spread of disease over time in a group of individuals exposed to a constant level of infection could therefore be used.

The model was constructed by assuming a two-stage infection process; individuals are exposed only to vertical and pseudo-vertical infection up until time t_0, when all vertical/pseudo-vertical infection ceases and they become exposed to horizontal infection by infected rabbits they are in contact with. The absence of sufficient data from pre-weaned individuals prevented using a detailed model of the vertical processes, so the combined effect of vertical and pseudo-vertical transmission was represented by a single probability P_v that individuals are infected at age t_0. The horizontal infection process was modelled as a homogeneous Poisson process (representing the simplest mathematical form for horizontal infection within a homogeneously mixing social group of rabbits, see below for group size) with a constant infection rate in which I is the (constant) number of infected individuals in the population as a whole and λ_0 is the per capita rate of infection.

$$\lambda_0 = \beta I \tag{3.17}$$

In a homogeneous Poisson process with rate parameter λ_0, the probability that an event occurs in the time interval $(0, \tau)$ is

$$1 - e^{-\lambda_0 \tau}. \tag{3.18}$$

Including the effect of vertical transmission there are two ways that an individual could be infected at time τ – by being infected vertically/pseudo-vertically from its mother, or horizontally, with a combined probability

$$P_v + (1 - P_v)\left(1 - e^{-\lambda_0 \tau}\right) \tag{3.19}$$

whereas, in order to escape infection up to time τ an individual must avoid infection through both routes, leading to a probability of being uninfected

$$(1 - P_v)\left(e^{-\lambda_0 \tau}\right). \tag{3.20}$$

Combining these probabilities with the data, the likelihood

$$\begin{aligned} L(\lambda_0, P_v, t_0) = \prod_{i=1}^{N} \Big\{ & I[y_i = 1]\left(\mathrm{P_n} + \mathrm{I}[\mathrm{t} > \mathrm{t_0}](1 - \mathrm{P_n})\left(1 - \mathrm{e}^{-l_0(t_i - t_0)}\right)\right) \\ & + I[y_i = 0](1 - P_v)\left(I[t \le t_0] + I[t > t_0]e^{-\lambda_0 (t_i - t_0)}\right)\Big\}, \end{aligned} \tag{3.21}$$

is formed which was maximised numerically in order to obtain maximum likelihood estimates of the parameters λ_0 and P_v. Note that in practice it is the negative of the logarithm of the likelihood that is minimized. The data consist

(continued)

Box 3.1 (continued)

of the infection status y_i ($y_i = 1$ corresponds to infection and $y_i = 0$ to susceptible) of $I = 1, N$ individuals and their estimated ages $t_{i.}$ $I[..]$ denotes the indicator function which is unity if the expression in square brackets is true and zero otherwise. The first line of Eq. (3.21) corresponds to the probability that susceptibles become infected, whilst the second line represents the probability that susceptibles escape infection.

Maximum likelihood estimates were $\lambda_0 = 0.037$ and $P_v = 0.14$ when using a weaning age of $t_0 = 1$ month. These values can be expressed in terms of the underlying transmission probabilities. This per capita rate of horizontal infection per month (λ_0) is specific to the study site and will vary depending on the number of infectious (I) and susceptible animals in regular contact. The generic horizontal transmission coefficient per month (β) can be estimated as

$$\begin{aligned} \lambda_0 &= \beta I = \beta N p \\ \beta &= \lambda_0 / Np, \end{aligned} \tag{3.22}$$

where p is the overall prevalence and N is the total population size.

Adult rabbit social group sizes at the study site were conservatively estimated at between 2 and 7 individuals, equating to a conservative β value range of 0.013 to 0.046. The proportion of individuals entering the population after weaning (at 1 month old), which were infected via vertical and/or pseudo-vertical transmission (P_v), estimated from the maximum likelihood procedure, was 0.14. As only offspring from infected does can be infected vertically or pseudo-vertically, the probability of transmission via these routes can be calculated from the proportion of infected juveniles entering the population after weaning and the proportion of infected females of reproductive age. There was no significant difference in the prevalence of *Map* between sexes at either site therefore it was assumed that equal percentages of males and females were infected with *Map*. For adults of reproductive age (i.e. >6 months), 42.9% (85/198) were *Map* positive. Assuming that there is no effect of *Map* infection on either reproductive output or juvenile survival, this gives a probability of infection via vertical and/or pseudo-vertical transmission of up to 0.326 (14% of young infected when entering the population at 1 month /42.9% of infected females of reproductive age). These estimates of rabbit-to-rabbit routes of *Map* transmission were subsequently used in a modelling study to show that infection is highly persistent in rabbit populations (Judge et al. 2007) a critical step in understanding the role of rabbits in the epidemiology of paratuberculosis within the host community as a whole.

distributions for the parameters of interest, and incorporate prior belief regarding parameters (if available). Such models can be compared using the Bayesian Information Criterion (BIC) or deviance information criterion (DIC) as appropriate.

Using Longitudinal Data Estimating the force of infection from prospective studies of individuals (i.e. susceptible individuals are followed and their time to infection

Box 3.2 *Mycobacterium bovis* (bTB) in wild pigs – testing for treatment effects

The study data under consideration (Table 3.2 and Fig. 3.2) come from the Northern Territory, Australia, and estimate the proportion of wild pigs (*Sus scrofa*) with visible lesions typical of bovine tuberculosis (caused by *M. bovis*) during two territory-wide surveys. The first survey during the early 1970s (Corner et al. 1981), occurred at a time when bovine tuberculosis was highly prevalent in sympatric populations of wild cattle (*Bos spp.*) and water buffalo (*Bubalus bubalis*). The high prevalence observed in pigs was hypothesised to be a result of their association with these infected bovid populations. Subsequently, the populations of cattle and buffalo were dramatically reduced as part of the Brucellosis & Tuberculosis Eradication Campaign (BTEC). The second survey was undertaken in 1992, with the aim of determining whether

Table 3.2 Prevalence of wild pigs with lesions resembling bovine tuberculosis by age (in years). Adapted from McInerney et al. (1995)

Survey	Age	Sampled	Lesioned
1	0.5	128	21
1	1.5	132	59
1	2.5	117	55
1	3.5	83	47
1	4.5	105	66
1	5.5	82	56
1	6.5	45	35
2	0.5	251	8
2	1.5	227	9
2	2.5	131	10
2	3.5	113	13
2	4.5	38	2
2	5.5	16	4
2	6.5	14	3

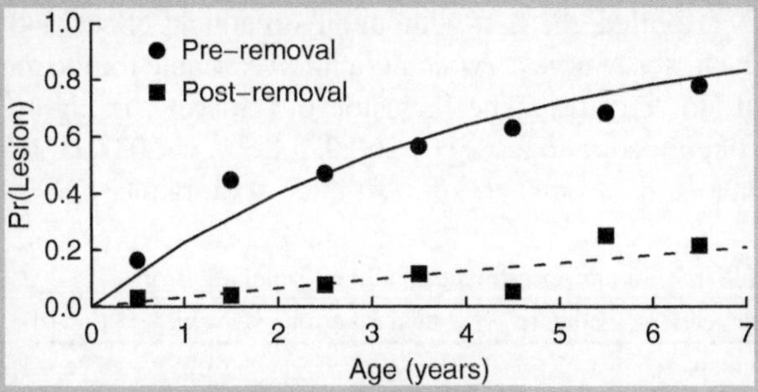

Fig. 3.2 Prevalence of lesions typical of bovine tuberculosis in wild pigs before (pre-removal) and after (post-removal) culling of sympatric cattle and buffalo populations known to be infected

(continued)

Box 3.2 (continued)

BTEC had indeed reduced the level of disease in wild pigs as predicted (McInerney et al. 1995).

There appears to be a difference in the age-specific prevalence of lesions between the two surveys (Table 3.2). How do we quantify this difference in terms of an underlying model that accounts for this data? This model is determined by how the age-specific prevalence relates to the force of infection. Assuming that animals are exposed to a constant force of infection from birth, then the prevalence (p) at a given age (a) is

$$p(a) = 1 - e^{-\lambda a} \tag{3.23}$$

This equation can be linearised with several simple algebraic operations:

$$\ln\left(-\ln\left(1 - p(a)\right)\right) = \ln(\lambda) + \ln(a) \tag{3.24}$$

This equation is straightforward to fit as a generalised linear model (see Crawley 2005 for details). The left hand side of this equation is what is known as a complementary log-log transformation of p. Hence the link function within the GLM is specified as complementary log-log. The age-dependent probability of infection ($p(a)$) is binomially distributed – so the error structure is specified as binomial. By specifying $\ln(a)$ as an offset (equivalent to fixing its slope to 1), we can directly assess the effect of factors and other explanatory variables on $\ln(\lambda)$. The steps to fitting this model are

1. Specify the proportion infected as the **response variable**.
2. Specify the **error structure** as binomial.
3. Specify the **link function** as complementary log-log.
4. Specify $\ln(a)$ as an **offset**.
5. Fit the model.

Two models are fitted, the first without any treatment effect, and the second including the factor "Survey" (which is a proxy variable for the removal of Tb positive buffalo & cattle). The inclusion of "Survey" is highly significant based on a likelihood ratio test ($\chi^2 = 266.9$, d.f. $=1$, $P < 0.001$). It is, however, the parameter estimates that are of most interest (Table 3.3).

Table 3.3 Parameter estimates and their standard errors

Parameter	Estimate	Standard error	Z value	Pr(>|z|)
Intercept	−1.37	0.06	−23.75	<2e − 16
Survey	−2.03	0.15	−13.19	<2e − 16

Note that the parameter estimates in Table 3.3 are on a logarithmic scale. That is

$$\lambda = e^{-1.37} = 0.26/\mathit{Year} \qquad \text{(Survey 1 - Pre-removal)}$$

$$\lambda = e^{-1.37-203} = 0.033/\mathit{Year} \qquad \text{(Survey 1 - Pre-removal)}$$

That is, force of infection post bovid removal was about 13% of that in pre bovid removal times – a substantial and statistically significant reduction (note though the lack of an experimental control in this simple before-after experimental design).

This basic exponential model fitted in this way can be expanded to include further factors and covariates. However, if for example, the mortality rate of animals increases as a result of being diseased, then the new expression for the prevalence of infection is intrinsically non-linear and unable to be fitted as a GLM. It can still, however, be fitted by standard maximum likelihood techniques (see Caley and Hone 2002 and Heisey et al. 2006 for details).

measured, or their infection status after a known length of time is determined) is similar to modelling age-prevalence data, only with the exposure time substituted for age (e.g. Ramsey 2007). A rich family of models exists for analysing this type of data centred on Cox's proportional hazards model (Cox 1972) and variants thereof. Note, however, that Cox's model is primarily concerned with estimating the proportional effects of different factors on the force of infection, rather than the baseline force of infection, which is sometimes the variable of intrinsic interest.

3.2.2 Estimating β

Since β is dependent on the underlying transmission function, for it to be estimated requires that the relevant variables (e.g. densities of the different infective classes) and/or parameters are also known or estimable.

Estimating β Directly from Trajectory of Prevalence or Cases There are several approaches to estimating transmission coefficients from such data, which typically includes additional data on temporal changes in the population size. Often enough simplifying assumptions can be made to enable the model to be fitted as a generalised linear model and coefficients estimated directly, with the response variables being either the prevalence of infection (e.g. Caley and Ramsey 2001) or the density of infectious classes (e.g. Begon et al. 1999). If the model cannot be solved analytically, then typically the series of differential equations that describe the host/pathogen dynamics will be solved numerically to yield the fitted number and/or density of animals in the relevant disease classes at the times of observation. If prevalence is chosen as the response variable then the model may be fitted by minimising the binomial negative log likelihood where estimable parameters enter into the likelihood through the fitted prevalence (e.g. Arthur et al. 2005). Likewise, Miller et al. (2006) modelled temporal changes in the number (or cumulative number) of cases of chronic wasting disease in elk (*Cervus elaphus*) as a means of estimating disease transmission rates.

Estimating β from the Force of Infection If the underlying transmission function is known (or assumed), then estimates of λ in conjunction with other variables enables estimation of β. For example, under density-dependent transmission for a single-species model, and assuming the area occupied by the study populations is constant over time (Begon et al. 2002), the rate of conversion from susceptibles to infecteds (di/dt) must equate with the term βsi, where i is the density of infectious animals. That is, $\lambda s = \beta si$, hence $\lambda = \beta i$ where β has the units "potentially infectious contacts per infectious individual per unit area per unit time". Under frequency dependent transmission, infecteds are created at rate $\beta si/n$ where n is the density of all individuals. The rationale is that there are βi potentially infectious contacts per unit area of which a proportion s/n will be with a susceptible individual and hence lead to transmission. Caley and Ramsey (2001) apply both transmission models to leptospirosis infection of brushtail possum (*Trichosurus vulpecula*) populations.

Where a host species may be infected from several sources, the observed force of infection is the summation of the contribution of the different sources of infection. In the case where both intra- and inter-species transmission is density-dependent, the force of infection experienced by the jth species is the sum of the products of the inter-specific transmission coefficients and their densities

$$\lambda_j = \sum_{k=1}^{n} \beta_{jk} i_k \tag{3.4}$$

In Eq. (3.4), n is the number of species, i_k is the density of infectious individuals of species k, and β_{jk} is the transmission coefficient from species k to species j (this follows the notation order of Dobson and Foufopoulos (2001)). Where there are independent estimates of λ_j and i_k, then estimates of β_{jk} can be obtained by regression. An application of this model to a two host (possum, ferret) one pathogen (*M. bovis*) system is given by Caley and Hone (2005). Clearly one could have a mix of frequency-dependent & density-dependent transmission processes occurring in a multi-host system.

3.2.3 Estimating R_0

Estimating R_0 from λ or β Anderson and May (1991) provide a number of steady-state solutions for the basic disease reproductive number. Under Type I mortality (death rate consistently low until the older age classes) and assuming a constant force of infection, they derive the following expression:

$$R_0 = \frac{\lambda L}{1 - e^{-\lambda L}}, \tag{3.5}$$

where L is the life expectancy (clearly disease-induced mortality is assumed to be negligible). However, under Type II mortality, where life expectancy declines exponentially with increasing age, they obtain (again under steady-state assumption and with negligible disease-induced mortality):

$$R_0 = 1 + \lambda L. \tag{3.6}$$

As under these conditions λ is simply the reciprocal of the mean age of first infection (A), Eq. (3.6) can be rewritten in terms of L and A:

$$R_0 = 1 + \frac{L}{A}. \tag{3.7}$$

Anderson and May (1991) also provide a general argument relating R_0 for a microparasite in a homogeneously mixed host population to the overall fraction who are susceptible at equilibrium (x^*) (Eq. (3.8)). The parameter p is the proportion of hosts that are infectious. Note that $x^* = S^*/N^*$, where S^* and N^* are equilibrium densities of the susceptible and total population respectively.

$$\begin{aligned} R_0 &= \frac{1}{x^*} \\ &= \frac{1}{1-p} \end{aligned} \tag{3.8}$$

Applications of these estimators for estimating R_0 in wildlife are hard to find, and note that they are inappropriate for making inference on host status as they are greater than or equal to one for all non-negative values of λ, L, A, or p. This is because these estimators for R_0 assume the system is in a steady-state with a non-zero prevalence – clearly the disease must be persisting, and hence R_0 must be unity or greater.

Assuming that the rate of conversion from the susceptible to the infected class is described by density-dependent transmission, βsi, with horizontal transmission only, a more general estimate of the basic reproduction number of the disease is given by Anderson et al. (1981):

$$R_0 = \frac{\beta S}{\delta + b + \gamma}, \tag{3.9}$$

where β is the transmission coefficient, b is the natural mortality rate, S equals the number of susceptible animals (that can be replaced by the density s), γ is the rate of disease recovery, and δ is the rate of disease-induced mortality. The latent period is assumed equal to zero. Host population dynamics assume exponential population growth, with the exponential rate of increase $r = a–b$, where a and b are the instantaneous *per capita* birth and death rates respectively. Many studies have estimated R_0 using Eq. (3.9) or variants of it. If host population growth follows the simple logistic model, the solution for R_0 is essentially the same, although S may be replaced by K (population carrying capacity), and a replaces b, and a disease latency period ($1/\sigma$) incorporated if required (e.g. Anderson et al. 1981; Pech and Hone 1988). Anderson and Trewhella (1985) used Eq. (3.9) to estimate the R_0 of *Mycobacterium bovis* infection in badgers (*Meles meles*) assuming generalised logistic growth. Equation (3.9) can be interpreted as one infected animal, equivalent to population density $I = 1/H$ (where H is the home-range area), making $\beta s/H$ infectious contacts per unit area per unit time for its infectious life expectancy $1/(\delta + b + \gamma)$, over an area H. This term for life expectancy (whilst diseased) assumes δ, b and γ are additive.

For the frequency-dependent approximation of the transmission process, the initial maintenance of disease is independent of the population size because the density of susceptibles is assumed to be equivalent to the population density, and occurs (May and Anderson 1979) when $\beta' > (\delta + b + \gamma)$. It follows that the basic reproduction number may be calculated (Roberts and Heesterbeek 1993; Heesterbeek and Roberts 1995) as:

$$R_0 = \frac{\beta'}{\delta + b + \gamma}. \tag{3.10}$$

A heuristic explanation of Eq. (3.10) is an infective individual meeting β' susceptible individuals per unit of time, and it does this for a period of $1/(\delta + b + \gamma)$ (Heesterbeek and Roberts 1995). Assuming local population density does not vary (and hence affect the contact rate), this expression for R_0 is considered to be independent of population size (De Jong et al. 1995). This is also the case if local population density does vary; however, individuals have a fixed number of infectious contacts per unit time (as may be the case for sexually transmitted diseases).

Estimating R_0 from Case Notifications If T_G, the mean serial interval between infections or the generation length is known and the rate of increase (r) of cases in the epizootic can be estimated, then the effective reproduction number during the course of the epizootic may be estimated as

$$R(t) \approx e^{r(t)T_G}, \tag{3.11}$$

providing there are not substantial heterogeneities in transmission. An estimate of R_0 can be obtained during the early phase of epidemic growth when depletion of susceptibles is insignificant. It is commonly assumed that T_G is simply the reciprocal of the recovery rate added to the latent period (defined as infected but not infectious). This assumes that infectivity is constant throughout the infectious period whose length is distributed exponentially – unlikely in practice but difficult to measure. A more realistic pattern, particularly of directly transmitted infectious diseases of animals, is for infectivity to peak early during the infectious period. Unfortunately, estimates of transmission rates and hence R_0 are highly sensitive to the assumed shape of the infectivity function and the associated serial interval – overestimating the serial interval overestimates R_0 and *vice versa*. If the form of the infectiousness function $\beta(x)$ is known then R_0 may be obtained by solving Lotka's equation (here modified to include the proportion of the population that is susceptible (s)

$$s\int_0^\infty e^{-ru}\beta(u)du = 1/R_0 \tag{3.12}$$

If the form of the infectiousness function is known or can be reasonably assumed, there have been recent advances in using case notifications to estimate the effective reproduction number of the course of a completely observed (Wallinga and Teunis 2004) or truncated (Cauchemez et al. 2006) epizootic. The method of Wallinga and Teunis (2004) is reasonably robust to substantial under-reporting of cases (see Caley et al. 2008), which will inevitably be the case except in captive populations of wildlife (e.g. see Miller et al. 2006). Where there is a prolonged though variable delay between infection and case diagnosis, methods of back-calculation may be used to reconstruct the infection curve and thus estimate disease transmission rates (e.g. Isham 1989).

Estimating R_0 from Epizootic Attack Rates If an epizootic occurs over a period of time short enough for births and deaths to be considered negligible and the population is reasonably well mixed, the final size equation (Diekmann and Heesterbeek 2000) describes the relationship between the attack rate (α – the overall proportion of the population infected), the initial proportion of the population that is susceptible (s_0) (not to be confused with the initial density of susceptibles $s(0)$), and R_0:

$$\alpha = s_0(1 - e^{-\alpha R_0}) \tag{3.13}$$

For known values of R_0 and s_0, an estimate of α is obtained numerically – predicting α may be of interest where a pathogen is being deliberately introduced into a population (e.g. bio-control). Alternatively, when estimation of transmission rates are of interest, rearranging Eq. (3.13) gives an expression for R_0:

$$R_0 = -\frac{\ln(1-\alpha/s_0)}{\alpha} \tag{3.14}$$

where α and s_0 are estimated with error (as will often be the case), the variance of the estimate can be approximated using the delta method. The final size equation for α has been shown to be robust to quite a range of spatial contact structures including spatially isolated patches of susceptibles (Ma and Earn 2006), although it breaks down if inter-patch coupling (movement of infectious individuals) is insufficient. A stochastic equivalent of the final size equation (Becker 1989; Yip 1989) has been applied to wildlife disease modelling (Hone et al. 1992), and has the added attraction of enabling straightforward estimation of the variance:

$$\hat{R}_0 = X\left[\frac{1}{X} + \frac{1}{X-1} + \ldots + \frac{1}{X-Z+1}\right]\Big/ Z \tag{3.15}$$

$$\text{var}\left(\hat{R}_0\right) = \left[X^2\left(\frac{1}{X^2} + \frac{1}{(X-1)^2} + \ldots + \frac{1}{(X-Z+1)^2}\right) + \hat{R}_0^{\,2} Z\right]\Big/ Z^2 \tag{3.16}$$

where X is the initial number of susceptibles and Z is the final number of individuals infected. The attraction of both approaches is their simplicity (see Box 3.3). The attack rate may be measured by the observed proportion of animals that die (in diseases with high case fatality rates or if the case fatality rate is known) or the proportion with serological or clinical (e.g. scars) evidence of infection at the completion of the epizootic. Estimating the proportion of animals that die is difficult in many situations as animals are cryptic at the best of times and carcasses are often difficult to locate. Where the attack rate is very high (near one), as was the case for some populations of European harbour seals (*Phoca vitulina*) during the phocine distemper virus epizootic in 1988 (Swinton et al. 1998), the precision of the estimated R_0 is poor.

Box 3.3 Classical swine fever (CSF) in wild boar – comparing estimators

The data set used (Inayatullah 1973) documents the number of wild boar (*Sus scrofa*) found dead on each day following the reported release of a single wild boar inoculated with CSF into a population inhabiting a forest plantation in Pakistan (Table 3.4). Prior to release, the number of wild boar in the population was estimated by a drive count at 465. In the days following the release, a total of 77 wild boar were found dead (Table 3.4). However, approximately 6 months later the population was estimated at 87. There is uncertainty as to whether as many as 379 (= 465 + 1 – 87) boar died from CSF during the epizootic, or whether the wild boar unaccounted for had simply moved out of the area (quite possible considering the forest plantation was only 44.6 km^2). Suitable methods for estimating R_0 using the data include the final size equation (Eq. (3.14)), the method of Wallinga and Teunis (2004) (assuming the time from infection to death has little variance) and trajectory matching.

Using the stochastic version of the final size equation, and assuming a case fatality rate of 90% (i.e. 9 in 10 wild boar that became infected died), then R_0 is estimated to be 1.1 ± 0.2 assuming 77 wild boar died, and 2.7 ± 0.2 assuming 379 boar died. In contrast, if we assume that the inoculated boar died 20 days following inoculation and the CSF infectiousness function is uniformly distributed between 5 and 20 days following infection (after Hone et al. 1992), then applying the method of Wallinga and Teunis (2004) estimates the effective reproduction number (R) of the early cases (what appears to be the1st generation) of the epizootic to be about 4 (Fig. 3.3). We would expect that at this stage the depletion of the susceptible population of wild boars would be minimal, and hence this estimate of the effective reproduction number would be close to R_0. The serial interval is uncertain, and if shortened would lead to a lower estimate of R_0, however, this would be inconsistent with the observed temporal distribution of cases.

All methods have strengths and weaknesses. The method of Wallinga and Teunis (2004) is independent of the epizootic attack rate and robust to consistent

Table 3.4 Observed deaths of wild boar in the days following the introduction of a single boar inoculated with classical swine fever. Adapted from Hone et al. (1992) (permission granted)

Days	Deaths
31	6
32	3
33	1
43	5
44	6
45	2
46	2
47	7
48	7
49	1
51	13
52	2
53	4
54	2
58	5
61	3
62	2
63	2
69	4

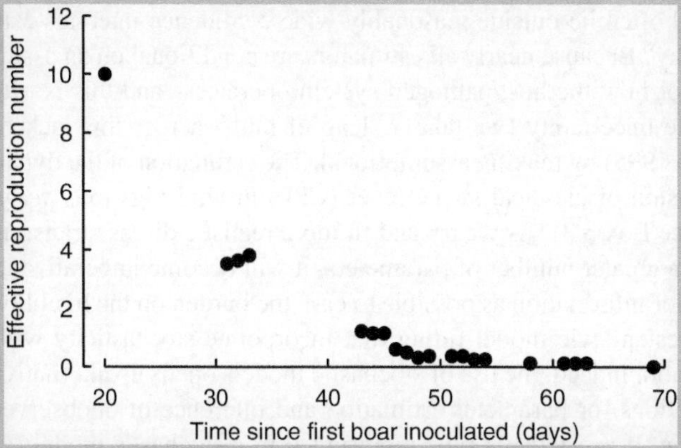

Fig. 3.3 Relationship between the estimated effective disease reproduction number and the day of carcass discovery for wild boars inoculated with classical swine fever

under-reporting of cases, but is heavily influenced by the assumed serial interval. Methods based on the attack rate are independent of nearly all parameters, but are strongly influenced by the assumption of the population being well mixed and the estimated attack rate being accurate. In this case the attack rate was based on the expected high case-fatality rate; doubts exist however,

(continued)

Box 3.3 (continued)

whether CSF causes uniformly high mortality in wild boar populations (Kramer-Schadt et al. 2007). Underestimating the attack rate will underestimate R and *vice versa*.

In summary, there is considerable uncertainty in our estimates of the R_0 of classical swine fever in wild boar – the method of Wallinga and Teunis (2004) strongly suggests a value of about 4, whereas methods based on the epizootic attack rate suggest an upper limit somewhere around 2. Can we reconcile the estimates? Yes, if we recognise that the wild boar population is structured into family groups with limited mixing between them, then it is quite possible for R_0 to be about 4, yet at a broader population level observe an attack rate consistent with a much lower value. It may well be that our assumption of homogeneous mixing is playing havoc with our estimation, reflecting uncertainty in how the system under study operates.

3.3 Dealing with Uncertainty

In a perfect world, different methods of estimating transmission rates should produce the same results. In practice this is rarely the case, and worse still, estimates using one method often lie outside reasonably wide confidence intervals estimated using another. Why? Because nearly all estimators are conditional on an assumed underlying model of how the host/pathogen system operates – and this is often subject to considerable uncertainty (we take a "leap of faith" across this lack of knowledge (McCallum 1995) by making assumptions). The estimation of the dynamics and rate of transmission of classical swine fever (CSF) in wild pigs (*Sus scrofa*) is a good example (see Box 3.3). As we try and fit more realistic disease transmission models containing a greater number of parameters, it will become imperative to incorporate as much prior information as possible to ease the burden on the likelihood functions. Hence Bayesian style model fitting that incorporate stochasticity will become the more common. Indeed, the use of stochastic models opens up alternative statistically rigorous options for parameter estimation and inference of unobserved features of the epidemic. If every event type represented in a stochastic model (e.g. infection, recovery etc.) were to be observed in a real epidemic, then it would be possible to construct a complete likelihood (based on this complete set of observations) from the definition of the model, and from which parameter values could be estimated as described in the examples above. However, in reality we typically have access to rather limited data; for example describing the prevalence or reported cases over time, and therefore we must infer not only the parameter values but also the missing infection (and other) events. Fortunately, it is possible to frame this problem in a Bayesian framework in such a way that the so-called posterior distribution of model parameters and missing events is known up to a normalising constant. Modern stochastic integration techniques such as Markov Chain Monte Carlo (MCMC) can then be employed to generate true samples from the posterior for increasingly complex models. The Bayesian framework also allows (and requires to some extent) prior information about the value, or possible range, of parameter values obtained from the literature or particular empirical studies, to be taken into account. The samples generated from the posterior allow the calculation of essentially any statistic of interest based on the parameters, and/or the missing events. For example, Streftaris and Gibson (2004) employed such methods to fit non-Markovian stochastic models for the transmission dynamics of a particular strain of foot-and-mouth disease (FMD) virus to data from a controlled experiment. In addition to transmission rates they inferred the hidden transmission history of the epidemic. Cook et al. (2007) used such techniques to estimate multiple transmissions rates within and between crop species in a spatial context, using information theoretic measures of deviance to show that the best–fitting model requires a fully parameterized transmission rate matrix; that is different transmission rates from species A to B and *vice versa.*

3.4 Assessing Host Status

Once the known host range of a disease has been established or extended there is a need to assess the role of these new hosts in the wider epidemiology of the disease. Assessing the host status in the epidemiology of a disease is crucial to its control (Caley and Hone 2005). Top of the agenda is determining whether the disease persists within a host population since all self-sustaining/persistent sources of infection (e.g. reservoirs) should be considered as part of a disease control strategy. Identification and quantification of transmission routes is often central to characterising the persistence of infection in wildlife populations. For example, the known host range of *M. avium* subspecies *paratuberculosis* has recently been extended to include a number of non-ruminant wildlife species (Daniels et al. 2003b). Of these new host species the European rabbit (*Oryctolagus cuniculus*) was identified as posing the greatest risk to sympatric livestock as rabbits are often abundant on livestock farms, they excrete high numbers of bacteria in their faeces and grazing livestock show no avoidance of rabbit faeces resulting in high exposure rates (Judge et al. 2005a). Given that paratuberculosis is a widespread and difficult disease to control in livestock populations and also has possible links to Crohn's disease in humans, the identification of a persistent wildlife source of infection would greatly impact on our understanding of current livestock control strategies. Judge et al. (2007) used a combination of field studies to quantify the rates of rabbit-to-rabbit transmission of paratuberculosis and mathematical modelling to show that infection can persist in rabbit populations for extended periods (see Box 3.4). This finding may go some way to explaining the persistent nature of the disease in livestock populations, and rabbits are now included in farmer led disease control strategies in the UK (e.g. The Premium Cattle Health Scheme).

Box 3.4 Persistence of *Mycobacterium avium* subspecies *paratuberculosis (Map)* in rabbit populations

European rabbits (*Oryctolagus cuniculus*) have recently been identified as a key wildlife species in terms of paratuberculosis transmission to the wider host community. Judge et al. (2007) tested the hypothesis that *Map* can persist in rabbit populations for extended periods of time. A spatially-explicit stochastic simulation model of a generic host-disease interaction was developed to quantify the inter-play between vertical and horizontal routes of transmission, needed for the persistence of *Map* in rabbit populations and to test the hypothesis. The model was parameterised based on empirical studies on rabbit population dynamics and on rabbit-to-rabbit routes of *Map* transmission. Predictions from the model suggest that *Map* persists in rabbit populations at all values of the horizontal and vertical transmission parameters in the range estimated from the field data (taken from Judge et al. 2006; see Box 3.1), and in many cases at all values within 95% confidence intervals around this range (Fig. 3.4). The persistence of *Map* infection in rabbit populations in the absence of external sources of infection suggests that they may act as a reservoir of infection for sympatric livestock. These findings, in combination with the ubiquitous distribution of rabbits in the United Kingdom and elsewhere, suggests that if *Map* becomes established in rabbit populations, they are likely

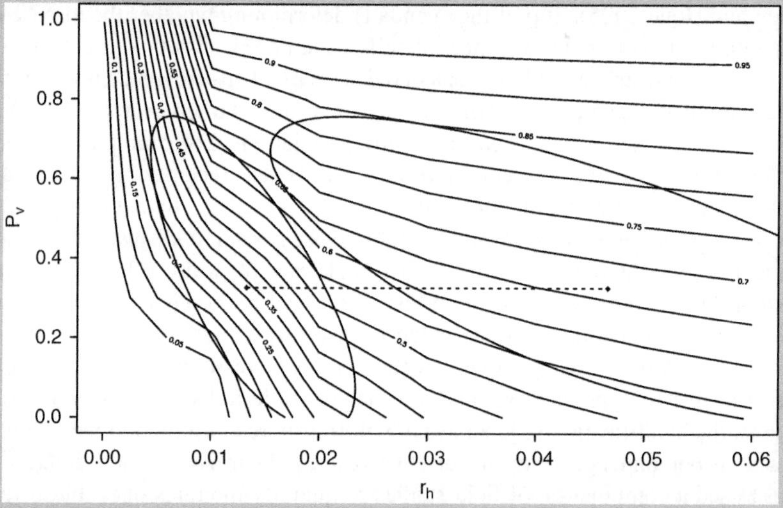

Fig. 3.4 Isopleths of equilibrium prevalence at differing vertical (P_v) and horizontal (r_h) transmission rates for a rabbit population approaching carrying capacity (dotted horizontal line is the estimated range of r_h values along the vertical transmission rate from field data). Ellipses represent 95% confidence intervals around the lower, and upper β estimates (from Judge et al. 2007)

to provide widespread and persistent environmental sources of infection, and thus a disease risk to livestock and potentially humans. Judge et al. (2007) conclude that where local rabbit populations are infected they should be included in any future paratuberculosis control strategies.

3.5 Implications for Management

Being able to quantify disease transmission and identify contributing factors is critical to not only evaluating management, but also designing management actions in the first place. Estimates of disease transmission coefficients are critically dependent on the assumed underlying model of transmission, and it is here that the greatest uncertainty is introduced. Where the mechanisms of transmission cannot be observed, or reasonably inferred by alternative means (e.g. disease pathology), data-based inference on the underlying mechanisms of transmission will need to be employed. This could take the form of critical experiments to identify routes of transmission. For example, Ramsey (2007) clearly demonstrated the importance of sexual transmission of leptospirosis (caused by *Leptospira interrogans*) in brushtail possums in a longitudinal experiment entailing castrating male possums to stop their sexual contacts. Likewise, Palmer et al. (2004) demonstrated the ability of *M. bovis* to be transmitted between white-tailed deer (*Odocoileus virginianus*) via contaminated feed. In doing so they overthrew the respiratory only paradigm of tuberculosis transmission in true reservoir hosts (Caley 2006).

Critical experiments needed to quantify the routes of transmission of wildlife diseases are typically difficult to undertake once let alone adequately replicate. Where critical experiments have not been undertaken, or are difficult to do, model selection techniques as applied to observational "experiments" may be the only way of (1) making inference on the underlying mechanism of transmission, and (2) estimating transmission parameters given a chosen model of transmission. Caley and Hone (2002) demonstrated how information-theoretic model selection techniques may be used to make inference on transmission routes by identifying how age-specific prevalence will vary as a function of age under different hypotheses. Miller et al. (2006) similarly used model selection techniques to demonstrate that a model that included indirect transmission of chronic wasting disease (CWD) amongst mule deer (*Odocoileus hemionus*) was the most supported model of transmission. These and other similar studies have greatly increased our understanding of the transmission of wildlife disease.

Chapter 4
Modelling Disease Dynamics and Management Scenarios

Graham C. Smith, Glenn Marion, Steve Rushton, Dirk Pfeiffer, Hans H. Thulke, Dirk Eisinger, and Michael R. Hutchings

4.1 Introduction

Mathematical modelling now plays an important role in developing scientific understanding of complex biological processes such as epidemics. Model-based risk assessments make such studies relevant to policy makers and resource managers. However, in providing such advice it is important to ensure that model predictions are robust to alternative plausible assumptions, and also that any predictions arising from such models correctly reflect the uncertainty in current knowledge and any intrinsic variability of the system under study. To see why this is so, contrast a point estimate of the efficacy of a given disease control measure with a prediction which gives the probability associated with varying degrees of success, and crucially, failure. The former gives a false sense of confidence, whilst the latter allows the decision maker to carry out a more complete risk assessment of the proposed strategy. In all cases, model predictions should be interpreted in the light of model structure and assumptions.

G.C. Smith
Central Science Laboratory, Sand Hutton, York, YO41 1LZ, UK

G. Marion
Biomathematics Statistics Scotland, The Kings Buildings, Edinburgh, EH9 3JG, UK

S. Rushton
University of Newcastle upon Tyne, Newcastle upon Tyne, NE1 7RU, UK

D. Pfeiffer
The Royal Veterinary College, London, Hawkshead Lane, North Mymms, Hatfield, Hertfordshire, AL9 7TA, UK

H.H. Thulke and D. Eisinger
Helmholtz Centre for Environmental Research – UFZ, Permoserstraße 15, D-04318 Leipzig, Germany

M.R. Hutchings
SAC, West Mains Road, Edinburgh, EH9 3JG, UK

R.J. Delahay et al. (eds.), *Management of Disease in Wild Mammals*,
DOI:10.1007/978-4-431-77134-0_4, © Springer 2009

In this chapter we are going to look at how modelling should be used to investigate disease in wildlife, with a strong focus on using models to make management decisions. We will largely avoid the vast area of purely theoretical modelling and concentrate on finding practical solutions to real world problems. Nevertheless, it is worth noting that theoretical analysis can, and has, provided profound insights into key aspects of system behaviour. One notable and particularly relevant example is the extensive range of theoretical results showing the importance of the basic reproduction number (R_0) as a threshold parameter in epidemics, starting with the work of Kermack and McKendrick in the 1920s (see Kermack and McKendrick 1991). However, this chapter is not about complex mathematics, but about defining the types of models available, describing the pros and cons of different approaches, and helping managers to determine the strengths and weaknesses of each in particular circumstances.

The objective of this chapter is not to turn the reader into a modeller, so there is no need to have a high level of understanding of mathematics, but to appreciate how models should, and should not, be used and interpreted. We will not give an exhaustive description of types of models, but concentrate on more commonly used approaches. Any model that is used to propose a management decision needs to be critically examined and our objective here is to give an understanding of modelling terminology, and the tools with which to question the model and the modeller.

4.1.1 What Is a Model?

As a matter of definition, no model is right in the philosophical sense of representing truth, but some models may be useful. All models are a simplification of reality. George Box (1979) stated "*All models are wrong – but some are useful*" and Oreskes et al. (1994) wrote "*... the establishment that a model accurately represents the 'actual processes occurring in a real system' is not even a theoretical possibility*". Models are simplified logical constructs of what we believe to be true, and in the context of this chapter are used to explain disease patterns in space or time, and to predict their future patterns. We construct models in our minds all the time, for example to assess the likely traffic flow of alternate routes on our way home or which queue to join in the supermarket. We know that these models only work in limited circumstances, and we should be equally willing to accept this as true for mathematical models, which are the focus of this chapter. The only real difference with these conceptual models is that mathematical models are a *formal* abstraction of our thought processes expressed in terms of a series of equations. Indeed, the act of constructing such a model, forces us to consider the problem in detail, and in a logical fashion. A mathematical model is simply an extension of a conceptual model into a mathematical framework.

Quantitative modelling activities can be broadly categorised into statistical (data driven) and mathematical (knowledge driven) models. An assessment of the strength of the inferences that can be drawn from these approaches needs to take this into account. Data-driven modelling uses statistical approaches to derive quantitative relationships from datasets. One strength of these models is that they are based on directly measurable factors, but this is also a potential weakness as these factors are usually proxies for underlying biological processes that cannot be directly measured in a field study, although statistical approaches can be used to infer the value of unobserved factors indirectly from the available observations. Such models can be used to generate knowledge in relation to cause–effect relationships (see Box 3.2), but they are not usually dynamic and only predictive in a limited domain determined by the range of the data used to construct them; extrapolation of statistical models is perilous indeed. Mathematical (knowledge driven) models can be analytically tractable or simulation-based. It is often advantageous to express even simulation models in terms of a formal mathematical description (e.g. differential equations or stochastic processes i.e. processes with a random element) for a number of reasons, including clarity in model definition and independence of the model from a particular implementation (i.e. easier translation to different simulation software, which is useful for model verification). Such models are based on existing understanding of the biological relationships within a system, and in principal, to the extent that such knowledge is correct, can be used to extrapolate predictions to novel situations. The strength of knowledge-driven modelling lies in the ability to represent the dynamics of complex biological systems. For an infectious disease this is particularly important as propagation of infection is inherently a time-dependent phenomenon, in that the number of new infections at a particular time depends on the number of infectious and susceptible individuals at preceding points in time. It is often possible to identify key factors within such a system. This group of models may include factors or quantities that cannot be directly measured, and are defined on the basis of measured proxy variables or hypothesised relationships. Such models may also have 'emergent properties' that are the unexpected result of the interactions between multiple effects represented in the model. Knowledge-based models can also be used to test the impact of changes in a system. Some of the information included in developing knowledge-driven models is generated through data-driven models, and in some cases the distinction is further blurred by the use of statistical methods to infer parameters in knowledge-driven models, but such models may also include expert opinion. In the rest of this chapter we will concentrate on these knowledge-driven (mathematical) models.

4.1.2 Why Model?

The question 'why model?' is important to address. If models are simplifications of reality and are always wrong then why bother? The answer hinges on the extent to which we understand the disease we are looking at. Models are a reflection of our understanding of the 'real world' in that they provide a structure in which to consider the complex biological interactions within a disease system. They allow us to explain or predict effects, whether these are the result of interactions of numerous factors or emergent properties of a system that evolve over time. Different approaches can be used to categorise them, one being to group them into qualitative and quantitative models. The models discussed here are all quantitative (knowledge-driven) models. Models are needed where mental simulation is not able to represent multiple causal links within a system (Lempert et al. 2003). According to Klein (1998) the limit is usually reached with three key variables and six transitions from one state to another. If we do not have full understanding of the underlying mechanisms and processes of disease, then modelling can allow us to investigate how the disease system as a whole functions. It can also reveal how weak our understanding is. This can be used to direct research to gain knowledge on the disease and the ecology of the host. Conversely, when we have more knowledge it can also be used to reveal how the disease system will respond to management interventions and to compare different approaches. Modelling thus provides either a strategic tool for increasing our understanding of disease or a tactical way of dealing with it. From this it should be apparent that we do not subscribe to the view that models should not be constructed until all suitable data are available. Rather the act of model construction itself forces us to formalise our ideas on the processes and mechanisms that we believe occur in the system under study; a process which often yields valuable insights.

Here we focus on the development of models that can be used to explore disease control strategies. As such, the models should capture the biological processes driving the disease and be able to simulate some management intervention whether that be at the host population level (e.g. population reduction) or the individual host level (e.g. restrict animal behaviour to prevent exposure to disease).

We can also use the output of models to inform field research. From the results of sensitivity analysis (see Section 4.4) we can extract those parameters that have a large influence on the output. In particular, we should differentiate between parameters that have known variability, and those that have uncertainty. Once we have identified the important uncertain parameters with significant influence on the system, we can then use this list to decide on research priorities for further data collection. This is discussed further in Section 4.4.

4.2 Basic Approaches

The modelling process usually starts with a question and is followed by the development of the underlying biological framework, mathematical model development, model testing and ultimately, predictive modelling to answer the question.

4.2.1 Approaches to Modelling

If a model is analogous to a scientific experiment, then the original question to be asked of the model is analogous to the hypothesis to be tested. There are three basic steps in constructing any model. These steps are so basic we often overlook the first two. But only after answering all three can we start to choose what sort of model would be most useful.

1. What is the question we wish to answer and are we aiming to increase our understanding of disease or develop methods for control?
2. What is the scope of the problem we wish to include?
3. What is our understanding about the mechanisms under study?

The first question we have to answer is what are we modelling for? Do we wish to use modelling conceptually or strategically, to increase our understanding about the underlying disease, or do we want a practical or tactical model for disease management. Emerging diseases such as the recent introduction of bluetongue virus to Northern Europe are a good example of where we might want to use modelling to investigate the potential spread of disease among wild and domestic ruminants, or identify key factors in the disease process that we do not understand in its 'new context'. Investigating the dynamics of bluetongue and its interaction with its vector species in the north European countryside with a strategic model would be a first step to understanding the magnitude of the problem for Northern European countries. Recent models have highlighted the possibility of bluetongue spread in northern Europe by climatic matching of vector species (*Culicoides* midges) with recent records of disease (Purse et al. 2007), demonstrating the potential involvement of novel midge vectors. Where we have greater understanding of the disease process, such as in rabies in wildlife, or bovine tuberculosis, then a tactical modelling approach would be more appropriate. In most cases the strategic versus tactical argument is easily defined at the outset, usually on the basis of our underlying knowledge of the disease.

It is impossible to undertake modelling without a context. This is fundamental to producing a useful model, and we will look at rabies in wildlife to illustrate the process. A request to "model the dynamics of rabies in wildlife" may be interpreted differently by almost every modeller. In this simplified example, no particular output is requested; so one person may model genetic changes to the virus over each viral generation, while another may construct a multi-species model of the evolution of rabies over centuries. The question needs to be carefully constructed, and a good modeller will help the manager to define the question. Even extending the question to "what is the best way to eradicate rabies from a focal outbreak in a naive population of red foxes?" does not precisely define the question. Do we mean the quickest, or most cost effective? Clearly by now we must realise that the question needs to be asked in such a way that it specifies the answer we expect to get from the model.

We move now to the second point, which is one of scale. In an isolated population, such as red foxes (*Vulpes vulpes*) in Britain, geographical scale has a maximum bound imposed by the surrounding sea. But should we model all the foxes in Britain, or just those within some distance of the outbreak? There are an estimated 240,000 foxes in Britain (Harris et al. 1995), so at this scale we could probably assume the population is infinite. But if in the case of an outbreak of rabies in the Ethiopian wolf (*Canis simensis*) with an estimated total world population of 500 individuals (Randall et al. 2004), then the issue of low numbers and chance events arises. Also, we need to examine the temporal scope of a question. For example, we may wish to ensure the survival of the Ethiopian wolf over decades or centuries. Consequently, with respect to the threat of rabies, the only way to ensure this is to eliminate the disease completely from the domestic dogs (*Canis lupus familiaris*) in the surrounding area.

Lastly we need to specify the mechanism we are interested in, and indeed what mechanisms we are not interested in. Biological systems are effectively hierarchical. At the lowest level we have underpinning biochemical processes such as the level at which individual drugs act in a veterinary context (e.g. Acetylcholinesterase inhibitors). Above this we have whole organ responses (e.g. renal failure) and above this whole animal responses at the level of the individual (e.g. alterations in behaviour, morbidity and mortality). Going higher still we come to inter-individual interactions (e.g. disease transmission) and to population level behaviour where we come into the realms of epidemiology. At even higher levels we have multi-species epidemiology and then the effect on food webs and biodiversity. Clearly, it is not practical to model at all levels in this biological hierarchy, so whilst we might be interested in looking at developing a model system to investigate the spread of bluetongue or rabies, we would not want to be involved with modelling the biochemistry of the immunological response to exposure to the infective agent. One could argue that a key component in the historic success of modelling in epidemiology is the assumption that the complex processes occurring within an individual whose immune system has been challenged by a pathogen can often be adequately (e.g. for the purpose of population and community epidemiology) summarised by a series of transitions between a small number of distinct states (e.g. Susceptible, Exposed, Infected and Recovered: SEIR), despite the true internal state of the individual being more precisely described by something closer to a continuous range. However, if the key concern is the behaviour of a diagnostic test applied to individuals then a model of within-organism response may be more appropriate. In the context of rabies spread we are interested in inter-animal transmission and subsequent impacts on disease spread. In most cases, we would not model individual animal behaviour and we might not need to consider age or the sex of the animal (or just consider females). But how sure are we on issues such as sex- or age-biased infection? Generally we should only add model components (e.g. age or sex) when there is evidence that they impact on disease dynamics either directly, or indirectly. Note however, that they may impact on the management of disease even if they have no significant impact on (unmanaged) disease dynamics.

When constructing a model it is necessary to choose what to incorporate, and crucially what to leave out. Some elements of the model are dictated by the goal of the study, for example understanding the dynamics of sexually transmitted diseases is likely to require modelling both sexes! However, decisions whether or not to treat certain aspects come down to pragmatic considerations including current knowledge, resources and the data available for the project. As a general rule, models should be as simple as possible to describe the phenomena of interest (but not too simple). A commonly encountered problem when modelling biological systems is the explosion of model complexity, leading to poorly understood model behaviour and potentially low predictive ability due to over-fitting of the available data. In statistical models information theoretic approaches (e.g. AIC) are widely used to formally control the growth of model complexity. Pragmatically, model complexity can be limited by developing models for a particular purpose and incorporating only those features that are critical to that end. Like a map, models are an abstraction of reality and are at their best when they incorporate the appropriate level of detail, as too much detail can obscure the most important features. Of course, not enough detail means that the model may not be able to achieve the original goal, but this may just be an accurate reflection of current knowledge.

Thus, having defined the question, the scope, and our understanding of the system, we now need to decide how to model it. In mechanistic terms models can be classified in different ways on the basis of how they are constructed mathematically. Again we have three decisions to make. Should the model be continuous or discrete in time, spatial or non-spatial, and deterministic or stochastic.

The first is often a matter of personal choice. Continuous time models are usually differential models or stochastic processes, which are generally preferred by mathematicians, while discrete time models are difference equation models, which are generally preferred by biologists. Indeed there are often clear biological reasons for choosing a discrete-time model, for example in modelling populations with highly synchronous (e.g. annual) reproduction. However, it is important to note that time related parameter values (e.g. for birth and survival: are rates in continuous models and probabilities in discrete time models) need to change between these two models. A mortality rate/probability (s) in a discrete time model of interval length t, is related to the continuous-time differential equation mortality rate (μ) by

$$\mu = -\ln(s)/t \tag{4.1}$$

where discrete time models are chosen it is important to consider the time step used. This is generally set to one of the shortest events that occur in the system. With simple models (e.g. numerical solutions to differential equations) it may be possible to check that the time step is adequately short by making it shorter and checking that no differences occur in the output. However, this is not practical for most models. A simple way to determine if the time step is too long is if too many competing events occur within one step (e.g. if both primary and secondary infections could occur within one step). For example, in discrete-time models of rabies dynamics, the time step is often about one month (the average incubation period of rabies) on the assumption that the period of infectiousness (a few days) is regarded as an instantaneous event. However, in this case the number of individuals which are rabid on any one date will not be recorded since infectiousness is always followed by death and thus these individuals will be removed from the model. Such considerations typically do not arise in continuous time models, although algorithms used for numerical solution of differential equations will typically determine a short time-step to be used, this is relatively automatic and does not require any reformulation of the model. It is often the case that continuous-time models are structurally simpler, and thus conceptual whilst discrete event models are more complex and designed to be more practical. However, even most moderately realistic continuous-time models are not solvable using present-day mathematics, but nonetheless there are a range of approximate mathematical approaches that can provide valuable insights into model behaviour.

Another crucial, yet often over looked aspect of modelling disease dynamics is the issue of waiting time distributions. Consider an individual that becomes infected with say a virus, the *latent period* is the time it takes for the virus to become established and the individual to become infective i.e. to start shedding the virus. The latent period will vary between individuals, but across the population is described by a latent (waiting) time distribution. Particular waiting time distributions typically describe other transitions e.g. from infected to recovered or susceptible. The details of such waiting times are crucial in determining disease dynamics, for example at the start of an epidemic, particularly for emerging diseases, uncertainty about the latent period can result in large uncertainty in the predicted size of an outbreak: HIV-AIDS and variant CJD being two notable examples in humans. Models may fail to account correctly for waiting times due to a lack of information, but also because widely used mathematical formalisms such as deterministic ordinary differential equations, and stochastic Markov processes are based on exponential waiting times. However, such shortcomings can be addressed and, for example using individual-based stochastic methods, it is relatively easy to account for any required waiting time distribution.

It is essential to be aware that models should strive to capture the key ecological processes that drive disease dynamics, and be capable of including proposed management options.

4.2.2 Deterministic or Stochastic

Most of the early mathematical modelling of disease in human, livestock and wildlife populations was undertaken with continuous time – differential models. These modelling approaches were based on calculus, originally developed by Isaac Newton. This approach is deterministic in the sense that, for any set of inputs to the series of equations used in the model, the output is determined and fixed. This approach was used to predict the motions of the planets around the sun. However, one of the most obvious features of biological systems is that they show inherent, but often unexplained variation. Many biological systems, including epidemics, exhibit a high degree of variability. For example, the introduction of a single infected animal into a population may or may not result in a disease outbreak, and the size of any resultant outbreak will likely vary between populations. This leads to the adoption of stochastic modelling methods in which there is randomly induced variation between different model-runs (even where all parameters are held constant). In theory, the output of such models is represented by a probability distribution, which can be estimated from simulation, as a histogram across many model runs. Therefore stochastic models are more computationally demanding than deterministic models. However, a key advantage of this approach is that it represents variability parsimoniously with relatively few parameters and without necessarily increasing the number of variables needed to represent the state of the system. Since for every iteration, model output will be different, many (hundreds or thousands of) iterations are required for stochastic models to produce a representative distribution of possible outcomes. A major advantage of stochastic models is that they are able to capture emergent properties of a system arising from stochastic or rare events. This ability may be considered especially useful when quantifying the effects of population reduction as a means of disease control. Box 4.1 describes an example where the use of population reduction is explored as a means of paratuberculosis control in rabbits (*Oryctolagus cuniculus*). The observation that the success of single one-off population reduction events comes largely, not from the probability of removing all of the infected individuals in the population, but from the failure of the disease to spread from the infected animals that remain, highlights one advantage of stochastic over deterministic models.

Deterministic models are therefore best used early in any biological investigation, to improve our understanding of the processes being modelled. They can lead to useful insights, but generally stochastic models are more valuable if the objective is to make a management decision. It is important to realise that the introduction of stochastic effects into a previously deterministic model can alter predicted outcomes both quantitatively and qualitatively. For example, with low levels of stochasticity model outputs are typically distributed around a mean value corresponding to the deterministic model prediction (the stochastic model predicts system mean *and* variability). At intermediate levels of stochasticity the mean prediction of the model is typically different to the deterministic case, and where stochastic effects dominate they can drive a transition not observed in the deterministic model (e.g. stochastically induced disease extinction where the deterministic model predicts disease persistence). Stochastic effects are typically most important for relatively small populations (or sub-populations), however heterogeneity can also amplify stochastic effects.

4.2.3 Non-Spatial Models

Non-spatial models were the first to be developed, and generally treat the whole population as homogeneous: without having to consider space or any social interactions, they are relatively simple. Such homogeneous mixing models based on differential equations are relatively amenable to mathematical analysis, although typically most recent models are not solvable mathematically. Nonetheless, mathematical analysis of such models has led to important insights into system dynamics and this is where many of the theoretical developments in epidemiology have been produced. These developments have included insights based on R_0, the average number of new infections that a single infectious animal will produce during its "infectious lifetime" when placed in a completely susceptible population (see Box 3.3 on estimating R_0). This ratio depends on the density (and other factors, such as spatial organisation and behaviour which are essentially ignored in non-spatial

Box 4.1 Modelling population reduction to control wildlife disease: rabbits and paratuberculosis

Reduction of wildlife population density is a common method used to control wildlife disease. Given the financial and logistical difficulties in experimentally testing the efficacy of wildlife control programmes, modelling is often employed to explore wildlife population reduction as a means of disease control. Here, stochastic modelling offers a significant advantage over deterministic modelling as it can capture both the likelihood of the population reduction event removing all the infected animals and also the probability that stochastic fluctuations prevent the persistence and subsequent recovery of the infected population following population reduction. This was demonstrated by Davidson et al. (2008) when modelling the control of paratuberculosis (*Mycobacterium avium* subsp. *paratuberculosis*; *Map*) in rabbit (*Oryctolagus cuniculus*) populations. They used a spatially-explicit stochastic simulation model of *Map* dynamics in rabbit populations to quantify the effects of rabbit population control on disease persistence. The model was parameterized from empirical studies on rabbit population dynamics and on rabbit-to-rabbit *Map* transmission. Single population reduction events targeting up to 96% of all individual animals did not result in any noticeable chance of disease extinction, while culls at the (even more) unrealistically high levels of 98% and 99%

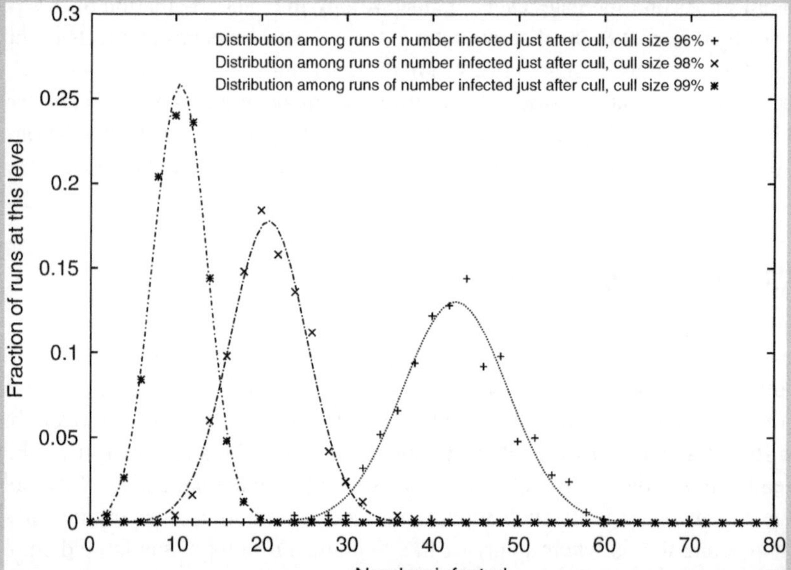

Fig. 4.1 The distribution among realizations of the number of infected individuals immediately after a cull event. Gaussian curves have been fitted to the data points. Three levels of culling are shown – 96%, 98% and 99% (from Davidson et al. 2008)

of the population yield disease extinction probabilities of just 0.08 and 0.34 respectively. These results can also be seen in Fig. 4.1, where the distribution of the number of infected animals among simulations *immediately* after a population reduction event is shown. Although no simulations were disease-free straight away, even with the 98% and 99% culls, many simulations were left with a small number of infected individuals (e.g. a mean of 11 infected individuals in the case of a 99% cull), which resulted in chance eradication in the subsequent recovery period due to small populations being more susceptible to stochastic effects.

The study demonstrated that high rabbit population reduction levels (greater than 96%) are necessary if a one-off rabbit cull strategy is to have even a small probability of eradicating the disease. At these high reduction levels the main contribution to this small eradication probability emerges not from the probability of removing all infected individuals at the cull (which is highly unlikely), but from subsequent fluctuations while the disease remains for a short time with a reduced incidence brought about by the cull i.e. the failure of the infection to spread post-cull. This effect can only be captured with a stochastic model.

homogeneous-mixing models) of the host population. If the value is above unity then the disease will *probably* produce an epizootic, whereas a value of less than one means that the disease will die out, although stochastic factors may mean that a relatively large number of animals will become infected beforehand.

Other important insights include estimating the threshold density at which the disease will die out (i.e. when *R* drops below unity), or the proportion of a population that needs to be vaccinated to eliminate the disease (i.e. an alternative way for *R* to drop below unity). A large literature exists which presents the mathematics of disease dynamics (see for example Anderson and May 1991) and simple disease modelling is mentioned in most ecological texts. Whilst these models have generic value in understanding disease dynamics in systems that are adequately captured by the free-mixing assumption, they are less useful where the animal-pathogen system shows heterogeneity. This heterogeneity can occur at different levels which range from differences in the susceptibility of individual animals to disease, variations in the pathogen, or more commonly, heterogeneity that arises from the distribution of hosts in time and space and determines levels of population mixing (e.g. territoriality in some wild mammals). In addition, animal behaviour may change as population density is reduced through interventions such as culling (Chapter 7), which gives rise to the more correct concept of a threshold contact rate, rather than a threshold density (Sterner and Smith 2006). Early non-spatial homogeneous mixing models assumed linear density dependence in transmission as density is reduced (often referred to as βSI). A refinement assumed instead a fixed contact rate between individuals regardless of density (referred to as $\beta S(I/N)$), such as may occur with sexual contacts, since

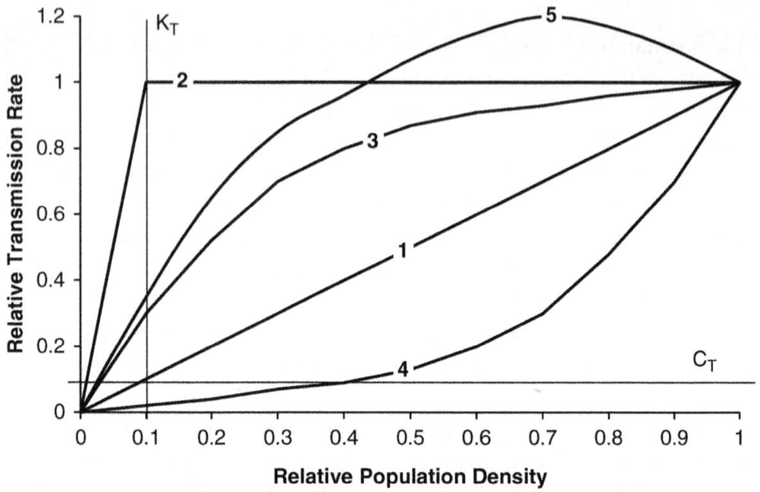

Fig. 4.2 A variety of relationships have been hypothesised between (susceptible) host density and the transmission rate. The earliest (line 1) is linear density dependence (βSI), which gave rise to the idea of a threshold density (K_T) below which the disease will die out since R_0 declines to below unity. Alternative relationships include frequency dependence (line 2), which technically cannot be a constant since the relationship must go through the origin, convex up (line 3) and convex down (line 4). Line 5 represents social perturbation discussed in detail in Chapter 7. From these lines it can be seen that K_T should be replaced by a critical contact threshold (C_T), although this is much harder to measure

these do not normally scale with density. However, empirical evidence often suggests a non-linear response with density (Caley et al. 1998; Ramsey et al. 2002). These relationships are shown in Fig. 4.2.

Non-spatial models have come a long way since simple linear relationships, and can now adequately model spatial heterogeneity (e.g. Barlow 2000), although the mechanism causing this non-linear relationship is not specified. However, by using a simple model, Keeling et al. (2000) showed that such non-linear relationships could be interpreted as the effects of spatial heterogeneity. Box 4.2 illustrates a widely used approach in which a deterministic model is constructed (using a technique known as closure) as an approximation to a fully stochastic and spatial model in a manner that captures some spatial effects. The accuracy of such pseudo-spatial models should be tested, but they are typically an improvement on non-spatial homogeneous-mixing models also illustrated in Box 4.2. Closure-type approximations of this type have been used to capture spatial effects in a computationally efficient manner, for example to model the UK 2001 Foot-and-mouth epidemic (Ferguson et al. 2001b). It is also worth noting that it is often possible to identify cases, e.g. high rates of migration or cross-infection, where deterministic homogeneous mixing models, moment-closure type approximations, and stochastic spatial models coincide. Such limits are useful both in developing understanding and in verifying correct implementation of models.

Box 4.2 Assessing the importance of stochastic and spatial effects in determining disease risk exposure in grazing systems

Many of the most pervasive disease challenges to livestock, and other herbivores, are transmitted via the faecal-oral route, from mycobacterial pathogens such as *Mycobacterium avium* subspecies *paratuberculosis* (the causative agent of Johne's disease) (Judge et al. 2005a), to nematode parasite infections such as *Haemonchus contortus* and *Teladorsagia circumcincta* (Hutchings et al. 2003). Marion et al. (2005, 2008) developed an agent-based modelling framework, based on a series of empirically observed rules of thumb, governing the grazing and faecal avoidance behaviour of grazing animals, which can be used to assess disease risk to livestock from faecal contacts. The key features captured by the model are (i) animals only have limited local knowledge e.g. they can visually assess swards from some distance but only smell faecal contamination at short ranges; and (ii) there is a trade-off between faecal avoidance and the desire to maximise intake which controls the risk of exposure to faecally transmitted disease.

Marion et al. (2005) demonstrated how to develop an analogous non-spatial deterministic model which ignores both spatial and stochastic effects. In the limiting case of large movement rates these models give equivalent predictions. However, their comparison is useful in quantifying the importance of stochastic and spatial aspects of the model. Not only is this useful in developing a better understanding of the system at hand but for example, if the two models agree then it would make more sense to use the deterministic version which is simpler, quicker to run and potentially more amenable to analysis. In general, deterministic models can be thought of as differential equations for the mean value of quantities in the stochastic case. However, a formal mathematical derivation of equations for such mean values shows that they depend on higher-order statistics of the stochastic model like variances and co-variances, which are simply ignored in deterministic models. The only exception to this is a completely linear model, but biologically plausible models will usually contain some non-linearity in which case the deterministic model is not guaranteed to match the mean of the stochastic model. Unfortunately, although it can usually be easily simulated on a computer it is typically not possible to solve the stochastic model analytically, although various approximate methods are available. For example, so-called closure approximations that attempt to model both mean values and some higher-order statistics, such as variances and co-variances, have been widely applied in epidemiology. Figure 4.3 shows the total intake rate across all animals for continuously occupied pasture versus the density of animals (stocking rate) for each of the three model formulations: deterministic; stochastic spatial; and moment-closure based. The peak-value identifies the optimal stocking density and comparison of the three curves shows that both the deterministic and moment-closure models underestimate the optimal stocking density and overestimate the associated intake rate, in comparison with the stochastic spatial model.

(continued)

Box 4.2 (continued)

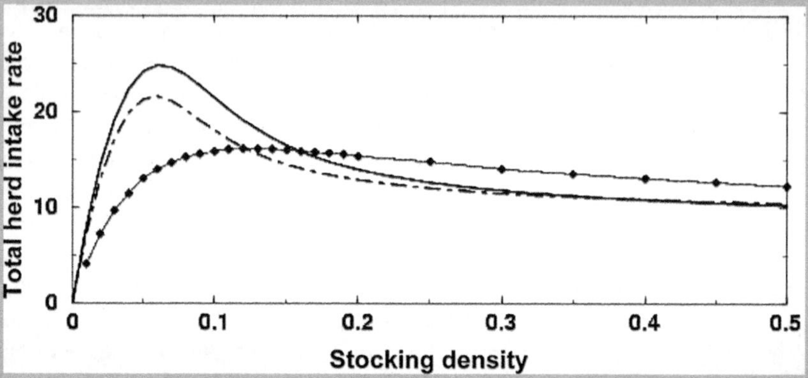

Fig. 4.3 Total intake rate across all animals versus stocking density for non-spatial deterministic (solid curve), stochastic and spatial (black dots), and moment-closure based (dot-dashed curve) models (taken from Marion et al. 2005)

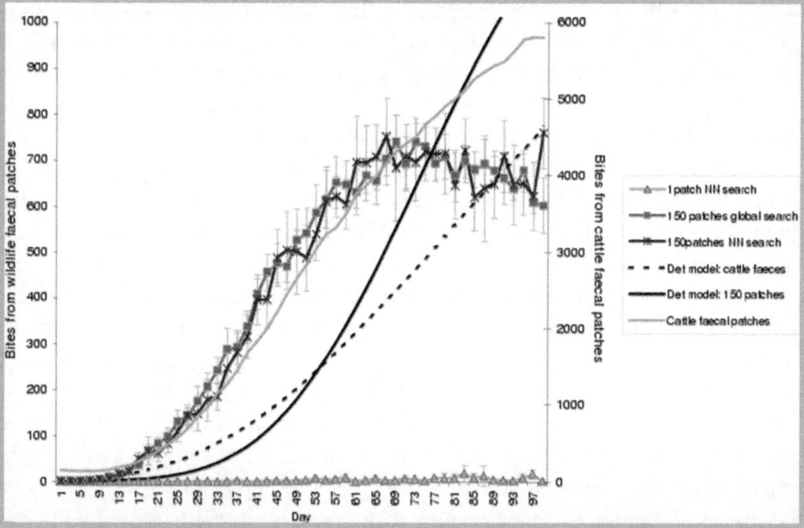

Fig. 4.4 Effect of wildlife faecal defecation pattern ('1 patch' corresponding to a latrine & '150 patches' to dispersed faecal distributions) and search distance (global and local, or NN) on the number of bites taken by cattle from wildlife faecal contaminated patches (mean over 10 realisations +/−1 standard deviation for the stochastic model). The deterministic model outputs relate to the case of dispersed wildlife faeces as indicated. The scale on the right refers to the daily bite rate on livestock faecally contaminated patches (with standard errors omitted for the stochastic outputs) (taken from Marion et al. 2008)

Marion et al. (2008) extended the original model to explore the risk of exposure to faecally mediated disease, now comparing the spatial stochastic version of the model to a non-spatial deterministic model parameterised to represent a set-stocked scenario in a temperate beef herd. Figure 4.4 shows the bite rate from dispersed and highly clumped (representative of wildlife latrines) distributions of wildlife faeces, and from a dispersed distribution of livestock faeces for the first 100 days after cattle are turned out into the pasture. The search distance of herbivores is currently unknown and difficult to measure (Phillips 1993), and therefore the robustness of their conclusion to this poorly determined parameter was assessed. For the case of dispersed patterns of wildlife faeces, Fig. 4.4 contrasts strictly local searching, and global searching in which moves are simulated (at least potentially) across the entire pasture. Similar robustness to search distance was observed for both wildlife latrines and livestock faeces. In addition the results show that the non-spatial deterministic version of the model initially underestimated disease risk and crucially predicts the peak in risk much too late (the deterministic model predicts a peak well after the 100 days shown in the figure).

4.2.4 Spatial and Network Models

It has been a useful starting point to model disease spread within a population by assuming that all individuals within it mix evenly with each other (variously known as mean-field, mass action, homogeneous or complete mixing). However, spatial heterogeneity in wild mammal populations is an important determinant of contact patterns between individuals, with potentially profound implications for disease dynamics (Chapter 2). Hence, the introduction of heterogeneity in modelled contact structures typically produces a more realistic model. The impact of spatial heterogeneity, where contact patterns between individuals is a function of the distance that separates them, has been studied extensively. Another scenario of spatially heterogeneous contacts is where individuals mix uniformly with others in some localities (e.g. within a group territory), but only occasionally make longer-range contacts (e.g. with residents of other territories). Where further information is available it is sometimes possible to develop more detailed models of the network of contacts between individuals (e.g. social and sexual contacts: see Box 2.3), and indeed spatial contact processes themselves can also be thought of as a special class of contact networks. This duality is most simply demonstrated by a lattice where hosts are placed at the nodes and connections are allowed only to the four nearest neighbours; a structure can be viewed as a spatial neighbourhood model and a network. The mathematical study of networks is a rapidly developing field with recent results demonstrating the profound impact that different contact network structures have on disease dynamics (Chapter 2). For example, Pastor-Satorras and Vespignani (2001, 2002) showed that for a contact network with a power-law distribution (i.e. a few individuals have very many contacts) there is no critical threshold density below which the disease will be eradicated. However, in such circumstances targeting highly connected individuals is an effective strategy. Durrett (2007) however, suggested that such extreme power-law contact networks are rarely observed in practice. Nonetheless, given the possibility for such profound effects, it is important not only to study the effect of observed network structures, but also to explore the robustness of any results obtained with respect to uncertainties in such contact structures. There is a growing literature on the estimation of complete contact networks from (inevitably) partial observations. Where data such as the mean observed number of contacts, or higher-order network statistics, are available, likelihoods can be constructed for so-called exponential random graph, or p^*, models (Frank and Straus 1986), and then computational statistical methods can then be used to generate complete networks consistent with available data (Handcock and Jones 2004). Closure type approximations have also been applied to epidemics on networks (Boots and Sasaki 1999; Keeling 1999) and can make direct use of measured network statistics without the need to model the network directly (Keeling and Eames 2005).

Models can be further refined by introducing variation between individuals, or across space, for example by specifying variation in the susceptibility of individuals. Intuitively, such heterogeneity and variability tends to sub-divide the population into smaller, partially connected clusters, enhancing the importance of stochastic effects in the spread of disease. Even in cases where there is no intrinsic or initial difference between sites or individuals, such heterogeneity in contacts interacts with stochasticity to generate heterogeneity within the system, often resulting in qualitative differences compared to the analogous homogeneous deterministic model. Such differences are typically smaller in situations where the contact network is highly connected and the population relatively homogeneous in nature.

The extent to which space is important as a modelled feature depends to a large extent on the mechanisms of disease transmission, how animals are distributed, or interact within it and the scale at which you model. If we are interested in investigating the dynamics of disease in a herd of dairy cattle then the animals can be considered to be homogeneously mixed, and to frequently come into close contact. In these cases the time domain for inter-animal contact is short relative to that of the transmission process. Disease spread under these conditions would not need to incorporate a spatial component. On the other hand, if we were interested in the spread of infection between herds and across farm holdings then space might be important. However, if two competing strains of disease were present, then space may even be important in the first example, since the homogeneity may be disrupted by the presence of the second outbreak. However, it is not certain if herds of wild mammals will mix homogeneously within any particular time frame.

Spatial models often provide a tactical context for managing disease. They can be used to simulate explicit/hierarchical contact networks between individuals, groups or sub-populations, even in specific geographical regions. In these models specific spatial locations are required, which may or may not be linked through Geographical Information System (GIS) software.

The use of spatial models has led to a number of findings that illustrate the importance of space in disease dynamics. Non-spatial badger (*Meles meles*)-TB models predicted substantial population reduction when the disease was present (Anderson and Trewhella 1985). The development of spatial models, along with a thorough analysis of the data (Delahay et al. 2000a), has allowed the simulations to become more accurate with a relatively small population reduction and a poor relationship between population size and prevalence (Smith et al. 1997). These effects appear to arise because the disease moves slowly between social groups, which themselves change in size, and thus the disease never reaches equilibrium in all groups. This has demonstrated the importance of territoriality for disease spread in wildlife. Early non-spatial models of rabies in foxes suggested that approximately 70% of the population needed to be vaccinated to eliminate rabies (Anderson et al. 1981), and this has been taken as the target level ever since. However, spatial modelling indicated that this threshold density may be too high (Eisinger and Thulke 2008), because local remnants of rabies infection are unable to spread in the vaccinated population and so die out. For infected foxes the probability of encountering a susceptible host is less than that predicted from the overall density because spatial structure results in less susceptible animals in their contact neighbourhood. Since disease spread is a local phenomenon, many of the local foxes will already have been infected.

Spatial models are often run on a grid (x, y coordinates) with each cell representing a unit area, or a territory. Animals are then assigned to each cell as required. Such grids have received some discussion, since in a square grid each cell may be considered to have either four (called a Von Neumann neighbourhood) or eight (a Moore neighbourhood) neighbours. In reality, territorial animals often have an average of about six neighbouring territories. The original reason for using simple grids was related to the limitations in computing power. However, it is now easy to combine small cells to produce territories (e.g. Smith and Harris 1991), or to utilise a modelling framework based on the underlying geographical structure (using a GIS). This not only allows spatial heterogeneity in territory size and shape, but has recently been shown to remove significant potential bias related to movement of individuals, disease or other information in regular model landscapes (Holland et al. 2007).

4.3 Parameterising Models

An area of considerable importance is the parameterisation of dynamic knowledge-driven models (see McCallum 2000). A given dynamic model often exhibits a range of interesting and plausible behaviours, which can be explored via analysis where possible, but more commonly via numerical simulation and sensitivity analysis. It is usually necessary to determine parameter values in order to apply the model to a particular system, and this is critical if the model is to be used tactically for quantitative risk analysis or management purposes.

The parameters in dynamic models are typically biologically meaningful and therefore have often been measured directly or inferred from empirical studies. The values of such parameters are often quoted as a mean and standard deviation, or in some cases as a distribution of values. If the parameter distribution is skewed (i.e. not symmetrical: e.g. lognormal) then quoting a mean and standard deviation can be misleading. In addition, inherent variability, for example between regions, sites, or even between times at the same site, often leads to apparently inconsistent parameter estimates between different studies. It is important to differentiate between inherent variation and uncertainty in parameter estimates. Mortality rates may be regarded as variable (between years or places), if a number of measurements are available. Disease transmission rates are often uncertain, since few studies have attempted to estimate or measure them, and they are often inferred.

Thus, there is usually uncertainty in our knowledge about some parameter values, which can be expressed as a range or probability distribution of possible values. In order for model outputs to reflect variation and uncertainty it is possible to randomly sample values for each parameter (independently) and then run the model for each set of parameter values thus generated. This builds up a histogram of model outputs reflecting the uncertainty. The computational cost of such a scheme can be reduced by employing an intelligent sampling scheme such as Latin Hypercube sampling (Vose 2001). However, since we typically know nothing about the correlation between model parameters, many parameter sets generated contain combinations of values that are unlikely. However, if system response data (e.g. the number of clinical cases observed over time during an epidemic) are available, a relatively limited, but rapidly developing, set of tools allows statistical parameter inference (i.e. unknown or poorly determined parameters can be estimated from data). Methods of Bayesian statistics can be used to combine both top-down system level data and bottom-up data on parameters, in which the distributions for each parameter are used. This process results in parameter distributions from which means, variances and *correlations* between different parameters can be deduced. Parameter combinations for which the model predictions are far from the observed data receive a correspondingly small probability.

In order to apply such methods it is necessary to define a *likelihood*, which, conditional on the model itself, determines the probability that the observed data was generated for each possible set of parameter values. In the case of stochastic models the true likelihood follows from the definition of the model and any assumptions about the accuracy of the observations e.g. Gaussian errors. Such an approach is arguably the most statistically rigorous, but is often difficult to implement and requires computationally intensive methods such as Markov chain Monte Carlo (MCMC). For example, Streftaris and Gibson (2004) use MCMC in a Bayesian framework to fit stochastic models for the transmission dynamics of a certain type of Foot and Mouth disease (FMD) virus to data from an experimental setting. These authors illustrate a key benefit of such an approach by not only inferring parameter values, but also missing data in the form of the hidden transmission history of the epidemic. Such techniques can be used to infer contacts in a real epidemic (Demiris and O'Neill 2005). In the case of deterministic models, parameters are typically determined by least-squares which implicitly assumes Gaussian measurement errors and ignores correlations in time.

What happens if we cannot agree on a single structure for a model? For many exotic diseases we may not even have enough information to decide what the mode of transmission is (for exotic vector-borne diseases we may have no evidence for how effective local vector species are). In such cases we can build two or more alternative models. Then, as more data becomes available we may be able to exclude (invalidate) some models. Where the models are relatively simple we may be able to choose between them using a multi-model inference approach (Burnham and Anderson 2002). Alternatively information theoretic criteria can be used to select between models sequentially (Spiegelhalter et al. 2002). If the alternative models are more complex we can still use any data ("patterns") at hand to qualify the good and the bad representations according to their ability to reproduce all the information simultaneously (for example after specifying the latent phase in our model, one could compare the resulting temporal epidemic curve emerging from different assumptions (fixed time vs. fixed rate) to field data on an outbreak. While the data may only show the temporal trend of virus positives, in the model we can manipulate the inputs and observe the outputs to compare it with the available data. Another approach is to produce one model that includes all structures and iterate the model repeatedly, with the number of iterations of each structure depending on the weight of evidence for that structure (see Smith et al. 2008). However, this approach is generally not possible if different research groups produce the models. Where no one model can be identified as significantly better than all others it is appropriate to employ a model averaging approach; each model is run and a combined output is formed by weighting the output of each model by some measure of our prior belief in the model combined with a measure of the extent to which it accounts for the available data. In such cases, uncertainty in model structure is combined with uncertainty in parameters and any intrinsic variation in the model to produce probabilistic outputs reflecting all these sources of uncertainty.

4.4 Quality Control

After constructing a model we then need to interrogate it. We should bear in mind that strategic models should be used to answer questions on improving our understanding, and tactical models should be used to answer questions of the form "what if". When deciding whether to consider model outcomes for policy development, the aim should be to determine whether a model is good enough rather than whether it is correct. But how do we assess the quality of the output and how do we deal with individual objections? There are three main objections used against models: (1) the "I don't believe it" approach, (2) "the model is not validated!" and (3) the model is "too complex" or "does not include some critical component", for example it only includes one sex. The first objection stems from a lack of understanding of the formal structure of modelling. In some cases unbelievable results will stem from errors in the coding, or structure, of a model. However, since the objective of a model is to gain *new* insights, we should not be surprised by unexpected results.

If the criticism can be more specific than "I don't like the answer", then it moves into the second objection category. If not, then this objection is irrational.

When considering the use of model outputs for informing policy development, it is crucial to evaluate their uncertainty (precision, random error) and validity (bias, systematic error). For knowledge-driven models, assessing the validity of outputs can be complex, since models are typically based on quantitative relationships, which have usually been derived from different studies, or may even be based on expert opinion. Validity is usually assessed by comparing model behaviour with the behaviour of the observed 'real world'. Since often 'real world' data does not exist, or the available study has been used to parameterise the model, it is often necessary to consider the plausibility of quantitative outcomes resulting from varying inputs. It should be emphasised that this is not the same as a full sensitivity analysis.

Usually the validity will have to be assessed in a qualitative fashion, whereas uncertainty can be quantified. Validity is therefore particularly difficult to assess, requires a good understanding of the biological system being studied and the methods used to study it. The uncertainty is a reflection of the natural variability in the 'real world' system and of lack of knowledge with respect to causal relationships. Both uncertainty effects are often difficult to separate or measure, but clearly any model prediction should also include an estimate of the uncertainty associated with the predicted outcomes. Policy makers may perceive scientific enquiry as a means for reducing the uncertainty associated with decision-making. However, this may not always be the case as research leading to the advancement of knowledge often results in the realisation that uncertainty has actually increased. Pielke (2003) wrote: "*Ignorance is bliss because it is accompanied by a lack of uncertainty*".

The question of model validity is a important one. By definition, no model can be valid for all circumstances. As we stated at the beginning of this chapter, all models are wrong. Equally, no model can be truly validated – like hypothesis testing, model validation can continue until a model is falsified, and even then it could remain useful in some circumstances. A model that predicts well in the short-term may predict badly in the medium to long term, but we should still consider it valid for short-term predictions. However, some important steps can and should be taken before using the results of a model. Firstly, a model should be verified. This means that its structure should be tested to ensure that the processes are modelled logically and that the output of interest changes in a plausible manner when input values are adjusted. A model should not generate results that are unfeasible, although judging what is feasible is not always straightforward. Nevertheless, there is clearly a problem if model outputs are negative when they should be positive or if it generates numbers that are larger than the total number of atoms in the universe! It is important to distinguish errors of logic from errors of coding. For this purpose it is useful to have a hierarchy of models based on different mathematical approaches e.g. deterministic non-spatial homogeneous mixing to stochastic and spatial, within which results can be compared. For nationally important management decisions, two similar models could be constructed by different teams using the same data and agreed mechanistic processes. From this, verification can occur by cross-model analysis of output.

For discrete-time models there is an additional verification step; examination of all processes (subroutines) that can occur within a single time step. The order of all subroutines within a time step should be clearly stated. One misguided approach that has been taken is to randomise the order of subroutines in each time step. However, all subroutines are either continuous (e.g. mortality) or effectively instantaneous (e.g. changing age categories, disease transmission), and the latter should usually all occur together either at the start, or the end, of each time step, and the order of all subroutines should be checked for logical consistency. If we can assume that the model is logical and verified, and there are no coding errors, then there are two other key processes that have to be considered before it can be used in anger, namely sensitivity analysis and model validation.

Sensitivity analysis assesses how sensitive the model is to its input parameters. There are a number of ways of assessing model sensitivity. The most common approaches include adjusting each parameter by a fixed value (say ±10%), or adjusting each parameter to its maximum and minimum bound, and re-running the model. The former approach is often called local sensitivity analysis and is a form of model verification that tests the sensitivity of the model structure to change. The latter, tests the sensitivity of the uncertainty or variation in the system under study and is used to determine the most important drivers of a system. As an example, a model of population control of a fossorial mammal may indicate that a 10% increase in mortality of young (i.e. pre-emergence from their underground lair) reduces the population more than a 10% increase in mortality of adults. However, the cost of increasing juvenile mortality by 10% may include finding and digging into the lair, whereas a 10% increase in adult mortality could be achieved for far less effort. Thus, from a management perspective the 'best' management option may be to increase adult mortality rates. Thus, local sensitivity analysis by changing values by a fixed percentage should not be used to inform management decisions. It is also important to distinguish between parameters that have large natural variability (e.g. juvenile mortality) and those that are relatively unknown (e.g. disease transmission rates or their surrogate, individual contact rates). A key feature of sensitivity analysis is to provide insights into which features of the model have the greatest effect on the output. This is important particularly if there is any imprecision or over-simplification of fundamental processes within the model. The sensitivity analysis is then used to identify areas where the model requires more precise data. This can be of considerable significance in modelling disease spread because key processes such as transmission are often poorly understood or quantified. This approach can be used to identify those parameters or processes that have the greatest influence on the outcome, and if amenable to human influence, provides insight into management and control. Sensitivity analysis is also useful for identifying parameters or processes that have limited impact on the model outputs. If a model is insensitive to a variable or a process that is incorporated in the model then it is quite legitimate to remove the process from the model completely.

Model validation is the next step in model assessment. During this process, the outputs of the model are compared with real data. These data should come from a system different to that which was used to create the parameters used as inputs in the model. In effect the modeller is attempting to replicate what happened in another system. Models may also be validated against secondary predictions, in other words with data not used during model parameterisation, but nevertheless taken from the same system. If the verification, sensitivity analysis or validation fails in some respect, then the model has to be redesigned, or refined, until it passes the tests. Once it has 'passed' the tests it could then be used to inform management decisions. Ideally, the requirements for model validation should be specified in advance, since it is often easy to find some aspect of model output that does not fit well with the available data, or belief.

4.5 Using Models to Inform Policy Decisions

If we therefore assume that we have a model that has been subjected to verification and validation, and has not failed these tests, it can be used to help make management decisions. However, it should be noted that models do not produce decisions, but simply extrapolate under a number of "what if" scenarios. The consequences of these management options need to be considered in a wider context. Pielke et al. (1999) stated "*Predictions must be generated primarily with the needs of the user in mind. For stakeholders to participate usefully in this process, they must work closely and persistently with the scientists to communicate their needs and problems*". Thus there should ideally be constant dialogue between the modeller(s) and the user (decision maker), although in our experience this is rare.

Models generally assume that all parameters remain constant (or for stochastic models that the variation does not change with time), except for the parameter being changed (e.g. the management option). In many cases however we expect that some aspects of the model assumptions may change with time (e.g. landowner behaviour), thus, model 'predictions' need to be interpreted in the light of our expected changes in the system.

It is also critical that the user understands the uncertainties in the model, and how they may affect the outcome of different policy options. Lempert et al. (2003) describe the use of quantitative models in policy development as follows:

> *Under conditions of deep uncertainty, we suggest that analysts use computer simulations to generate a large ensemble of plausible scenarios about the future. ... The goal is to discover near-term policy options that are robust over a wide range of futures when assessed with a wide range of values.*

When using models to provide management advice it is desirable to take account of uncertainty in our knowledge of, as well as the intrinsic variability in, the system under study. As discussed above, variability can be accounted for by introducing a stochastic element into the behaviour of the model, and uncertainty in knowledge may be accounted for by using statistical approaches. For example, the estimation of parameter values from data is uncertain, and statistical methods provide a distribution of estimates or at the least a mean and standard deviation. It is often the case that a range of models are available and there is therefore uncertainty associated with the choice of model. Accounting for either, or both, system variability and uncertainty in our knowledge about the system leads to a probability distribution across the predicted response of the system. This profoundly changes the advice arising from the model from unequivocal, to for example the relative probabilities that a given disease reduction strategy will be successful or fail (by some criteria). In addition to such quantifiable uncertainty it is also critical to communicate the qualitative limitations of different model formulations as these are likely to be critical to interpreting results.

A recent UK government review into infectious diseases concluded that useful models would in the future need to include stochasticity, individual and population level heterogeneity and spatial structure. It stated "*Combining these refinements into ever more complex ... models provides a better chance of accurate prediction. This will be invaluable in ... deciding on control options.*" Further, the report suggested how models could be used to assess new technologies: "*The development of rapid tests to detect infection earlier could, in theory, help isolate infectious individuals and stop disease spread. However, a model is needed to estimate the potential of such diagnostics and to show* [how] *they might best be deployed*" and also pointed to a new area for consideration in modelling "*To be really useful, however, future models must embody more understanding of human behaviour*".

Modellers need to understand that the results of their model will not be used without being put into a policy context, and users need to understand that model results should not be used without critical interpretation. How models have developed over time, for the UK wildlife rabies contingency plans, are shown in Box 4.3. This development is also instructive in informing policy makers, or budget holders, where to direct further research.

4.6 Model Limitations

All models are subject to a number of assumptions. Much like for statistical tests, some assumptions can be broken without affecting the validity of the answer, and there is no clear definition of which assumptions are of over-riding importance. For example, many models that do not include sex, or age, or season, can result in robust results. In any written presentation, a list of model assumptions should be given, including those that seem obvious to the modeller. Indeed it is worth noting that important caveats concerning the model can easily be lost when crossing between one discipline (epidemiology) and another (policy-making). It is only by examining a list of model assumptions that model output can be interpreted correctly. The failure to provide an adequate list of assumptions often leads to 'overselling' the model.

One of the most important limitations for wildlife disease models is that the disease transmission rate can rarely be measured directly. This is such an important parameter

Box 4.3 The development of rabies modelling for contingency planning in the UK

The simplest mathematical model of rabies spread in red foxes (*Vulpes vulpes*) was a deterministic non-spatial model (Anderson et al. 1981). Although not designed or parameterised for the UK, this simple model could be used to calculate the level of culling (or vaccination) required to eliminate a rabies outbreak, as a function of fox density. Although structurally overly simple, the model relied on assumptions about the threshold density below which fox rabies does not persist (i.e. $R_0 < 1$), which was estimated at one fox per square kilometre. However, rabies is known to persist in foxes in Canada at far lower densities (Voigt and Macdonald 1984), so the generality of this assumption is uncertain. This model was then parameterised for the UK and a spatial dimension was added in the form of a diffusion term, which assumed that disease spread was caused only by the itinerant movement of rabid foxes (Murray et al. 1986). A map of fox density in England and Wales was then used, on which the equations were numerically solved, producing time series pictures of rabies spread in England and Wales. In a first attempt to utilise a model to evaluate the local introduction, spread and control of rabies, an existing simulation model (the 'Ontario Rabies Model': Voigt et al. 1985) utilised a grid where each cell represented a fox home range (Smith and Harris 1989). This model simulated a large range of fox densities, and being stochastic, could evaluate the probability of disease elimination for any given level of control. It permitted disease spread by neighbour-to-neighbour contact and fox dispersal, and the threshold for disease persistence was an emerging function of biological parameters, and was not pre-determined. This latter model also demonstrated that low levels of fox culling would result in extending the duration of the epidemic. However, it was known that fox density in local areas of English cities could exceed nine foxes per square kilometre (Harris 1981), whereas in nearby rural areas it was likely to be less than one fox per square kilometre (Macdonald et al. 1981). Therefore, a revised simulation model was constructed based on 500× 500 m grid cells, which were combined to form fox home ranges of different sizes (Smith and Harris 1991). This not only permitted the incorporation of heterogeneity in fox density, but had the added advantage of removing the bias inherent in regular geometric simulation models (Holland et al. 2007). Refinements of this model were used to investigate fox vaccination (Smith 1995) and fertility control (Smith and Wilkinson 2003), and the model was integrated within the UK contingency plan for dealing with an outbreak of wildlife rabies (Smith and Fooks 2006).

in the models that we dedicated a whole chapter to estimating it (Chapter 3). However, this is not an insurmountable problem, since, if the structure of the model is correct, and a prevalence estimate has been measured in the field, a limited range of transmission rates will produce this output. This is similar to having one unknown in an equation – only one value (of the transmission rate) will make the equation balance. In two-species models there are four transmission rates (two within-species and two between-species rates), which makes estimating these values with limited field data difficult. Few attempts have been made to formally parameterise two-species disease models for practical use (but see Morgan et al. 2006). Some theoretical work has been done in this area (Dobson 2004), particularly with parasite/parasitoid models (e.g. Preedy et al. 2007). With three-species models there will be nine transmission rates, thus making accurate estimation nigh impossible.

Given that modelling outcomes will always be associated with varying degrees of uncertainty and validity, the decision to use them for informing policy making will have to be based on opinion and judgment. One recent advance is the inclusion of the economic dimension within computer models to help managers to make informed decisions.

Chapter 5
An Economic Perspective on Wildlife Disease Management

Richard Bennett, Graham C. Smith, and Ken Willis

5.1 Introduction

This chapter considers what an economic perspective can bring to the management of disease in wildlife. The chapter starts by outlining the scope of economics as a discipline and its relevance to wildlife and disease management. It goes on to identify the main economic impacts of disease in wildlife and the economic dimensions of wildlife disease management issues. An economic framework for the consideration of wildlife disease management is explored, as is the relevance and use of the cost–benefit analysis approach, which is fundamental to policy analysis and project appraisal. The economic literature relating to wildlife disease management is briefly reviewed, and problems and issues for economic analyses in this area are discussed.

5.2 Importance of an Economic Perspective

An economic perspective is required to adequately address wildlife disease management issues alongside natural and physical science, ethical and political perspectives. Indeed, the discipline of economics can provide a useful framework within which to consider wildlife disease and its management.

Economics is a social science, which has close links to moral philosophy and political science, as well as psychology. It provides a discipline for thinking about social issues and management problems. Economics has been described as concerned with decisions about 'how we should live' (Sen 1987). Alfred Marshall, the

---

R. Bennett
School of Agriculture, Policy and Development, University of Reading, Reading, RG6 6AR, UK

G.C. Smith
Central Science Laboratory, Sand Hutton, York, YO41 1LZ, UK

K. Willis
School of Architecture, Planning and Landscape, University of Newcastle, NE1 7RU, UK

R.J. Delahay et al. (eds.), *Management of Disease in Wild Mammals,*
DOI:10.1007/978-4-431-77134-0_5, © Springer 2009

notable Victorian economist, wrote that 'economics is a study of mankind in the ordinary business of life' (Marshall 1947).

Management of wildlife, whether due to disease problems or for other reasons, involves decisions that affect the allocation of resources (i.e. to undertake various management and disease control actions). Economics is concerned with the allocation of scarce or limited resources to the achievement of alternative ends in the pursuit of greater human welfare. Thus, wildlife management and the control of disease in wildlife are economic activities as much as they are scientific ones. Indeed, economists would view wildlife disease management as primarily an economic problem that has scientific dimensions. They are economic activities because resources are scarce in the sense that we never have enough (land, labour, time, raw materials, etc.) to do or produce everything that we would like to, and their allocation to any one area necessarily deprives another area of their use. For example, resources devoted to the management of a particular disease in wildlife may mean that fewer resources are available to control other diseases in wildlife, livestock or humans. This means that resource allocation decisions, such as those involved in wildlife disease management, have an 'opportunity cost' (i.e. the benefits that could have been derived from doing something else with those resources). Thus, when making a resource allocation decision, (opportunity) costs have to be weighed against the expected benefits of the decision. This is the basic philosophy behind economic cost–benefit analysis, which is considered further in Section 5.5.

The 'economic system' can be simply described as the transformation of inputs into outputs (see Fig. 5.1). Inputs are resources that are combined and transformed by the production process into outputs in the form of goods and services to satisfy human wants and increase human welfare. Wildlife are largely external to this simple production model, since they are rarely used as inputs to production (although game hunting and bushmeat are economically important in some countries) and are rarely the outputs (results) of economic activity. However, wildlife may affect the production process, for example by acting as disease vectors passing infection on to farm animals (which are inputs to livestock production and whose productive performance will therefore be affected) and may be affected by it (for example by

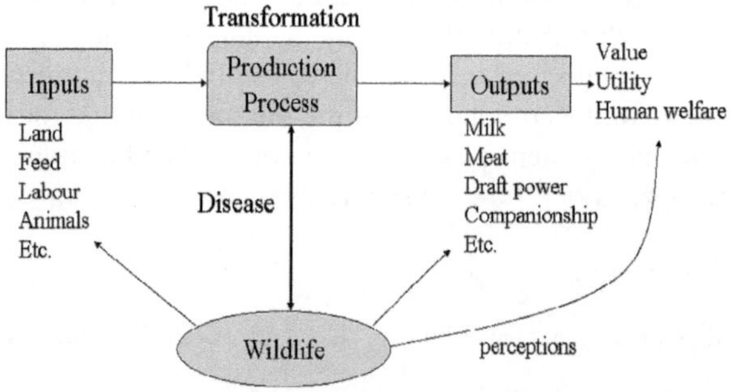

Fig. 5.1 An economic framework for wildlife and their diseases. Generally wildlife disease can be considered as an externality, impacting negatively on livestock production for bovine tuberculosis in badgers, but positively on cereal production for myxomatosis in rabbits

loss of habitat/feed sources due to the use of land for farming or by disease transmitted from farm animals). The former clearly impacts on the economics of livestock production, by reducing output and/or increasing inputs, which has a cost both for livestock producers and for society more widely (see McInerney 1996; Bennett 2003) whilst the latter only has an economic dimension if human welfare is affected. For example, if people value wildlife populations, consider them important to environmental sustainability, derive pleasure from seeing wildlife populations (or from knowing that they exist) or are concerned about the welfare of wild animals, then the impact of production on wildlife has an economic dimension. Thus, in this case, if wildlife populations are adversely affected by livestock production or by wildlife management then there is an economic cost involved that needs to be acknowledged and taken into account.

Without an economic perspective on wildlife disease management issues it is likely that resources will not be used efficiently which will impose unnecessary costs on society.

5.3 The Economic Dimension

It is worth noting from the outset that, from an economic perspective, disease in wildlife only has an economic impact if it affects human welfare, either directly or indirectly.

When considering the management of disease in wildlife there are three key questions that need to be addressed, which are:

1. Why should disease in wildlife be managed?
2. What are the objectives of the management of wildlife disease?
3. How should wildlife disease be managed?

The economic dimensions of wildlife disease management can be numerous. To help address the questions posed above, it is worth considering the main categories of economic impacts of disease in wildlife. These are:

1. Direct negative impacts on the host species, such as mortality, population reduction, animal suffering and threats to species survival
2. Impacts on ecosystems and the environment, such as infection risks for other wildlife species (through spillover infection), disruption to ecosystems, and impacts on environmental stability and sustainability
3. Impacts of disease on domesticated species, including companion, zoo and farm animals
4. Risks to human health if the disease is zoonotic
5. The resource costs of prevention and control of disease in wildlife (including monitoring and surveillance).

The effects of disease on host mortality and the size of wildlife populations may not have an economic impact unless at least some people feel worse (or better) off as a result. For example, people may not like to think of wildlife species suffering and so their (human) welfare would be diminished if this were to happen. This was clearly the case for the phocine distemper outbreak in seals in the North Sea in 1998 and again in 2002 (the large and highly visible numbers of dead and dying seals stranded on beaches prompted an extensive response, including rehabilitation). This means that selective culling, for example, of diseased animals might be perceived as beneficial to the wildlife population but unselective culling might not. Also, people may value the opportunity of seeing wildlife in its natural habitat and the chance of doing this may be reduced if wildlife populations decline due to disease. People may also value the existence of the species and thus any threat of extinction of a species (even if only at a local level) would cause them concern and hence reduce their (human) welfare. There is a growing economic literature on the valuation of wildlife species and of environmental resources more generally. Notable examples include a consideration of the economic value of marine biodiversity in UK waters (Beaumont et al. 2006), and an estimation of the Australian public's willingness to pay to conserve various wildlife species (Tisdel and Wilson 2006). The latter study found that, on average, people were willing to pay between Aus$1.36 and Aus$1.73 per species per week for their conservation. Wildlife species may be valued for recreation or sport. A tangible effect of disease in wildlife and reduced wildlife populations (and/or increased likelihood of seeing wildlife suffering) might be reduced visitor numbers to a wildlife area – which in turn could impact on tourism and the local economy. However, the impact on tourism and the local economy is likely to be mainly a distributional issue rather than an efficiency issue. Visitors would probably spend their money in another area of the country; thus there would be no loss to the national economy as a whole, unless a disease outbreak reduced the number of international tourists.

Similarly, disease in one wildlife species may have implications (whether real or perceived) for other wildlife species. For example bovine tuberculosis (bTB) infection in buffalo (*Syncerus caffer*) may spillover into other species such as lions (*Panthero leo*) (Michel et al. 2006). Disease in a host population may also result in changes to habitats (e.g. myxomatosis-induced mortality reduced rabbit density with consequent ecological effects: Thompson 1994), ecosystems (e.g. disease induced mortality in rabbits reduced prey availability for Spanish imperial eagles: Moreno et al. 2004) or landscapes that are valued by people. In all these instances, the wildlife disease in question is an economic problem with associated economic costs.

The risk of wildlife disease to domesticated species is an area of particular concern and where substantial tangible economic impacts can be apparent. For a number of diseases, wildlife may be able to infect domesticated species, either companion animals (e.g. rabies) or farm animals (e.g. bovine tuberculosis). In addition, wildlife disease may also threaten animals in zoos, stables and other establishments (Bengis et al. 2002; Simpson 2002). The economic impacts of disease on farm animals may include (Bennett 2003):

(i) A reduction in the level of marketable outputs
(ii) A reduction in (perceived or actual) output quality
(iii) A waste (or higher level of use) of inputs
(iv) Resource costs associated with disease prevention and control

(v) Human health costs associated with zoonoses or with disease control (e.g. organo-phosphates in sheep dip)
(vi) Negative animal welfare impacts
(vii) International trade restrictions.

These can be substantial costs for farmers, livestock industries and for economies (Bennett and IJpelaar 2005). For example, the 2001 outbreak of Foot and Mouth Disease in the UK was estimated to cost up to £10 billion to the economy (Barclay 2001). A study of the risks of bTB in New Zealand, where the possum is an important wildlife vector, estimated that international trade losses due to TB could potentially be $NZ1.3 billion per year but that there was only a 2% risk of such losses occurring (Clough and Nixon 2000).

Risks to human health can be a concern, either through direct infection from wildlife or their waste, or via infection of companion or farm animals. Potential human health economic impacts include loss of human life, reduction in quality of life, loss of productivity (of workers), and resource costs of medical treatment and care. A study of canine rabies in Africa and Asia (Knobel et al. 2005) estimated that human mortalities of 55,000 per year with 1.7 million disability-adjusted life years (DALYs are a composite measure of years of life lost due to premature mortality and years of life lived with a disability caused by a condition) lost each year and treatment costs, resulted in an annual cost of around US$ 583.5 million. Box 5.1 provides an example of economic analyses of wildlife rabies management, taking account of the potential costs in terms of human health, livestock and companion animals, and control.

Box 5.1 Rabies: cost-effectiveness of control options in wildlife

The UK has been free of terrestrial rabies since 1922, and the government contingency plan aims to re-establish rabies freedom as soon as practical should an outbreak occur. Rabies freedom is obtained 24 months after the last reported case, assuming that suitable surveillance is in place to detect new cases. Here, we look at a cost-effectiveness analysis of two options to eliminate a new focal outbreak of wildlife rabies. The UK government policy is to follow the European Union fox *(Vulpes vulpes)* vaccination plan where applicable (European Commission 2002). However, in areas of high density, vaccination is expected to be less successful and take longer than focal poisoning with a ring of vaccination (Smith and Fooks 2006). Focal culling with a ring of vaccination has worked to eliminate focal outbreaks of rabies in raccoons in Canada (Rosatte et al. 2001), and could be considered in the UK if the red fox density was medium to high and the initial case was identified.

Focal poisoning, using baits distributed by hand and checked daily, would be more expensive per unit area than oral vaccination, where baits are distributed by hand, or by air, and left in situ. Focal poisoning would involve defining locations for 15–30 bait stations per square kilometre, laying pre-bait for 7–10 consecutive days and monitoring bait uptake daily (taken baits are replaced). When bait uptake levels asymptote, the pre-bait is replaced by poison

(continued)

Box 5.1 (continued)

bait for 3–5 days; again monitored, and taken bait replaced daily. All baits would then be removed. This whole procedure could be repeated at two-monthly intervals until a suitable population reduction is achieved (2–5 campaigns). Computer models are used to determine the minimum area for focal culling that does not compromise the success of the strategy. Generally this is only an area with a radius of 3–6 km. Outside the focal culling, oral vaccination is performed out to a radius of some 20 km. Following focal culling and ring vaccination, oral vaccination is performed in the entire area in late spring and autumn of each year until the area is declared rabies-free.

Oral vaccination would be performed by distributing baits up to a radius of 20–50 km from the outbreak. One emergency vaccination would be performed immediately on confirmation of wildlife rabies, and then again in late spring and autumn of each year until the area is declared rabies-free.

Computer models were used to simulate a rabies outbreak in a rural area of medium fox-density. A 3 km radius of poison baiting was sufficient, with vaccine baits being laid out to 20 km. For oral vaccination a 20 km radius was sufficient to ensure disease elimination.

Costs of the control campaign were calculated based on man-power requirements, equipment and baits. Based on other studies we assumed that 70% of the domestic dog (*Canis lupus familiaris*) population and 50% of the cats (*Felis catus*) would be vaccinated during the outbreak and that 50 people out of every 100,000 would receive post-exposure treatment for rabies each year. We would expected there to be about 140,000 people, 20,000 cats and 16,000 dogs in the rural area. The total cost of the rabies outbreak is thus the sum of the costs of: pet vaccination, livestock vaccination, replacement of pets, replacement of livestock, quarantine of suspect animals, human pre- and post-exposure prophylaxis, adverse reaction charges, public health charges, animal control costs, insured human death claims and resource loss of rare or threatened animals (Sterner and Smith 2006). However, we assume that we are dealing with an initial focal outbreak in an otherwise rabies-free country and that the disease is rapidly controlled. Thus, we expect no human deaths, very limited replacement of pets and livestock, and no effect of rabies, or its control, on rare or threatened species.

Most previous studies have found that the majority of the cost is associated with human or pet vaccination, rather than wildlife control (Uhaa et al. 1992; Meltzer 1996). However, given the low number of rabies cases expected, the relatively short duration until disease freedom is obtained (about 30–36 months), and the high cost of the initial stages of wildlife control, this does not hold in the scenarios above. The cost of disease elimination for a 20 km radius oral vaccination breaks down as follows: human *and* pet vaccination £1 million, laboratory testing £50,000, wildlife control £9.5 million. The total cost of focal culling was just £2 million more, and took four months less to achieve rabies freedom. Thus, despite the higher cost of poison baiting, the costs are identical if the focal culling could reduce the radius of control to 77% that of the oral vaccination. Simulations in urban areas suggest that the area of vaccination would have to be much larger than that of focal culling due to the higher fox density. Thus, it is likely that focal culling could not only be more effective, but also more cost-effective in some high-density areas.

Substantial resource costs may be incurred in preventing and controlling disease in wildlife. For example, some \$230 million to \$1 billion are spent per year in the USA on the detection, prevention and control of rabies, where over 90% of all cases in animals occur in wildlife (Rupprecht et al. 1995). Resources may be employed to cull or reduce wildlife populations in some other way (Chapter 7), to treat them with vaccines or medication (Chapter 6) or to manipulate the environment so as to reduce disease risks (Chapter 8). These activities may not only involve resource costs but may also involve human welfare loss. For example, people may not like the idea of wildlife being culled (particularly if the method is perceived as inhumane), or of wildlife populations being reduced in magnitude, and if this happens they will feel a welfare loss that they would be willing to pay to avoid. This issue is addressed further with a practical example in Box 5.2. Resources may be employed to monitor disease in wildlife populations to help identify and manage potential disease problems (i.e. those that could result in substantial economic loss). Disease surveillance in wildlife (Chapter 10) may be undertaken by various government funded bodies and research establishments, but is rarely the responsibility of a single organisation in any country. We are not aware of any situation where the costs of disease surveillance have been evaluated against the cost of later disease detection in any formal way.

It is clear that avoiding, reducing or mitigating the effects of the economic impacts of disease in wildlife is good reason for intervention, and these provide the broad aims of any management policy. The question that then remains is precisely how wildlife disease should be managed. This relates to the nature and magnitude of resources that should be devoted to addressing the problem, how these resources should be used and the relative costs and benefits of their use. A further issue is who pays, or should pay, for wildlife disease management.

5.4 An Economic Framework

This section provides a more formal and analytical exploration of the economic impacts of wildlife disease and its management identified above.

Figure 5.2 shows a possible relationship between resources devoted to control of disease in wildlife and the level of disease. The curve shows that as more and more

Box 5.2 Bovine tuberculosis: valuing cattle and badger management

There is evidence to suggest that the European badger (*Meles meles*) is instrumental in infecting cattle with bovine tuberculosis (bTB) in the UK (Krebs et al. 1997). The incidence of bTB in cattle in England and Wales has increased substantially over the last 10 years particularly in the south west of England, where badger populations are most dense. The government operates a compulsory national 'test and slaughter' policy for cattle, so all cattle herds are routinely tested for bTB and cattle that test positive are slaughtered with the farm being under cattle movement restrictions until all animals test negative. In 2004, around 24,000 cattle were slaughtered due to bTB infection. Of these, around 95% came from the bTB 'hotspot' areas in the south west of England and south west Wales. Farmers received compensation payments (over £36 million in 2004/5) for cattle slaughtered in this way. Government expenditure on tackling the bTB problem was estimated at some £92 million in 2004/5.

The utility or welfare to society from conserving badgers and reducing cattle slaughtered through bTB in England and Wales was assessed using a stated choice experiment. The experiment comprised a stratified random sample of 400 general public households; stratified to generate a roughly equal number of observations in both badger/TB 'hot-spot' and 'non-hotspot' areas. Each respondent was presented with sets of alternatives, each associated with a cost: badger population size, number of cattle slaughtered because of bTB, management strategy, and tax cost, and asked to choose their most preferred alternative. Repeated choices by respondents from different sets of alternatives reveals the trade-offs respondents are willing to make between badger numbers, bTB infections, management strategy, and increases in household taxation. The choice of one alternative from the choice set was modelled as a function of the attribute levels plus an error term. The individual's utility function can be specified as

$$U_{ij} = V_i + \varepsilon_{ij}$$

where U_{ij} is the utility individual i obtains from alternative choice set j. This utility is known to the individual but not the researcher. The individual is assumed to choose alternative j over alternative k if $U_{ij} > U_{ik}$. The researcher observes attributes of the alternatives considered by the individual, and specifies a function, V_{ij}, relating these observed factors to the individual's utility. Since there are aspects of utility the researcher does not observe, ε_{ij} captures the factors that affect utility but are not included in V_{ij} (Train 2003).

The attribute and attribute levels used in the choice experiment (CE) were badger population size (100,000, 200,000, 300,000 or 400,000), badger management strategy (trap and shoot, application of contraceptive, no intervention or badger friendly road construction), cattle with TB slaughtered per year (0, 10,000, 20,000 or 50,000), and increase in tax per household per year (£5,

£20, £50 or £100). The levels of the badger population reflected both the estimated size of the population in England and Wales at the time of the study (300,000), the likely 'biological maximum' population (400,000) and the likely lowest size of the population if a wide-scale badger-culling programme were to be implemented (100,000). The badger management strategies were chosen to include no intervention (badgers as a protected species), a culling policy (as has been employed by the UK government) and progressively more 'badger friendly' policies to test the importance of management strategy to people's preferences. The levels of cattle slaughtered due to bTB reflected both the current level (just over 20,000), and possible future levels depending on whether the bTB problem worsened (50,000) or improved (10,000), with zero (bTB eradication in cattle) being the most desired outcome.

Relative to no intervention (badgers as a protected species), the utility obtained by people decreased with a management strategy of culling badgers, but increased for a strategy of controlling badger populations with contraception. People's willingness-to-pay (WTP) was estimated at £0.10 per household per year for every additional 100,000 badgers, and £1.52 per household per year for every reduction of 10,000 cattle slaughtered, £68.31 per household per year not to have badger culling and £13.58 per household per year to have badger contraception (in 2004 prices). These values were then extrapolated for the whole of England and Wales (an estimated 21.7 million households). This produced a WTP per additional badger of around £22 and WTP for a reduction in cattle slaughtered due to bTB of £3,298 per animal. This estimate only relates to the range of badger population size of 100,000 to 400,000, but the results suggested that changes in the size of the badger population within the limits considered are not of great importance to many people. The results implied a WTP of £4.4 million per annum to retain what was the current badger population of 300,000 rather than a reduced one of 100,000. The WTP for reducing cattle slaughtered due to bTB included people's total valuation of their perceptions of the importance of the disease and its impact on farming, wildlife, cattle welfare, food quality, risks to human health and so on. The cattle valuation suggested a WTP of over £80 million per annum amongst people in England and Wales, to have no cattle slaughtered due to bTB (i.e. to eradicate bovine TB), which is close to the current government expenditure on tackling the bTB problem.

In contrast to WTP for changes in the size of the badger population, badger management policy appeared to be very important to people, with a high WTP, in particular, not to have badger culling. Extrapolation of values to England and Wales gave a WTP of £1,480 million not to have a badger culling policy rather than no intervention and £294 million for a badger contraception policy. Clearly, these values are high, particularly for not having a policy that intentionally kills tens of thousands of badgers by trapping and shooting (73% of respondents in the survey said that they objected to badgers being intentionally killed).

resources are allocated to control, the level of disease in wildlife falls (and for a particular disease could reach zero). The exact shape of the curve will depend on the nature of disease and of the wildlife population affected as well as the control measures used. For example, if a technically superior method of control were developed the curve would be shifted downwards. Figure 5.2 also shows a relative price/value line (with slope vR/vDW). The line shows the relative value (i.e. cost of) disease in wildlife (vDW) relative to the value (cost) of resources used to control disease (vR). If this line is moved until it is tangential to the curve, the point of tangency (i.e. where vR/vDW = dR/dDW) shows the optimum level of resource use (from an economic perspective) that should be devoted to wildlife disease control (shown by point R with an associated amount of disease in wildlife DW). At this point, the total combined value of the cost of disease in wildlife and the cost of resources for control of that disease is at a minimum. Figure 5.3 shows that change in the relative values or costs of resources and of disease in wildlife will change this position. For example, if the cost/value of disease in wildlife is actually higher than first thought (represented by an additional price line in Fig. 5.3) then there is a new higher, optimum level of resource use (R′) and a resultant lower level of disease in wildlife (DW′).

Figure 5.4 shows this in terms of the costs and benefits associated with wildlife disease control. As more and more resources are devoted to the control of disease in wildlife, the benefits of lower disease in wildlife (i.e. in terms of less disease in domesticated animals and humans) increase. However, the costs associated with using more resources also increase. The optimum point of resource use, where the net benefit of control, shown by the greatest distance between the benefits curve and the costs curve (labelled Total cost1), is maximised, is shown at point R1. However, if for example, the costs of control of disease in wildlife have been underestimated, perhaps because of public sentiments concerning culling of wildlife species (for which they have a willingness to pay to avoid) or due to damage to

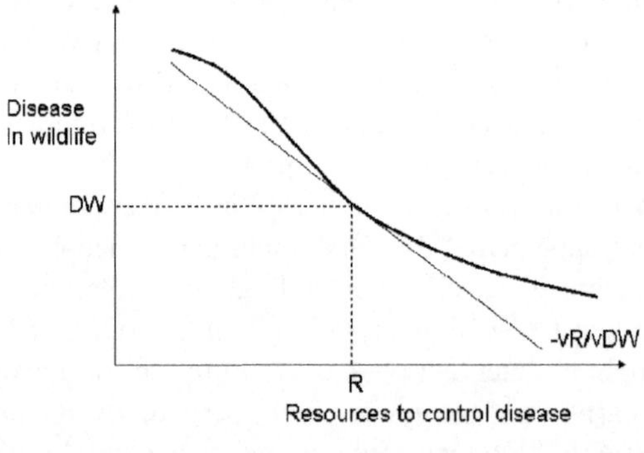

Fig. 5.2 Economic optimum resource use to control disease. The curve shows the 'true' relationship between the amount of disease in wildlife and the amount of resources used to reduce it. The straight line shows the cost of disease (vDW) relative to the resources used to control it (-vR). Where the line is tangential to the curve, the economic optimum is determined

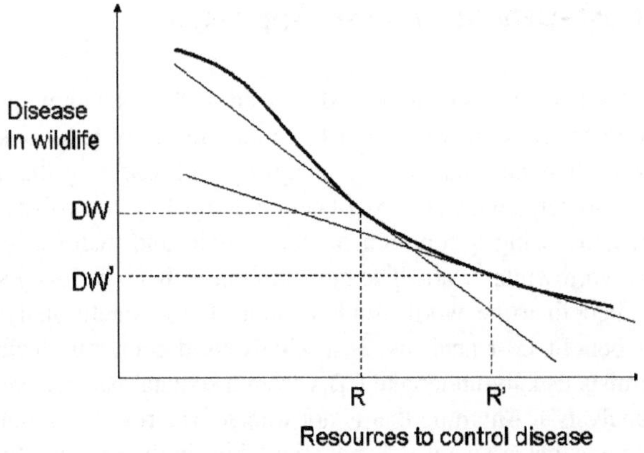

Fig. 5.3 Economic optimum resource use to control disease when resource costs and/or costs of impacts of wildlife disease change

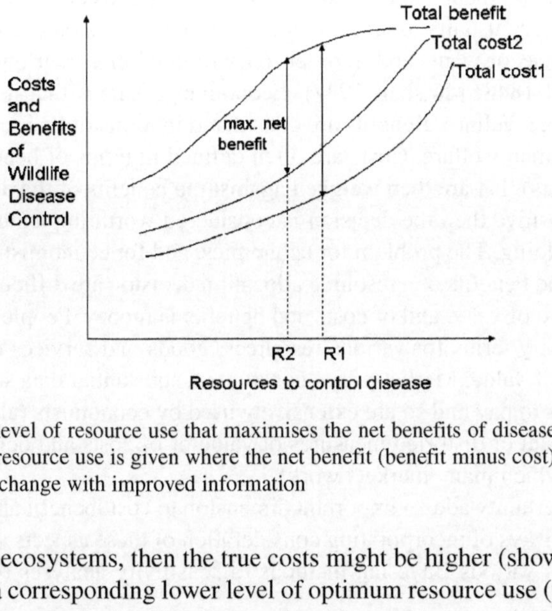

Fig. 5.4 The level of resource use that maximises the net benefits of disease control in wildlife. The optimum resource use is given where the net benefit (benefit minus cost) is maximised. This optimum may change with improved information

habitats and ecosystems, then the true costs might be higher (shown by curve Total cost2) with a corresponding lower level of optimum resource use (where net benefit is once more maximised) of R2. In practice this would mean a lower level of culling of the wildlife species or a different form of control.

The benefits of wildlife disease management are considered largely in the form of the lower level of negative economic impacts of wildlife disease that would result (i.e. the costs associated with wildlife disease that are prevented). In Section 5.3, disease in livestock (arising from wildlife) was identified as an important potential economic cost. Similar diagrams to those shown in Figs. 5.2 and 5.3 could be produced showing the relationship between wildlife disease control and livestock disease.

5.5 The Cost–Benefit Analysis Approach

Cost–Benefit Analysis (CBA) has been described as the 'economic ethic' (Boulding 1969) and the techniques are well described in a number of text books (e.g. Mishan and Quah 2007). It is an approach for analysing and assessing the advantages and disadvantages associated with a course of action or a policy. It involves weighing costs against the benefits using a common monetary unit, and then assessing whether a policy appears worthwhile or not. The presumption is that any policy should result in a positive net benefit to be worth implementing. Cost–benefit analysis (sometimes referred to as benefit–cost analysis) is a widely used economic technique (Mishan 1998). Economists usually undertake CBA from a societal perspective (called 'social cost–benefit analysis'). Anything that is not wanted as a result of a policy is regarded as a cost and things that are wanted as benefits. Thus, in the context of wildlife disease management, the reduction of infection in the wildlife population and any reductions in disease risks to domestic animals and human populations would be regarded as benefits. Adverse impacts on the wildlife population (e.g. in terms of its size, location, etc.) or on the environment and the resources used in disease control would be costs.

Economic theory usually assumes that people make decision choices according to their preferences or wants and in order to try to maximise their own satisfaction or 'utility' (Mill 1848; Marshall 1947). Economics assumes that it is utility that determines human welfare. Benefits are considered in terms of their contribution to utility and to human welfare. Costs are often defined in terms of benefits foregone (opportunity costs) that are then weighed against the benefits of the decision. If the net benefit is positive then the decision is considered worth implementing and the action is worth doing. The problem for economics, and for economists, is that often both the costs and benefits of a resource allocation decision are difficult to measure. A useful measure of value and of costs and benefits is money. People's willingness to pay in monetary terms for various resources, goods and services can provide a useful measure of value. Market prices constitute a substantial data source on people's willingness to pay and so are extensively used by economists (although prices are often imperfect or distorted measures of value or of costs and benefits because of the ways in which many markets work).

Risk and uncertainty add an important dimension to cost–benefit analyses. There are a number of ways of incorporating consideration of these aspects into the analysis ranging from various Bayesian methods to sensitivity analyses (Gollier 2001; Meyer 2003).

5.6 Economics (and CBA) of Wildlife Disease

There is a relatively limited literature on the economics of wildlife disease management. A review of articles in the Journal of Wildlife Diseases between 1965 and 2007 revealed no papers with the words 'economic' (or 'economics') in their titles and just one with the word 'costs' (or 'cost'). This highlights the relative absence of economic considerations to the management of wildlife disease. However, there are a small number of 'economic' studies reported in the published literature generally. The use of an economic framework for cost–benefit analysis has been advocated for wildlife damage studies (Shwiff and Sterner 2002), with a recommendation to use decision tree analysis to explore the costs and benefits of different management options using probabilities of outcomes and valuations of their physical impacts. The decision tree approach has been previously applied to cost–benefit analyses of livestock disease control at farm level (Bennett 1992). Indeed, most studies reported in the literature have taken a cost–benefit analysis approach to economic analysis of wildlife disease management problems. Such studies include a number relating to rabies in various mammals (Meltzer and Rupprecht 1998; Aubert 1999; Kemere et al. 2000; Gordon et al. 2005; Sterner and Smith 2006) and the control of bTB in badgers (Bennett et al. 2004; Smith et al. 2007b) and brushtail possums (Bicknell et al. 1999; Clough and Nixon 2000). These studies highlight the consideration of benefits in the analyses in the form of costs saved. Some studies do not take such a disease-oriented focus but consider more broadly the benefits of wildlife control, especially where the species concerned are considered as pests (Hone 1995; Vere et al. 2004; Jones et al. 2006; Smith et al. 2007a). These benefits include both disease risk/reduction in domesticated species and humans and environmental benefits. One study (Vercauteren et al. 2002) used dynamic systems simulation modelling to undertake cost–benefit analyses of rodent control on pig farms. The analyses considered a range of potential rodent impacts including damage to structures, loss of stored feed and transmission of disease to pigs.

Where public policy is concerned, it is important to consider wildlife disease control in terms of socially optimal management strategies (i.e. using the concept of social cost–benefit analysis). One such study (Horan and Wolf 2005) used a two-state linear control model to examine the socially optimal management of disease in a valuable wildlife population (in the context that diseased animals could not be selectively culled). The case study used was bovine tuberculosis in white-tailed deer *(Odocoileus virginianus)* in Michigan, USA. The researchers considered disease control from a societal perspective, which includes the external benefits of individual farm control for other farmers, landowners and hunters and the economic value of the wildlife population being culled. The analysis also included consideration of possible costs associated with trade restrictions. The study found that the optimal policy was for the disease to remain endemic in the area but at low levels, although if the State had trade and testing restrictions imposed upon it by other states or by federal government then disease eradication became the preferred policy.

5.7 Stages of Economic Analysis

There are a number of stages involved in any economic analysis of wildlife disease management and the main ones are listed below:

1. Specification of the problem
2. Identification of impacts and effects
3. Specification of possible technical solutions
4. Identification of impacts and effects of solutions
5. Measurement/estimation of impacts
6. Valuation of impacts and effects
7. Calculation of estimated/expected costs and benefits taking account of time (discounting) and risk/uncertainty
8. Net Present Value (NPV) estimates
9. Sensitivity analysis
10. Policy implications.

Understanding the nature of the problem to be managed is clearly important, both in terms of its technical and economic specifications. For example, the rising incidence of bTB in cattle in Great Britain has been identified as a problem because of the effects on cattle farmers, on animal welfare and potential risks to international trade (e.g. in cattle, beef and dairy products). There are also additional risks to human health, mainly those in contact with infected cattle, since milk pasteurisation and meat inspection practices largely protect the consumer. Potential strategies for addressing the problem include testing and slaughter of infected cattle, cattle vaccination, culling or vaccination of wildlife, amongst others. There is also a political dimension, as farmers and their representatives apply political pressure on the government to take action. In economic terms, the implications of the disease for cattle productivity and farm profitability and competitiveness are important, as are potential impacts on human health, international trade and animal welfare.

Economic analysis is also concerned with distributional issues of policy, or who wins and who loses economically from a policy option. For example, government funded action may benefit primarily cattle farmers at the cost of taxpayers. Decisions to cull wildlife to control transmission of disease to cattle may impose costs both on the taxpayer (to the extent that government is involved), on farmers (who might be licensed to cull badgers for example) and on society more widely where citizens place a value on the wildlife concerned and would prefer that they are not killed by farmers. There are therefore, issues concerning whether the disease control strategy is a national one, whether it is voluntary or mandatory, who pays for control and other costs, whether government should be involved on behalf of society (and the extent to which it should be involved) or whether the problem should be 'left to the market' for cattle producers to solve. The UK national test and slaughter policy for bovine tuberculosis in cattle imposes costs on cattle producers (time involved in testing, movement restrictions, disruption to the farm business, etc.) and on the government and taxpayer (the cost of state veterinary inspection, slaughter compensation paid to farmers, etc.).

The basic economic question is whether the cost of control is less than the benefits of control in terms of reduction in disease incidence in cattle and the costs of disease impacts. In terms of badger culling, this is the question addressed in Box 5.3. Sometimes impacts are difficult to measure and particularly to value. This is true where a wildlife species has a value to society (or a group within society), which is not represented by a market and therefore has no market price attached to it (for example, the value placed on the badger population). Where a specific outcome is specified, for example a standard or level of disease incidence, a cost effectiveness analysis might be appropriate which merely assesses the costs of alternative policies to achieve the outcome.

Sometimes impacts of disease or of policy to manage disease have uncertain (unknown risk/probability) or risky outcomes. These considerations need to be incorporated into the analyses, and simulation modelling and sensitivity analysis are two ways of doing this. Disease and its effects are dynamic as are the effects of

Box 5.3 Bovine TB: cost–benefit analysis of badger culling

A full cost–benefit analysis would involve calculating all direct, indirect and marginal costs involved in any management decision. As indicated in Box 5.2 the value of badgers (*Meles meles*) (and indeed cattle) to human welfare is difficult to measure, particularly since the value will change depending on how the badger population would be reduced (e.g. culling or fertility control). In this example we will ignore welfare valuations and illustrate a partial cost–benefit analysis based only on the financial costs of badger control and the financial savings achieved by any reduction in the number of cattle herds affect by bovine tuberculosis (bTB).

This is best achieved with an individual-based simulation model, which includes the badger population, cattle herds, and transmission of bTB between them (for details of the badger model see Smith et al. 2001). Cattle testing and slaughter, and badger removal are also simulated. Costs can be attributed to each of these activities (e.g. cattle testing, herd breakdown and restriction of cattle movement, badger culling) by randomly choosing a value from measured distributions for each event that occurs within each iteration of the model. Costs and benefits are calculated by comparing the costs of a simulation with badger control, against the costs of an identical simulation, but without badger control. Culling is known to affect the behaviour, and disease epidemiology of bTB within the badger population (see Chapter 7). This perturbation effect increases the disease transmission rate from wildlife to cattle, particularly in the areas immediately surrounding the area culled. For illustrative purposes we will ignore this spatial effect of social perturbation on the system, by modelling an isolated badger population.

As badger culling efficacy increases, the badger population is further reduced. Although transmission rates may increase as a result of culling, overall there are less infectious badgers and the frequency of cattle-herd breakdowns decreases (Fig. 5.5). This causes an economic saving in terms of herd breakdowns. The average herd breakdown imposes a cost of about £14,000 to society (Smith et al. 2007b). We choose here a method of badger

(continued)

Box 5.3 (continued)

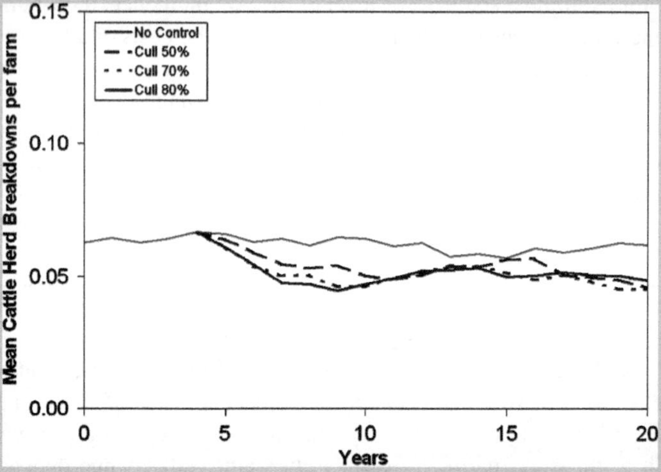

Fig. 5.5 Simulated reduction in mean cattle-herd breakdown rate with increasing levels of badger culling (no control, 50%, 70% and 80% of badgers removed each year for 5 years) in an isolated badger population. Note, that only those herd breakdowns caused by badgers can be affected by badger culling, so no amount of badger culling will reduce the herd breakdown rate to zero

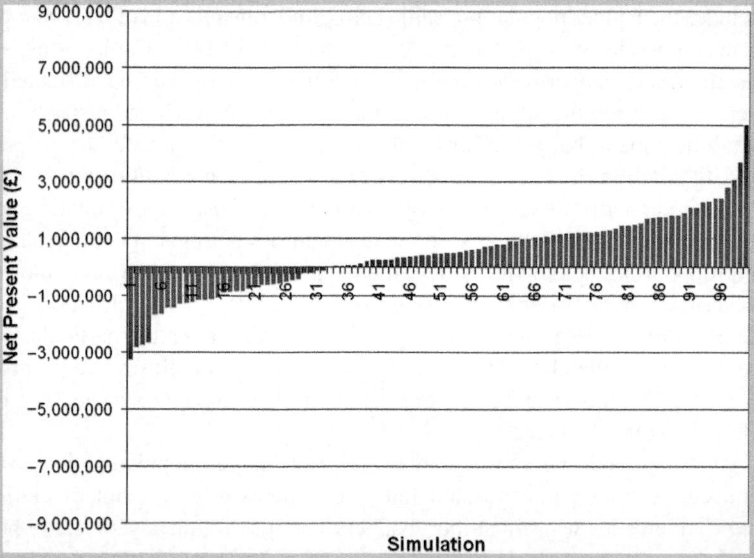

Fig. 5.6 Distribution of Net Present Values (discounted at 3.5% over 15 years) for 100 iterations of the model for badger gassing (assumed to be a relatively cheap and effective option killing 80% of all badgers), if performed for five consecutive years over 100 km^2

control that is expected to be relatively cheap and effective (gassing of badger setts). We are not making any ethical or value judgement about the method of culling, since we are not including the human welfare aspect. Gassing would be expected to cost approximately £1000 per square kilometre treated, if performed by individual landowners, including the cost of government licensing and monitoring. Results from 100 simulations show that in two-thirds of the cases an economic benefit is achieved (Fig. 5.6). This also means that the costs outweigh the benefits in one out of three cases. Clearly this is a policy that has an economic risk associated with it, and if badger control is less efficacious than simulated, or public demand makes it more costly by using an alternative method of control, this would increase the risk of an economic loss. This risk would be further increased if the population were not isolated, but resulted in social perturbation of surrounding badger social groups. This example not only demonstrates the utility of CBA, but also how stochastic models can be used to help decision-making (see Chapter 4).

wildlife disease management policies. In economic terms, this means that there are usually a stream of costs and benefits associated with disease and its control over time. This has to be taken into account and in economics this usually involves 'discounting', whereby a future cost or benefit is given a lower present value than a current cost or benefit, using a discount rate. The sum of current and future values (discounted benefit minus cost) is then the discounted Net Present Value (NPV).

From a policy perspective, this is likely to mean that economic analyses of wildlife disease management problems result in estimates showing a range of possible outcomes, depending on various assumptions and states of nature. It is then left to the decision maker to use these analyses as information on which to base a decision, taking account of other considerations, such as those of a social and political nature. It is acknowledged that economic analyses cannot sensibly place a money value on all considerations and so can never take the place of the decision maker.

5.8 Issues in Economic Analysis

The main issues for economic analysis of wildlife disease management common to most problems are:

1. Specification of the technical relationship between disease in wildlife and its impacts such as disease in domesticated animals and humans
   The extent of effects of wildlife disease is often uncertain or difficult to estimate with any precision due to a lack of empirical data – for example, the contribution

of wildlife to disease in livestock (such as badger to cattle disease transmission rates for bovine tuberculosis).

2. Efficacy of disease control methods in wildlife
   The extent to which alternative disease management methods will reduce wildlife disease and the problems associated with it are often equally uncertain. Sometimes, the efficacy of a control method will depend on both animal and human behaviours, which may be difficult to predict. For example, cattle farmers may react to increased cattle compensation by failing to adopt good farm biosecurity; hence increasing the likelihood of cattle infection from wildlife.
3. Economic benefits and costs – the full story
   All economic costs and benefits need to be identified and included in the analyses if possible. Reliable valuations need to be attached to impacts, even where non-market aspects might be involved. Market prices might not reflect true values or opportunity costs (for example, they may be distorted by government policy or some other factor which needs to be taken into account).
4. Economic optimum
   Because of problems associated with 1–3 above, the estimation of economic optima is difficult and it may not be realistic to seek such estimation.
5. Decision-making
   As stated above, economic analysis is one input to the decision-making process and needs to be considered alongside other types of information. Economic analyses cannot consider everything and are usually limited in a number of regards. The merits and limitations of any analysis need to be clearly recognised. Notwithstanding this, making resource use decisions without economic analyses is unlikely to result in well-informed decisions.

We therefore argue that an economic perspective should be included within any wildlife disease management strategy. We also note that there are many areas where costs are difficult to ascertain. The introduction of myxomatosis into Great Britain caused significant environmental changes as a result of the dramatic decline in rabbit numbers (Thompson 1994), and this could be viewed as a cost or a benefit! Some costs of intervention only become apparent afterwards. The large-scale elimination of rabies from foxes in Western Europe led to increases in fox density and an associated increase in the range and prevalence of other zoonotic parasites (e.g. *Echinococcus*, Pleydell et al. 2004). Thus, similar to modelling, economic analysis should be performed, but should not be used in isolation to make decisions.

Chapter 6
Options for the Control of Disease 1: Targeting the Infectious or Parasitic Agent

Jean Blancou, Marc Artois, Emmanuelle Gilot-Fromont, Volker Kaden, Sophie Rossi, Graham C. Smith, Michael R. Hutchings, Mark A. Chambers, Steve Houghton, and Richard J. Delahay

6.1 Introduction

There are three basic approaches to managing diseases: directly reduce the reproductive rate of the pathogen, reduce host (or infected host) density, or manipulate the environment to reduce contact between diseased and susceptible animals. In this chapter we will look at the first of these approaches. Since disease transmission results from direct or indirect contact between infectious and susceptible individuals,

J. Blancou
11, rue Descombes 75 017 Paris, France

M. Artois
Université J. Fourier, Laboratoire TIMC-IMAG, Unité Environnement et Prévision de la Santé des Populations F-38000 Grenoble; Ecole Nationale Vétérinaire de Lyon, France

E. Gilot-Fromont
Université de Lyon; Université Lyon 1; CNRS; UMR5558, Laboratoire de Biométrie et Biologie Evolutive, 43 Boulevard du 11 Novembre 1918, Villeurbanne F-69622, France

V. Kaden
Friedrich-Loeffler-Institut, Bundesforschungsinstitut für Tiergesundheit, Institut für Infektionsmedizin, 17493 Greifswald-Insel Riems, Germany

S. Rossi
Unité Sanitaire de la Faune, Micropolis, la Bérardie, Belle Aureille, 05000 GAP, France

G.C. Smith and R.J. Delahay
Central Science Laboratory, Sand Hutton, York, YO41 1LZ, UK

M.R. Hutchings
SAC, West Mains Road, Edinburgh, EH9 3JG, UK

M.A. Chambers
TB Research Group, Department of Statutory and Exotic Bacterial Diseases, Veterinary Laboratories Agency, Weybridge, New Haw, Addlestone, Surrey, KT15 3NB, UK

S. Houghton
Veterinary Vaccines Consultancy, Coles Cottage, Plum Park Lane, Paulerspury, Northants, NN12 7NN, UK

R.J. Delahay et al. (eds.), *Management of Disease in Wild Mammals*,
DOI:10.1007/978-4-431-77134-0_6, © Springer 2009

there are two ways to target an infectious agent: either limit the number of susceptible individuals by vaccinating them, or treat infected individuals in order to reduce the duration or intensity of the infectious period and the number of infectious individuals present at any given time. The overall aim of this chapter is to consider the conditions under which vaccination and treatment may make a valuable contribution to the control of infectious diseases in wild mammal populations. Both field research and mathematical modelling approaches have been used to address this question. For vaccination, early mathematical models of infectious disease dynamics suggested a simple answer: vaccination is useful as soon as the rate of control ensures that a sufficient proportion of the population is immune for a sufficient period of time (Bailey 1957). At the individual level, this herd immunity means that any given infectious individual has a low probability of encountering a susceptible animal. If the disease is introduced into a vaccinated population, the mean number of secondary infections caused by each infected case will be lower than unity, thus preventing further outbreaks from occurring ($R < 1$: see Chapter 3). However, this generalised scenario may be considered overly simplistic, as the practicalities of vaccination campaigns often complicate matters. For example, modelling studies often include assumptions about perfect vaccine efficacy, and the efficiency of delivering the vaccine to a population that may or may not reflect the situation in the field.

Red fox (*Vulpes vulpes*) rabies in Europe provided the earliest example of a disease of wildlife where vaccination appeared a realistic possibility, thanks to the pioneering work of Frantz Steck, Alexander Wandeler and their co-workers (see Section 6.2). Owing to the inadequacy of fox culling as a method of rabies control (see Chapter 7), European countries pursued the development of oral vaccination. As soon as appropriate baits for oral vaccination and safe vaccines were available for use on a large scale, the relative merits of vaccination and culling were investigated. The ensuing studies showed that vaccination of foxes was more efficient at halting epizootics than culling, it was less costly in the long-term, and importantly it could be rapidly deployed in response to the re-emergence of disease (Aubert 2003). Vaccination also had the added benefit that unlike culling it did not destabilise the social structure of fox populations, and so avoided the potential for perturbation to enhance transmission rates (Macdonald 1995) (see Chapter 2). The eradication of rabies from Western Europe at the end of the 20th century, following a period of intensive oral vaccination of foxes, finally paved the way for other vaccination strategies to combat wildlife diseases (Pastoret and Brochier 1999).

Contrary to the assumptions of most early models of wildlife disease dynamics, wild mammal populations are not homogeneous. Under the assumption of population homogeneity, all determinants of disease propagation are identical over space, and control measures are not expected to affect the spatial and social behaviour of individuals. These are clearly unrealistic expectations given the potential social complexity of mammal populations and the profound influence this can exert on disease dynamics and management efforts (see Chapter 2).

Hence, vaccination programmes may require heterogeneous effort over space and time (May and Anderson 1984) in order to optimally deploy resources for disease control. Vaccination is predicted to be the most efficient method in populations where rates of host birth and death, and disease propagation are relatively low. Elsewhere, culling or combined strategies may be more efficient (Barlow 1996). Nevertheless, the appropriate approach may vary widely between different mammal species, depending on their particular ecological and behavioural characteristics and the epidemiology of infection in the target population. Therefore, obtaining basic information on host population processes and disease dynamics at the appropriate spatial scale is an essential first step in determining the most appropriate control plan.

Financial cost is clearly a consideration when developing vaccination strategies (see Chapter 5) and will vary widely depending on how long vaccination is predicted to be necessary in order to achieve eradication or some other stated aim. The potential occurrence of multiple disease outbreaks or failure of early confinement may have dramatic effects on limited resources, and so contingencies for such events should be built into any vaccination campaign. Moreover, vaccinating or treating may have other indirect costs. Vaccinating a reservoir population may potentially lead to an increase in host density, and so enhance the risk of transmission of other diseases or amplify any other problems associated with the host (e.g. damage to crops or livestock). Human population growth in parts of Africa has been accompanied by a dramatic rise in domestic dog (*Canis lupus familiaris*) populations, which act as reservoirs of rabies and distemper viruses with spillover into wild canid populations. Vaccination of domestic dogs offers the potential to benefit endangered wild canids by reducing the risks of spillover, but could also have the potentially undesirable effect of increasing dog populations further (Cleaveland et al. 2002). Vaccination and direct medication of wild mammals may also have undesirable evolutionary consequences. For example, the long-term use of vaccines will impede any selection in the host population for resistance, or could lead to the selection of non-vaccinal strains of the pathogen, depending on the mode of action of vaccine induced immunity. Similarly, the inappropriate application of direct medication (e.g. antibiotics) may give rise to the emergence of antibiotic resistance in the pathogen.

Whether vaccination or treatment are useful options in any given situation will be guided in part by an evaluation of the population status of the host and the role of the pathogen in host dynamics. The extent to which pathogens have significant long-term effects on host population dynamics remains unclear. When the host is perceived to be 'over-abundant', the initial response is often to cull. In contrast, in small populations of endangered species, the value of each animal takes on a greater significance, such that treating individuals may become a viable option (see Chapter 11). In the future vaccination and treatment strategies may become more attractive as disease control options in a wider range of circumstances, particularly as more candidate vaccines are developed for use in wildlife, and if practical methods of deployment can be improved.

6.2 Early History

6.2.1 Medication

Large-scale use of direct medication has rarely been contemplated for wild mammals except in exceptional circumstances where there has been a serious threat to human health, or to highly valued wildlife. Although there are a few examples of the direct medication of wild mammals on a limited scale (see Section 6.3), this approach has remained largely restricted to those individual cases treated in veterinary surgeries, and during rehabilitation (e.g. at marine mammal centres) and translocation exercises. This is understandable, given the substantial practical difficulties in handling wild mammals, and the cost of veterinary drugs.

6.2.2 Vaccination

Vaccination is a term based on Edward Jenner's use of cowpox virus administered to humans to protect them against smallpox (*vacca* means cow in Latin). This practice worked because the immune response to cowpox protected individuals from the more virulent smallpox. Early vaccines were produced by one of two methods. Firstly, the repeated passage of the microorganism through rabbits or mice can create a weakened strain, which when injected into a patient may protect them from the virulent strain. Alternatively, microorganisms may be killed using chemicals or heat treatment and then used as a vaccine. In both instances the vaccine contains antigens associated with the pathogen and so invokes an immune response in the immunised host, but without causing disease.

In 1971 foxes were first immunised against rabies in the USA by the oral administration of the SAD ("Street Alabama Dufferin") vaccine, which is a live attenuated strain of the rabies virus (Baer et al. 1971). The research programme was prematurely terminated following the accidental infection of one of the scientists. However, work continued in Western Europe, and the first field trials using SAD virus in chicken-head baits were conducted in an Alpine valley in Switzerland by Frantz Steck, who unfortunately died when his helicopter crashed during the distribution of vaccine baits in the mountains. In 1983, trials using the SAD-B19 strain in chicken-head baits were conducted in Germany, followed in 1985 by the same strain in manufactured baits, and two years later by the use of the SAD-Bern strain in a field trial in southern Ontario (Canada). The first national vaccination campaigns using the SAD-B19 strain began in France and Luxembourg in 1986.

A new approach to rabies vaccination was developed in the 1980s, involving the delivery of the rabies virus surface antigen (glycoprotein) in a genetically modified vaccinia (pox) virus (VRG, or Vaccinia-rabies glycoprotein recombinant virus), which replicates within the vaccinated host (Thomas et al. 1990). The VRG vaccine was first used in Belgium in 1987 (Pastoret et al. 1988). A large-scale field trial of raccoon

(*Procyon lotor*) vaccination commenced in New England, USA in 1992 using VRG in oral baits distributed by hand and helicopter over 559 km^2. From 1995 to 2000, successful vaccination campaigns continued in Western Europe using the SAG2 strain of live attenuated virus and VRG, until the eradication of the disease. During this period large scale VRG vaccination campaigns were also conducted in coyotes (*Canis latrans*) and gray foxes (*Urocyon cinereoargenteus*) in Texas, USA (Black and Lawson 1980; Blancou et al. 1991; Brochier et al. 1996; Fearneyhough et al. 1998; Hanlon et al. 2002). An investigation of the relative cost-effectiveness of oral vaccination versus fox culling concluded that the former became economically beneficial after four years, and that culling had only ever resulted in a transient lull in the occurrence of the disease, while oral vaccination resulted in elimination (Aubert 1999).

Oral vaccination against classical swine fever (CSF) has been investigated in Europe since the 1960s, both under laboratory conditions and in the field. Initially, the vaccination of wild boar (*Sus scrofa*) was carried out in the former Soviet Union by adding liquid or freeze-dried CSF vaccine to cereal-based feed in heaps or troughs. Vaccination was performed in response to CSF epidemics in what are now parts of Russia, Byelorussia, Moldavia and Ukraine during the periods 1975–1976 and 1990–1991. The efficacy of vaccination varied depending, amongst other things, on the course of the epidemic and the size of the area affected. During the 1990s a freeze-dried vaccine was developed in Russia and was administered to boar in food provided exclusively in winter (Kolomitsev et al. 1998). In the mid-1980s in Romania, oral vaccination of wild boar against CSF was investigated by feeding them the hind legs of rabbits (*Oryctolagus cuniculus*) used for the production of rabbit-passaged CSF vaccine, or with hen's eggs that had been inoculated with live virus vaccine immediately before distribution. At the same time, oral immunisation of wild boar was carried out under laboratory conditions in the former German Democratic Republic using parts of rabbit carcasses derived from animals prepared for the production of rabbit-passaged C-strain (China-strain) vaccine. In Italy (Rutili et al. 1987) and in France (Chenut et al. 1999), oral vaccination of wild boar was also investigated experimentally, using rabbit-passaged Chinese CSF virus. The results of these studies were generally encouraging, and subsequent work led to the development of a commercial CSF bait vaccine in Germany (Kaden et al. 2000), and culminated in the introduction of oral immunisation of wild boar to control CSF in several parts of Europe.

In 1990, oral immunisation of wild boar against CSF was again studied in Germany under laboratory conditions using the live attenuated C-strain virus, in baits similar to those previously used to deliver rabies vaccine to foxes. From 1993 to 1995 a field study was carried out in an area of approximately 270 km^2 in Lower Saxony. The C-strain virus used for oral vaccination was incorporated into a cereal-based bait matrix. Vaccination campaigns took place in spring and autumn, and each consisted of two bouts of vaccine deployment 14 days apart. In a subsequent study the prevalence of CSF antibodies in boar varied from 49% to 60% in the vaccinated population, with over 50% of the young animals failing to ingest bait and become immunised. Consequently, intensive hunting of the young animals was deemed necessary as an adjunct to oral vaccination, and after the third immunisation campaign, no virus was detected in the treated areas (Kaden et al. 2000). The subsequent deployment of vaccine baits in other parts of Germany included the addition of a campaign in the summer (also consisting of two parts) and an extension of the interval between bouts of vaccine deployment to 4 weeks. In Baden-Württemberg, where this procedure was first applied, the prevalence of CSF antibodies in wild boar continued to increase until the third immunisation campaign when it peaked at 72%. The addition of a third campaign also succeeded in achieving higher levels of immunisation amongst young wild boar compared to previous campaigns (Kaden et al. 2005b) (see Box 6.3).

Although routine vaccination against CSF is prohibited in domestic pigs within the European Union, emergency vaccination is permitted (by oral immunisation) if, in its absence, the extensive spread of virus is considered to be likely. The vaccination procedure for wild boar, which has been used in several German Federal States, France (since 2004), Luxembourg (from 2003 to 2004), the Slovak Republic (since 2005) and Bulgaria (since 2006), has comprised of three vaccination campaigns in spring, summer and autumn (Kaden et al. 2006; Rossi et al. 2006).

6.3 Wildlife Medication

6.3.1 Basic Principles

Direct medication of free-ranging wild mammals is most likely to be seriously considered when there is no other way to control a disease that affects individuals of an endangered or valuable (e.g. as hunted game) species. Direct medication has however, also been used in some cases where the targeted disease threatened public health, or represented a threat to livestock or game animal production, or to international trade of animals and derived products. Nevertheless, the prohibitive costs and substantial practical difficulties of administering medication to wild mammals means that such examples are rare (see Chapter 11). In contrast direct medication is common practice in some countries during the rehabilitation of wild mammals, and is advisable during translocations that may incur a risk of spreading pathogens that are zoonotic or of significant potential economic or conservation concern.

6.3.2 Advantages and Disadvantages

Direct medication of wild mammals may allow the control of pathogens when the isolation or culling of hosts is not practical or acceptable, and is a more ethically attractive option. As individual animals may need to be captured or restrained, then this may provide an opportunity to carry out health checks, which may be particularly valuable in the case of endangered species. Also, where direct medication is aimed at specific individuals it is likely to incur negligible risks to non-target species.

Direct medication requires the use of regulated drugs, and official authorisation would be necessary for their use in wild animals. Such drugs may be expensive and may need to be deployed by qualified professionals, potentially causing costs to grow substantially if the disease is not rapidly eradicated. Undesirable potential side effects of direct medication include the persistence of harmful residues of veterinary drugs in the environment and non-target species (Green et al. 2006). Also, evolutionary effects such as the emergence of drug-resistant disease strains and inhibition of selection for resistant hosts, could potentially lead to more extreme epizootics in the future.

The effects of direct medication in wild mammal populations can be difficult to predict. Baits containing the anthelmintic (de-worming) drug praziquantel (Droncit®) have been deployed for the control of the cestode *Echinococcus multilocularis* in European red foxes (see below). However, since the drug is not an ovicide, it has been suggested that uptake by foxes could result in a mass release of *E. multilocularis* eggs into the environment, which could in turn increase the probability of host exposure to the parasite (Petavy 2008). Concern has also been expressed amongst some game managers over the use of salt stones (mineral licks) as bait for the distribution of anthelmintics (or other drugs) to free ranging deer, as this could potentially encourage aggregation and so facilitate the spread of contagious pathogens. Experimental use of an acaricide in rabbit burrows dramatically reduced the numbers of rabbit fleas, the vector of myxomatosis, and resulted in a two- to three-fold increase in rabbit density (Trout et al. 1992). This clearly demonstrated how much the disease was suppressing the population. Nevertheless, the ecological consequences of direct medication as a tool for controlling infectious disease in wild mammals remain poorly understood.

6.3.3 Approaches

An important and successful example of treating wild mammals was achieved in the National Wildlife Research Centre of Taif (Saudi Arabia), although this involved a prolonged period of captivity. Following an outbreak of bovine tuberculosis (caused by *Mycobacterium bovis*) in a herd of Arabian oryx (*Oryx leucoryx*), individuals were captured and treated in captivity with antimycobacterial combination therapy. This was successful in producing tuberculosis-free oryx for release into the wild (Greth et al. 1994). Similarly, capture followed by treatment was used to eradicate sarcoptic or notoedric mange in free-ranging Spanish ibex (*Capra ibex*), cheetah (*Acinonyx jubatus*) and other endangered or genetically compromised populations (Pence and Ueckermann 2002 and Section 11.2). Game species have also benefited from direct medication for the control of helminth infections. Successful examples include anthelmintic treatments in bighorn sheep (*Ovis canadensis*) (Schmidt et al. 1979), snowshoe hares (*Lepus americanus*) (Murray et al. 1996) and white-tailed deer (*Odocoileus virginianus*) (Qureshi et al. 1994). In these instances the anthelminthic drugs were administered orally, being mixed with food or salts in areas where the target animals were known to feed.

Finally, as mentioned previously, direct medication was employed to protect public health by controlling *E. multilocularis* in populations of wild red foxes in Europe through the delivery of a bait containing praziquantel (Droncit®). After six bait deployment campaigns the average prevalence of infected foxes had declined from 32% to 4%. However, re-infection is likely to occur, since the infection does not produce a strong immune response. This approach is therefore still under evaluation in southern and northern Germany (Eckert et al. 2001) (see Box 6.1).

6.4 Wildlife Vaccination

6.4.1 Basic Principles

The goal of vaccination in wild mammals may be to eliminate a disease (e.g. wildlife rabies), and therefore to remove the threat to human health or susceptible domestic species, to reduce the prevalence of a disease to an acceptable level, or to prevent the extinction of a valued population or species. The best known and most successful example of the application of vaccination to manage disease in wildlife is the immunisation of wild mammals against rabies. Targeting the European red fox has resulted in the near complete elimination of rabies from West and Central Europe. Similar strategies have subsequently been used to control rabies in other species, including raccoon dogs (*Nyctereutes procyonoides*) in Europe and coyotes, striped skunks (*Mephitis mephitis*) raccoons and arctic foxes (*Alopex lagopus*) in the USA and Canada. The oral vaccination of wild boar against CSF (see Section 6.2.2) has also met with some success. This approach achieved the elimination of CSF infection in wild boar in several German Landër (states) within one or two years (Kaden et al. 2003; Von Rüden et al. 2008), and subsequent successes have been anecdotally reported in other European countries.

Following these two success stories, vaccines have become more widely considered as potential options for the control of disease in wild mammals. Particular interest has focused on the potential vaccination of wild mammals against bovine tuberculosis caused by *Mycobacterium bovis*. Currently, the only tuberculosis vaccine available to investigate in wildlife is BCG (Bacille Calmette Guérin), which is a live attenuated strain of *M. bovis* used extensively in humans. BCG has been tested experimentally in Eurasian badgers (*Meles meles*) in the Republic of Ireland (Southey et al. 2001), in wild boar (*Sus scrofa*) in Spain (Ballesteros et al. 2007) in African buffalo (*Syncerus caffer*) in South-Africa, in white-tailed deer in the United States (Waters et al. 2004; Palmer et al. 2007), and in red deer (*Cervus elaphus*) (de Lisle et al. 2002), ferrets (*Mustela putorius furo*) (Qureshi et al. 1999), and brushtail possums (*Trichosurus vulpecula*) in New Zealand (Corner et al. 2002; Wedlock et al. 2005). These studies have usually involved administration by injection as proof of principle, but some work has also explored alternative means of delivery, such as by nasal, conjunctival

Box 6.1 Echinococcus treatment in foxes

Recent studies in Europe, Asia and North America have revealed that the zoonotic tapeworm, *Echinococcus multilocularis*, has a far wider geographic distribution in carnivores (predominantly foxes) than previously thought. In Europe, growing red fox *(Vulpes vulpes)* populations and their increasing colonisation of urban areas, may potentially represent an emerging hazard to public health. Therefore, the development and implementation of effective methods of disease control and prevention are required. *E. multilocularis* is typically perpetuated in a wildlife host community, which includes foxes (genera *Vulpes* and *Alopex*) as definitive hosts and various rodent species as intermediate hosts. Humans can accidentally ingest the eggs, which hatch, and the larval stages (metacestodes) then usually enter the liver but can spread to other organs, and can lead to potentially fatal alveolar disease. Risk factors for alveolar echinococcosis may include occupational and behavioural activities. Areas of eastern France with high water vole (*Arvicola terrestris*) densities yielded a ten-fold higher risk of human alveolar echinococcosis compared to those with low densities of this important intermediate host (Viel et al. 1999). In an area where *E. multilocularis* was endemic, as many as 39% of water voles and 7% of domestic dogs with free access to rodents were infected (Gottstein et al. 2001). Red foxes are likely to be the most important definitive hosts in many regions. In the past two decades, foxes have started to colonise cities around the world, and evidence is growing of a perpetual parasite life cycle in urban areas.

Few field studies focus on anthelmintic treatment of definitive hosts. In rural areas of Germany and Japan, baits laced with praziquantel (an anthelmintic) lowered the prevalence of *E. multilocularis* in foxes, although rapid recovery of the disease was also observed (Hansen et al. 2003), suggesting that prolonged repeated treatment may be necessary (Tsukada et al. 2002). Until recently no attempt had been made to evaluate the treatment of foxes in urban areas. In an experimental field study in Zurich, Switzerland, the effects of anthelmintic baits were investigated in urban areas where the organism was endemic (Hegglin et al. 2003). Over a 19-month period, 50 baits containing praziquantel were distributed per km^2 every month in six 1 km^2 areas and one 6 km^2 area. By the end of the trial, the proportion of fox faecal samples that were antibody positive to *E. multilocularis* had decreased significantly in all the baited areas. *E. multilocularis* prevalence in the intermediate host (water vole) also decreased significantly in treated areas. This experimentally controlled study suggests that a pronounced reduction in *E. multilocularis* egg contamination is achievable by treating foxes in urban areas where the organism is endemic.

(Corner et al. 2002; Corner and Buddle 2005) or oral (Aldwell et al. 1995b; Qureshi et al. 1999; Aldwell et al. 2003b; Wedlock et al. 2005; Buddle et al. 2006b) routes.

Vaccination of bison (*Bison bison*) and elk (*Cervus elaphus*) against brucellosis (*Brucella melitensis*) has been considered in the USA, using vaccines (either the S19 or the RB51 strains) administered by hand or contained in ballistic capsules. However, the S19 strain was not as effective in bison as it was in domestic cattle, and the RB51 strain caused inflammation of the placenta and spontaneous abortions. The release of live *Brucella* vaccine strains in wildlife is therefore of concern as it could lead to environmental contamination and infection of other wild species (Godfroid 2002; Olsen et al. 2002; Olsen et al. 2006).

Parenterally administered (injected) vaccines have been tested on a number of wild mammal species. These were either experimental studies or interventions with a follow up investigation, or actions performed for conservation reasons without any subsequent monitoring of individuals. Examples where useful data on the effects of vaccination were recorded include studies on anthrax in cheetahs and black rhinoceros (*Diceros bicornis*), (Turnbull et al. 2004), pasteurellosis in bighorn sheep (Kraabel et al. 1998), rabies in the Ethiopian wolf (*Canis simensis*) (Haydon et al. 2006) and myxomatosis and rabbit haemorrhagic disease (RHD) in European rabbits (Calvete et al. 2004b). Beneficial effects were reported in all these studies. In contrast, the control of anthrax (injection by hand, or by dart from an aircraft) in buffalo, black and white (*Ceratotherium simum*) rhinoceros, roan antelope (*Hippotragus equinus*) and hippopotamus (*Hippopotamus amphibius*) (Clegg et al. 2007), of morbillivirus infection in seals (*Phoca vitulina* and *Monachus monachus*), and of distemper in the endangered black-footed ferret (*Mustela nigriceps*) in the USA (Moutou, 1995) took place in the interests of conservation, but with no follow up of individuals.

Experimental oral vaccination by direct dosing has been demonstrated to protect black-tailed prairie dogs (*Cynomys ludovicianus*) against sylvatic plague (*Pasteurella pestis*) (Creekmore et al. 2002; Mencher et al. 2004; Morton et al. 2004) and wild rodent reservoirs of *Borrelia burgdorferi* against Lyme disease (Tsao et al. 2004; Gomes-Solecki et al. 2006; Scheckelhoff et al. 2006). Oral baits have been successfully employed in the experimental vaccination of feral pigs and wild boar against pseudorabies using a live recombinant vaccine. Recent work has indicated that abrasive agents in bait may enhance uptake of live vaccines by allowing them to penetrate the tissues of the buccal cavity (Edmonds et al. 2001).

The current body of evidence demonstrates the potential for vaccination to make significant contributions to the future management of disease in wild mammals. Although approaches involving the capture or darting of individuals may always be limited by the high levels of effort and costs involved, the delivery of vaccines in oral baits lends itself to larger-scale deployment. It is likely that in the near future vaccination will play an increasing role in the management of wildlife diseases other than rabies and CSF, with bovine tuberculosis looking to be at the top of that list.

6.4.2 Advantages and Disadvantages

If the practical challenges of delivering a vaccine to wild mammals can be overcome (see Section 6.4.4), and successful immunisation of the required number of animals is achieved, then this approach may offer a viable alternative to culling hosts. For this reason vaccination strategies are frequently advocated by conservationists, animal welfare groups and the general public. However, other considerations that need to be met include demonstration that the vaccine is either safe in, or unavailable to, non-target species (including humans) and is environmentally benign.

When evaluating candidate vaccines to be considered for use in wildlife, it is live vaccines that pose the most questions. Reversion to virulence has to be addressed as part of the licensing procedure by sequential passage in the target species (but see Section 6.4.3.1). Exposure of non-target species also needs to be considered. However, since experimental studies may be difficult to conduct it is important to understand the nature of the attenuation to inform an assessment of the likelihood of the vaccine strain posing any safety risk. Given the limitations of the analyses required to achieve marketing authorisation for a vaccine, the risks associated with its use in the field can only ever be minimised, not removed.

Vaccination of a wild host population may have significant long-term ecological consequences. By reducing the rate of disease-induced mortality for example, vaccination may have the effect of increasing host population size, and altering demographic structure and processes. This could have potential 'knock-on' effects for the wider ecological community, including for instance predator, prey and vegetation communities. A further concern regarding the long-term use of vaccines in wildlife populations is that protecting hosts from the selective pressure of infection, may remove selection for natural resistance to diseases (Woodroffe 2001). As a consequence, vaccinated populations could potentially become more susceptible to infection in the future, particularly after the vaccination campaign stops. Nevertheless, highly virulent infections such as rabies induce very low levels of natural immunity in most host species, and so in this instance the costs of vaccination in terms of loss of selection pressure, would be relatively small. On the other hand, these costs could be significant in the case of less virulent infections, which induce higher levels of natural immunity.

The financial costs of vaccination may increase substantially in the long term if the disease is not rapidly eradicated, although this would need to be weighed up against any benefits accruing from the level of disease control achieved (see Chapter 5). Rising costs may be a particular issue for chronic diseases such as tuberculosis, which may take many years of vaccination before they are eliminated. The sustainability of the long-term use of both medication and vaccines is therefore an important consideration that requires careful evaluation before any programme is implemented, particularly in the case of pathogens of low virulence.

The use of vaccines in wild mammal populations has not been without controversy. This has been most frequent where vaccination has required that animals are handled, which is known to incur risks of stress-induced mortality (Arnemo et al. 2006). After the disappearance of African wild dogs (*Lycaon pictus*) from the Serengeti National Park, it was argued that the stress associated with handling during anti-rabies vaccinations may have reactivated quiescent disease and caused increased mortality (Burrows 1992). However, subsequent field data and a review of the available evidence suggested that this was unlikely to have been a contributory factor (De Villiers et al. 1995; Woodroffe 2001). The handling of rabbits during vaccination campaigns against myxomatosis was also suggested to be detrimental, as young and sub-adult vaccinates exhibited enhanced rates of mortality during the first week after handling (Calvete et al. 2004a). Hence, the potential impact of capture-related stress and myopathy should be fully considered for any proposed vaccination campaign in which it is necessary to trap, restrain or handle the wild host.

6.4.3 Characteristics of Vaccines

The required properties of a vaccine will vary according to the characteristics of the pathogen and host. In the case of rabies, the vaccine must be delivered as a live modified virus or a live vector (e.g. vaccinia virus) because the immune reaction can only develop if the vaccine strain multiplies in the oral mucosa. In addition, the characteristics of the vaccine will be required to comply with prevailing legislation and guidelines for best practice. In Europe, vaccines intended for wildlife must fulfil all the requirements of the European Pharmacopoeia, which is a list of pharmaceutical substances and associated quality standards expected by the European Directorate for the Quality of Medicines (EDQM 2008). Recommendations on vaccine safety are also published by the World Health Organization (WHO 2008a).

To date, only three diseases have been targeted by vaccines that were either developed or adapted specifically for use in wild animal populations, the best known example being rabies. A number of different oral rabies vaccines, attenuated by repeated passage, have been produced. These may have one or more mutations that affect their virulence and pathogenicity. In general, the more mutations in a strain the less likely it is to revert to being pathogenic. These vaccines have been used in foxes and raccoon dogs in France, Belgium and Switzerland since 1985, with no major problems reported (Brochier et al. 1996; Aubert 2003; Cliquet et al. 2006). Since 1985 VRG has been used to vaccinate foxes and other carnivores against rabies in France, Belgium, Canada and the USA also without any problems (Blancou et al. 1986; Blancou et al. 1988; Blancou et al. 1992; Aubert 2003) (but see Section 6.4.3.1).

Classical Swine Fever vaccines consist of live attenuated strains of either CSF or another virus (e.g. bovine viral diarrhoea virus or adenovirus) that has been genetically engineered to carry the main immunogen (E2) of the CSF virus. Most conventional live attenuated CSF vaccines, including that contained within the German vaccine bait, are based on the C-strain. Recently, a chimeric Pestivirus (CP7_E2alf) has been developed and is being studied for oral vaccination against

CSF. Although no specific requirements have been defined for oral CSF vaccines in Europe, they must fulfil the general requirements of the European Pharmacopoeia regarding safety and efficacy, and of the relevant European Directives (European Directive 2001/82 as amended by Directive 2004/28).

The BCG vaccine against tuberculosis is a live attenuated bacterium that needs to be delivered to the host in a viable state in order to generate effective protection (Buddle et al. 1997; Skinner et al. 2005). This poses a substantial challenge for oral delivery in particular, as it requires that the immunising bacilli remain viable during formulation, storage and deployment in bait, as well as retaining viability in the host up to the point of immune induction, and ensuring that the consequent immune response is sufficient to confer protection (Cross et al. 2007b). If BCG is incorporated into a lipid matrix, it can be stored in a live state for weeks to allow distribution in the field (Aldwell et al. 2006). However, this matrix may require some modification to transform it into an attractive and palatable bait for the target species, and the time period needed for BCG stability has to take into account the time taken for batch testing and distribution. The steps being taken to evaluate the use of BCG in badgers in the UK and the Republic of Ireland are described in Box 6.2.

Other vaccines (e.g. anthrax, brucellosis, distemper, myxomatosis, pasteurellosis, RVHD) have been licensed for use in domestic animals and can be obtained from commercial sources, although their safety and efficacy in wild mammals cannot be guaranteed until the necessary studies have been performed in the target species. Novel types of vaccine that are considered to have potential for use in wildlife disease control include modified live bacterial vectors, plant-derived vaccines and DNA expressing protective antigens (Cross et al. 2007a). However, despite encouraging results obtained *in vitro* or in laboratory animals, none have been tested in wild mammals, and they are unlikely to be available in the near future.

6.4.3.1 Safety

In general the safety of candidate vaccines for use in wildlife is first assessed in laboratory animals and then in the target species in captivity, before being evaluated in the field. Further investigations may involve wild or domestic species that are likely to be exposed to the vaccine, particularly when it is to be delivered in bait. Safety studies will be required in order to obtain marketing authorisation for use of the vaccine. Establishing the safety of candidate vaccines in both target and non-target species is an essential early stage in the development of a vaccination strategy for wild mammals. Experimental studies on the safety of anti-brucellae vaccines in wild bison demonstrated viral shedding, chronic infection and vaccine-induced abortions (Godfroid 2002), which may indicate either that the correct dose is critical for protection, or that the vaccine is not suitable in bison. However, the licensing authority will take into consideration the risk-to-benefit ratio on a case-by-case basis when determining whether to grant a marketing authorisation. Consequently they may grant a licence subject to certain conditions and restrictions, and may

Box 6.2 Development of a BCG vaccine for badgers

The Eurasian badger (*Meles meles*) represents a wildlife source of recurrent *Mycobacterium bovis* infection to cattle in Great Britain (GB) and the Republic of Ireland and its vaccination against bovine tuberculosis (bTB) with BCG (Bacille Calmette-Guérin) is an attractive disease control option in both countries. BCG has the advantage of a long history of safety and efficacy in a variety of animal species (Murphy et al. 2008).

Safety of BCG (the Danish 1331 strain) was first demonstrated in captive badgers in a GB study (Lesellier et al. 2006a). Badgers were vaccinated with two consecutive doses of BCG via either the subcutaneous or intramuscular routes. The first dose was high ($16–22 \times 10^7$ colony forming units (CFU)), representing between 20 and 1,100 fold the actual target dose, and was followed 15 weeks later by a lower dose (of $4–7 \times 10^5$ CFU). The vaccine was tolerated well, with the only observed effect being localised swelling at the site of BCG injection, which disappeared 48 days after intramuscular vaccination but persisted at least three times longer in those vaccinated subcutaneously. Strong cellular immune responses were observed 13 days after the first vaccination, which persisted for at least 76 days. The lower dose induced a weaker and shorter-lived response.

There are active R&D programmes in both GB and the Republic of Ireland aimed at obtaining marketing authorisation for the use of BCG in badgers. As a starting point to both programmes, the Danish 1331 strain of BCG is being used as it is manufactured in an EU Good Medical Practice (GMP) facility, and is already licensed for use in humans. As such, essential quality and analytical data are already available for inclusion in a marketing authorisation.

Having been demonstrated as safe when administered to captive badgers, work in GB has progressed to evaluation of the vaccine in a small-scale ($55\,km^2$) field study. Permission to conduct the study was granted by the Veterinary Medicines Directorate (the UK veterinary medicines licensing body) in the form of an Animal Test Certificate, following submission of a summary of the quality data, a report on the GLP (Good Laboratory Practice) safety study and a detailed study protocol. The study started in 2006 and should be concluded by 2010. It is conducted according to the principles of Good Clinical Practice (GCP) (EMEA 2000), and has two specific aims: (a) to confirm the safety of BCG Danish 1331 previously demonstrated in the GLP safety study, when given intramuscularly to wild badgers at a dose of $2–8 \times 10^6$ CFU; and b) to investigate the immunogenicity and efficacy of BCG in wild badgers. These data will indicate the potential for investigating the likely benefits of widespread badger vaccination with BCG.

In parallel with the GB studies, protocols for the experimental infection of captive badgers by endobronchial instillation of *M. bovis* were developed in Ireland (Corner et al. 2007). These have been used to demonstrate the efficacy of BCG vaccine delivered via a number of routes, including subcutaneous, nasal/conjunctival, and oral (Buddle et al. 2006a; Lesellier et al. 2006b). Equivalent studies are underway in GB using either the intramuscular or oral routes of administration, in order to generate definitive efficacy data and define the lowest efficacious dose that might be used, thereby keeping the cost of the vaccine to a minimum.

A combination of the safety and efficacy data derived from studies with both captive and wild badgers, together with quality and analytical data on the vaccine, will form the bulk of the application to obtain a Marketing Authorisation for the intramuscular administration of BCG to badgers. Possible applications for the use of the injectable vaccine in the UK and Ireland are being considered by the respective Governments. However, it is broadly recognised that the application of an injectable vaccine will be significantly restricted by the cost and practicalities associated with its delivery in the field. Nonetheless, data obtained with an injectable form of BCG in badgers would build confidence in the possible performance of a future oral bait form of the vaccine.

Delivery of BCG in oral bait holds the best prospect for vaccinating badgers over a wide geographical area. However, as a live replicating vaccine, BCG has the limitation of little to no efficacy if delivered orally in a non-viable state (Skinner et al. 2005). This is exacerbated in the case of oral delivery, by inactivation in the low pH environment of the stomach (Aldwell et al. 1995a; Buddle et al. 1997; Skinner et al. 2005). Recent advances in the formulation of BCG for oral vaccination of possums *(Trichosurus vulpecula)* in New Zealand (Aldwell et al. 2003b) are being exploited for the vaccination of badgers. These studies are at an early stage but encouraging results are being obtained with BCG delivered in a lipid matrix (Lesellier et al. 2006b). An additional challenge remains in identifying a suitable bait that is compatible with the BCG formulation, and that has the optimal properties of attractiveness, palatability, and stability in the field, whilst complying with all pertinent legislation (see Section 6.4.4). Assuming all these criteria can be met, the method of delivery in the field may have an even greater impact on the success of any oral vaccination campaign than the choice of the bait itself (Cagnacci et al. 2007).

Alongside the vaccine development studies, supporting work in the UK and Ireland has resulted in a range of immunological tests for the badger (Goodger et al. 1994; Dalley et al. 1999; Southey et al. 2002; Greenwald et al. 2003; Kämpfer et al. 2003; Sawyer et al. 2007; Dalley et al. 2008), some of which are being used to monitor the responses of captive and wild badgers to vaccination and challenge. Whilst not strictly necessary for the monitoring of vaccine success, either during the development or implementation phases of a vaccine programme, a sensitive, non-invasive test (Dalley et al. 2008) has been instrumental in establishing the TB-free status of badgers brought from the wild into captivity. As well as the health and safety benefits associated with this screening, experimental efficacy data must be obtained from animals initially free of the disease of interest. A lack of suitable immunological or other tests for determining disease status of the target species may significantly hamper efforts to develop and license vaccines.

subsequently ask for more data to be generated in particular species, or may refuse the application altogether.

During the development of rabies vaccines for wildlife, safety was assessed in laboratory animals, and in both target and non-target wild species. In laboratory studies, the live attenuated SAD-B19 strain rabies vaccine was harmless in all but a few rodent species (Vos et al. 1999), which suffered from residual pathogenicity but no viral excretion. Safety in the field was first tested in small mammals on an island and subsequently in an isolated valley in Switzerland, without any evidence of rabies-induced cases or of uncontrolled spread of the attenuated virus. Since 1985, this and the related SAD-P5/88 strain have been used in several Western European countries with no reported adverse effects. Other strains (SAG1 and SAG2) have subsequently been derived from the SAD-Bern strain and their safety has been demonstrated in laboratory mice, wild rodents and monkeys, before deployment in the field (Coulon et al. 1992). In raccoon dogs direct instillation or delivery in oral bait of at least ten times the field dose of the SAG2 vaccine strain resulted in seroconversion and all animals remained healthy (Cliquet et al. 2006). However, the use of attenuated live rabies vaccines has in some instances resulted in disease in some vaccinates (e.g. Fehlner-Gardiner et al. 2008), and although there is no evidence that attenuated vaccines have reverted to a virulent strain and subsequently spread, this may remain a possibility.

The safety of VRG has been tested in laboratory animals, wild rodents and a wide variety of non-target species. Safety in wild non-target species was demonstrated in trials where baits containing the vaccine were deployed in fenced enclosures of varying sizes. Although VRG is now widely recognised as presenting no hazard to humans or non-target species, a reported instance of a mild pox infection in a pregnant woman after contact with VRG bait (Rupprecht et al. 2001) demonstrates that there will always be some, albeit small, residual risk.

In the case of CSF vaccines, the safety of the live attenuated C-strain virus used in bait, was experimentally assessed in the laboratory in mice, rabbits, foxes, domestic pigs, goats and cattle (Kaden and Lange 2008). Safety in wild boar was assessed in both laboratory and field studies (Kaden et al. 2003). A ten-fold vaccine dose was administered in safety tests that were carried out before the release of vaccine batches. The safety of the vaccine candidate CP7_E2alf has been experimentally evaluated in cattle, sheep and goats. As this vaccine candidate represents a genetically modified pathogen, substantial further safety studies are likely to be necessary in order to obtain a marketing authorisation for use in wildlife.

6.4.3.2 Efficacy

Demonstration of the efficacy (i.e. effectiveness in protecting against infection and/or the consequences of infection) of a vaccine destined for use in wild animals is a required element of the application for any marketing authorisation. The data are most frequently generated from studies using captive animals that are vaccinated and then subsequently challenged with the pathogen. The results of such studies may be supplemented with experimental data from field trials, and both form the basis for the claims made for the vaccine in the summary of product characteristics, the wording of which may be restricted and prescribed by licensing authorities. Unlike safety studies which must address the safety of an overdose of the vaccine (typically two times the field dose for non-living vaccines, and ten times for live vaccines), the efficacy of the vaccine should normally be demonstrated for the lowest possible dose, taking into account the potency or titre of the vaccine at the end of its shelf-life.

The immunogenicity and efficacy of rabies vaccines have been tested by antibody titration in target species (e.g. foxes, raccoons, raccoon dogs) and by the direct challenge of vaccinated animals and controls. In all instances the vaccinated animals resisted the challenge several months after vaccination (Brochier et al. 1996; Cliquet et al. 2006). Rabies vaccines have also been tested in susceptible non-target species, and shown to be less effective (e.g. VRG in badgers and striped skunks) (Brochier et al. 1989; Grosenbaugh et al. 2007).

The efficacy of CSF (C-strain) vaccine baits was investigated by challenging vaccinated domestic pigs and wild boar of different ages, and unvaccinated control animals. The studies demonstrated that animals that had received one dose of vaccine, whether in bait or by injection, were fully protected and did not develop clinical signs, viraemia (presence of virus in the bloodstream) or excrete virus (Kaden and Lange 2001). Oral vaccination of wild boar does not induce chronic infection, after either challenge of vaccinated pregnant sows, or infection of vaccinated non-pregnant animals. Efficacy of vaccination against CSF has been evaluated in relation to the prevalence of both antibodies and virus. Following the application of vaccination in a wild boar population, an increase in the proportion of antibody-positive animals (i.e. seroprevalence) in the hunting bag and a decrease in virus prevalence would be expected. However, the observed seroprevalence will not only depend on the performance of the vaccine but will also vary in relation to the composition of the hunting bag (see Box 6.3).

6.4.4 Vaccine Delivery

A variety of approaches have been considered for the delivery of vaccines to wild mammals. The most suitable mode of vaccine delivery will depend on the characteristics of the vaccine, the target species and the environment where it will be deployed. The two principle routes of vaccine administration are by injection (parenteral) and oral ingestion. Although rabies vaccines have primarily been delivered in oral bait, injection by hand has been used for vaccination of skunks and raccoons for a focal rabies outbreak in Canada (Rosatte et al. 1992). Similarly, intramuscular injection of BCG is likely to be the route of administration for the first licensed badger tuberculosis vaccine in the UK. As mentioned above (Section 6.4.1), there are now many examples of both parenteral and oral delivery of vaccines for species of conservation concern. During the early years of wildlife vaccination in the USA automatic injection devices were trialled, but have not been developed further

(Baer 1991). Dart guns have also been used to deliver vaccines to wild mammals in disposable darts, and in compressed pellets known as 'bio-bullets'. In the USA, vaccine darts have been used to immunise elk against brucellosis as they congregated on their feeding grounds (Wobeser 2002) and to deliver a vaccine against pasteurellosis to both bighorn sheep and elk (Cassirer et al. 2002). In Southern Africa, vaccine darts were delivered to antelopes from helicopter, to immunise them against anthrax (De Vos et al. 1973; Clegg et al. 2007) and in Canada bio-bullets were successfully employed for the vaccination of bison against brucellosis (Olsen et al. 2006).

Administration of vaccine to wild mammals via the oral route is usually achieved with an ingestible bait. The most successful examples are the rabies and CSF oral vaccine baits. Oral bait consists of two main components; the bait matrix, which is comprised of an attractive food, and the vaccine, which may be encapsulated within a protective capsule or substance. The bait matrix must obviously be attractive to the target species, and so a variety of imaginative formulations have been proposed as vehicles for vaccines, including eggs, meat, chocolate, polyurethane sponge and fishmeal. Flavourings and scented attractants can be used to enhance bait appeal. For example, synthetic fermented egg (the smell of rotting meat) appears to increase the rate of bait uptake by wild carnivores (Hunt et al. 2007). The ideal choice of bait matrix is usually determined by carrying out palatability studies, perhaps in captive animals initially, but should always be tested on the wild target species. For the main commercial rabies baits, there is little to choose between them in terms of bait acceptance (Smith and Woods 2007). Perhaps the most technically challenging aspect of bait formulation however, is ensuring that the vaccine remains stable during processing, storage and in the environment, and survives passage to its destination within the target animal. This may require encapsulation in some protective substance or structure. For example, large-scale rabies vaccination programmes have delivered liquid vaccine enclosed within a plastic capsule in either chicken heads, commercially produced tablets of ground meat (Blancou et al. 1991), or in blister packs. Rabies vaccine is believed to target the buccal cavity during mastication of the bait, and thus does not need to survive passage through the stomach.

A lipid-based formulation was developed in New Zealand for the oral delivery of BCG vaccine to brushtail possums, which permits survival of the vaccine through the stomach to the delivery site in the intestines. Use of the lipid-matrix allowed BCG to be retained in a viable, but static state for at least several weeks at ambient temperature (Aldwell et al. 2003a). In rodent models and brushtail possums, oral delivery of lipid formulations containing live BCG was shown to establish populations of viable, replicating BCG in the alimentary tract lymphatic system (Aldwell et al. 2005b; Wedlock et al. 2005), which in mice persisted for at least seven months post-vaccination (Aldwell et al. 2006). Voluntary uptake of the vaccine (which could be readily induced following flavouring of the lipid matrix) was shown to confer protection against virulent *M. bovis* or *M. tuberculosis* aerosol challenge in mice (Aldwell et al. 2003a; Aldwell et al. 2005a; Aldwell et al. 2006), and in possums and cattle against challenge to the respiratory tract with virulent *M. bovis* (Aldwell et al. 2003b; Buddle et al. 2005). The duration of protection after oral vaccination was maintained for at least seven months in mice and 12 months in possums (Aldwell et al. 2006; Buddle et al. 2006b).

Baits containing rabies vaccine were first distributed in the field by hand in the 1980's in Europe, and this is still the case for vaccination campaigns targeting specific populations, such as during the initial stages of an outbreak. Distribution of vaccine baits by hand is also the method of choice for CSF vaccination of wild boar. Rabies vaccine baits are distributed at an average rate of 15 baits per km^2. For CSF vaccine distribution, 20 to 40 baits per km^2 (Kaden et al. 2005a and Box 6.3). Vaccines can also be delivered from aircraft, as has been the case for most broad-scale vaccination campaigns against rabies in France, Belgium, Switzerland and Germany. Aerial distribution of baits may also be considered for vaccination of wild boar against CSF if distribution by hand is impractical or uneconomic, such as in challenging habitats like extensive coastal reed beds. The overall objective here should be to deliver the minimum number of baits per unit area, but to still achieve the objectives of disease management. Delivery systems are now increasingly subjected to economic evaluation to identify the most cost-effective solution (see Box 5.1).

One important consideration in the development of baits for use in wildlife is the potential for legal restrictions on the deployment of certain substances in the environment. This is especially likely to be a factor where exposure to non-target livestock cannot be ruled out. For example in the UK, current legislation relating to disease risks from animal by-products, significantly restricts the nature of the materials that can incorporated into any bait that will be deployed in an environment where livestock are present.

6.4.5 *Monitoring Success*

There are three distinctly different but complementary approaches that can be used to monitor the success of a vaccination campaign. These are quantification of the rate of bait uptake, quantification of the rate of vaccine-related immune response in the target population, and evaluation of the epidemiological consequences of vaccination.

At its simplest, the evaluation of vaccine bait uptake may involve observation of the rate of bait disappearance. However, this is generally not a sufficiently rigorous method for monitoring the success of a vaccination campaign, since many baits may be removed by a single animal, or by non-target species. More robust information on uptake may be gleaned by impregnating baits with a biomarker of some kind, and subsequently sampling the target population for its occurrence. Examples include the antibiotic tetracycline, which is detectable in bones and teeth, rhodamine dye, which can be detected in hair and whiskers, and analogues of iophenoxic acid, which are detectable in blood. Tetracycline biomarkers proved useful during early rabies vaccine trials, in which they demonstrated widespread acceptance of bait amongst target (and sometimes non-target) species. In European studies it was estimated that on average between 70% and 80% of baits were taken by red foxes (Blancou et al. 1988). A similar approach was taken to monitor bait uptake during the first field trial of oral baits for vaccinating wild boar against CSF in Germany. About 85% of baits disappeared within five days of deployment, and examination of the bones of shot wild boar identified the biomarker oxytetracycline (OTC) in 52 to 68% of individuals and indicated that uptake was high in areas where baits were buried but was low amongst juvenile boar (Kaden et al. 2000) (see Box 6.3).

Bait uptake rates may vary in response to external factors such as the availability of natural food or crops, and weather conditions. Also, in some populations it may take time for animals to become accustomed to taking a novel food source, so uptake rates may improve over time. The potential influence of such factors can be evaluated from appropriately designed field experiments using biomarkers.

The second way to evaluate the success of a vaccination campaign is to monitor the rate of immune response (e.g. presence of antibodies) resulting from vaccination. This is likely to require the collection of blood (or other body fluids) from a subsample of individuals, so that appropriate diagnostic tests can be performed to assess their immunological status. Following rabies vaccination campaigns, this approach identified rates of seroconversion in red foxes of 60% to 70% (Blancou et al. 1988). The deployment of CSF vaccine in bait in Germany was followed by the testing of all wild boar that had been shot, found dead or involved in road traffic accidents in the area for virus (by fluorescent antibody test, ELISA or Real Time-PCR) and antibodies (virus neutralising test or ELISA). This showed that rates of seroconversion differed considerably both spatially and between different age classes.

For some diseases, such as tuberculosis, the predominant immunological response is cellular rather than humoral, and thus, monitoring a serological antibody response is likely to miss a large proportion of vaccinated or infected animals. Although assays exist for measuring cellular immunity (Dalley et al. 2008), they are often considerably more time-consuming and expensive than antibody tests, which potentially limits their application for monitoring wildlife vaccination campaigns.

Immune responses induced by vaccines may potentially confound interpretation of the epidemiological situation, if no appropriate diagnostic tool is available with which to discriminate infected from vaccinated animals. So called DIVA (Differentiating Infected from Vaccinated Animals) methods have been successfully applied in the control and local eradication of Aujeszky's disease, infectious bovine rhinotracheitis, CSF, foot and mouth disease and avian influenza (Vannie et al. 2007). Such tests could potentially be applied more generally to the vaccination of wild mammals if sufficient information is known about the host immune response to vaccination and infection. DIVA tests however, are not required for all vaccination campaigns. For terrestrial rabies, animals generally only produce antibodies in the days immediately prior to death, so naturally seropositive animals are very rare in the population.

Finally, and perhaps most importantly, the success of any vaccination campaign can be assessed on the basis of its epidemiological consequences, and in particular the extent to which it reduces the incidence of disease in the target population. This requires that disease surveillance data is collected before, during and after vaccination campaigns, as was the case for red fox rabies in Europe. In 1989 the number of registered cases of fox rabies recorded in France peaked at 4,213. From the spring and autumn of 1992 onward, vaccine baits were distributed throughout the entire affected area (over 192,418 km^2) and as a result, the incidence of rabies diminished by about 60% each year until 1997, when it was finally eliminated (Aubert 2003). Similar figures were reported in other European countries following oral vaccination campaigns (Brochier et al. 1996).

As with any management intervention, monitoring is a vital component of a vaccination campaign. It not only provides hard evidence of success (or otherwise), but also permits a greater understanding of the epidemiology and logistics of disease control (see Box 6.3).

6.5 Conclusions

The effectiveness of any programme to vaccinate wild mammals will be a product of the proportion of animals that receive the vaccine and the proportion that become immunised. Hence, not only must a vaccine for wildlife be efficacious at the indi-

Box 6.3 CSF vaccination in wild boar

In Baden-Württemberg, Germany, vaccination of wild boar (*Sus scrofa*) against CSF began in 1999. The programme involved two deployments of baits containing vaccine every spring, summer and autumn until 2001. Hunting bags indicated that seroprevalence rates in wild boar were higher after these three seasonal campaigns than after vaccination in only spring and autumn. However, the hunting bags also revealed age-dependent variation in seroprevalence. During the first five seasonal vaccination campaigns, between 50% and 83% of adults (i.e. boar over 1 year old) were seropositive, compared to an average of only 45% of juveniles (i.e. those less than 1 year old) which decreased to approximately 30% thereafter. Further investigation showed that the proportion of antibody-positive young boar was less than 40% in the 3–5 month age group suggesting that these animals were largely responsible for the lower prevalence amongst juveniles (Kaden et al. 2005b). In further vaccination campaigns in Germany (e.g. in Saarland and Rhineland-Palatinate) and in France, post-vaccination seroprevalence was also lower amongst juvenile boar (Rossi et al. 2006).

Following three vaccination campaigns in France, CSF was still present in wild boar in the treated areas. This failure may have been related to poor uptake of vaccine baits, particularly amongst young animals (Rossi et al. 2006). Infected boar also remained following vaccination in North Rhine-Westphalia, although there was no indication of virus persistence in vaccinated individuals. One possible explanation was the vertical transmission of infection from sows to their offspring, although this phenomenon was not

(continued)

Box 6.3 (continued)

observed in a separate area with moderate infection pressure (North-Western Pomerania). Laboratory studies in which pregnant sows received a single oral vaccination, failed to demonstrate the transmission of virus to the foetus following experimental challenge. This indicates that transplacental virus transmission does not play a crucial role in the perpetuation of CSF virus in wild boar (Kaden et al. 2008). Rather, individual young wild boar that survive infection with moderately virulent virus, or partially protected piglets (e.g. animals with low maternal antibody titres), might exhibit a transient infection. The infrequent occurrence of persistently infected wild boar after post-natal infection, and the absence of infected foetuses in an experimental field study suggest that these are unlikely to be important routes of transmission. Rather, it seems likely that the high proportion of susceptible juvenile wild boar and population density are the crucial determinants of virus persistence.

The effectiveness of CSF vaccination and successful eradication of the disease in wild boar populations depend on several factors. Of principal importance is ensuring adequate provision of vaccine baits. Consistent results have been achieved employing 0.5–1 bait stations km^{-2}, each of which contained between 20 to 40 individual baits. This has been combined with population reduction achieved by hunting throughout the year, which is targeted at juveniles (i.e. <6 months old). To maximise the likelihood of local eradication of CSF in wild boar populations, vaccination should continue for at least one, if not two years after detection of the last CSF virus positive animal. During this period, all animals found dead, involved in road traffic accidents or shot should be the subject of virological and serological monitoring. Thereafter, surveillance for individuals at an early stage of infection should be carried out in wild boar populations.

vidual level, but it must also be delivered to a sufficient number of animals to impact on disease prevalence at the population level. This requires a clear understanding of the practical constraints that may be imposed by ecological factors. In addition, the vaccine must be safe for use in the host and in any non-target species that may be exposed to it. These issues represent considerable challenges to the development of effective vaccination programmes for wildlife, but the dramatic reduction in rabies incidence in Western Europe illustrates what is possible.

The most likely reason for the failure of an efficacious vaccine in wildlife is likely to be that it is not delivered to a sufficient proportion of the target population. For instance it has been suggested that an insufficient level of immunisation of the fox population against rabies allowed the infection to persist for longer in comparison with non-vaccinated areas (Smith and Harris 1989; Suppo et al. 2000), and may explain the resurgence of disease in suburban areas of Germany (Thulke et al. 2000). Sub-optimal vaccine coverage may arise if public or financial support for the campaign is inadequate, or if the treatment is not applied to a sufficiently large area.

It can also occur if the delivery of the vaccine fails to account for important aspects of host behaviour. For example, the social organisation and density of red foxes appears to have a key effect on the success of rabies control strategies involving culling, vaccination or fertility control (Smith and Wilkinson 2003). Thus optimum strategies can involve focal culling with ring vaccination in some circumstances (see Box 4.3). In order that mistakes are not perpetuated, and to enable vaccination strategies to be adapted when necessary, some form of monitoring is crucial during interventions. This should also be sufficient to identify the pre-determined conditions that will indicate that the objective of disease control or eradication has been achieved, and the campaign can end (see Chapter 9).

Whilst vaccination is often seen as one of the most attractive wildlife disease control options, it is not without its potentially undesirable side effects. Vaccines, bait compounds and methods of deployment can be potentially harmful to target or non-target species. Attenuated "live" vaccines can induce infection in species for which the vaccine has not been developed; LEP (low egg passage) rabies vaccine is known to induce rabies in several non-canids, and similarly live canine distemper vaccines can be problematic for highly susceptible mammal species (Griot et al. 2003). Consequently, although their use for the protection of small populations of African canids has been considered, these two vaccines should probably be disregarded (Laurenson et al. 2004). It is essential that the potentially negative effects of direct medication and vaccination are always thoroughly and systematically evaluated prior to their deployment in free ranging wildlife. In this regard there is no substitute for rigorous scientific investigation and economic evaluation (see Chapter 5) of any vaccine and proposed programme of deployment.

Intervention targeting the pathogen in the host population is aimed at achieving the ultimate goal of preventing inter-individual transmission, such that the pathogen eventually dies out. Basically, that goal is achieved when R (the effective reproduction number) is reduced below unity (see Chapter 3). There is evidence that R is influenced by risk factors related to host ecology, behaviour, density, phenotype (mass immunity), and host and parasite genotypes (Woolhouse et al. 2005). Hence we may speculate on the potential ecological and evolutionary consequences of reducing R below unity through the application of vaccines.

Parasites and pathogens can influence ecosystem structure and processes, and as a consequence, the control of pathogens in natural systems can have far-reaching consequences. For example, where a pathogen limits host abundance, then vaccination may lead to an increase in host population size which will in turn impact on other components of the ecological community (see also Section 7.3.2). It has been suggested that oral vaccination of foxes against rabies in Europe may have facilitated the spread of echinococcosis, although the role of rabies in limiting fox populations is not proven (Chautan et al. 2000). It is thought that some pathogens may effectively mediate competition between species (Hudson and Greenman 1998), in which case vaccination might theoretically enable a previously suppressed host species to become dominant.

Recent years have seen growing research interest in a new generation of vaccines. Genetic engineering techniques offer increasing opportunities to develop live vaccines with specific characteristics that can spread between hosts and so enhance coverage. However, these opportunities bring with them significant risks: transmissible vaccines could spread into non-target species or populations, with unexpected results. In Australia, an engineered myxomavirus has been proposed to control the fertility rate of introduced rabbits, but in parts of Europe rabbits are considered as desirable for hunting and biodiversity. Consequently, in Spain, hunters consider that rabbits should be immunised against RHD, and so a myxomavirus modified to express RHD genes has been developed as a vaccine (Angulo and Cooke 2002). Such opposing uses of the same technology demonstrate the potential risks of translocation of one such vaccine outside of its intended range. Enthusiasm for the development and application of such vaccines for the control of diseases in free-ranging animals should be tempered with a critical appraisal of the associated risks.

In conclusion it seems unlikely that direct medication of wild mammals will be an appropriate approach for the management of disease outbreaks in wild mammals in all but a minority of specific circumstances. Nevertheless, it will continue to be an important routine tool in the rehabilitation and translocation of mammals. Vaccination on the other hand, has already demonstrated its ability to eliminate disease from wild mammal populations over extensive areas. This is largely the result of the highly successful application of oral rabies vaccines. The results of current attempts to manage CSF in wild boar and the ongoing development of a TB vaccine for badgers will demonstrate over the coming years whether similar success can be achieved with other diseases of wild mammals.

Chapter 7
Options for the Control of Disease 2: Targeting Hosts

Stephen P. Carter, Sugoto S. Roy, Dave P. Cowan, Giovanna Massei, Graham C. Smith, Weihong Ji, Sophie Rossi, Rosie Woodroffe, Gavin J. Wilson, and Richard J. Delahay

7.1 Introduction

Targeting the host has been the most common approach to managing disease in wildlife. This has essentially involved some form of host population reduction, achieved through dispersing, culling, or controlling reproduction.

Dispersion of animals from the site of a disease outbreak has mainly been employed for birds (Wobeser 2007) but has also been attempted for some herding mammals such as bison (*Bison bison*) (Meagher 1989). This works best for non-infectious diseases; otherwise it requires that only susceptible individuals disperse, since the movement of infected animals will increase the geographic spread of disease. Unsurprisingly this method has had little success in practice, and is seldom likely to be of value in controlling infectious disease in wild mammals.

Culling is a long established method of population reduction, for both disease and pest control. This approach assumes that reducing the population size of the targeted species results in a concomitant decrease in the prevalence (and more importantly the absolute number) of infectious individuals. If the aim is to eradicate the pathogen then the number of infectious individuals must fall below a level at which infection can be maintained. However, it may often be sufficient that infection is reduced to a level below which spillover to other host species (e.g. humans, domestic animals, or endangered species) either ceases or is tolerable. Wild mammal populations have most commonly been subjected to culling because they have been perceived as agricultural pests, and less often because they may transmit diseases.

S.P. Carter, S.S. Roy, D.P. Cowan, G. Massei, G.C. Smith, G.J. Wilson, and R.J. Delahay
Central Science Laboratory, Sand Hutton, York, YO41 1LZ, UK

W. Ji
Massey University, Albany, Private Bag 102 904, North Shore Mail Centre, Auckland, New Zealand

S. Rossi
Unité Sanitaire de la Faune, Micropolis, la Bérardie, Belle Aureille, 05000 GAP, France

R. Woodroffe
Institute of Zoology, Zoological Society of London, Regent's Park, London, NW1 4RY, UK

R.J. Delahay et al. (eds.), *Management of Disease in Wild Mammals*,
DOI:10.1007/978-4-431-77134-0_7, © Springer 2009

Culling can be an effective means of controlling disease in livestock since, generally, all animals can be caught, tested, and if necessary dispatched. As techniques for the management of disease in livestock are well established, these approaches have often been regarded as the first choice for the management of disease in wildlife. Since each animal is only 'treated' once, the effect is immediate and permanent, so culling has also often been perceived as a simple and decisive approach with few complications. However, in practice this is seldom likely to be the case where wild mammal populations are concerned. Although there are many challenges associated with the development of non-lethal techniques (e.g. vaccination or fertility control), and their beneficial effects may not be felt immediately, their attractiveness is growing due to increasing evidence of the potential disadvantages, the ecological consequences of culling, and the pressure of public opinion.

Any disease control strategy must have clear objectives; although it is surprising how often this has been overlooked when culling has been employed to control disease in wildlife. The objective should be clearly defined at the outset, and the effectiveness of the intervention measured. Either host eradication or reduction may be considered necessary to achieve the goal of disease control, but neither should be considered as an objective in itself.

Fertility control is emerging as a useful technique for the non-lethal management of wildlife populations, although substantial research will be required before its full potential can be realised (Section 7.4). When used in isolation, fertility control has the potential to reduce population turnover or growth, but in combination with vaccination it could provide a powerful and publicly acceptable alternative to culling. In some circumstances culling is expected to be more effective in controlling disease than vaccination, because of the birth of new susceptible animals (which increases the density of the susceptible population above the threshold required for disease persistence: K_T; see Section 7.3). However, the effectiveness of a vaccination programme could be substantially increased by the addition of effective fertility control to curtail the recruitment of susceptible young animals (Smith and Wilkinson 2003). Furthermore, given the availability of appropriate diagnostic tests, either approach could be combined with selective culling to potentially enhance their effectiveness.

7.2 Host Eradication or Population Control?

There are two basic approaches to non-selective culling of wild mammal populations. One is to attempt to eradicate a species from a defined area, and the other is to maintain numbers below a certain (but seldom specified) level. Eradication may be the favoured option if the host is an exotic introduction, particularly if it is also an economic pest or threatens native ecosystems. A good example is the brushtail possum (*Trichosurus vulpecula*), which was introduced to New Zealand from Australia in 1858 to establish a fur industry. Following their intentional and accidental release, possums have spread throughout New Zealand, severely damaging native ecosystems

(Payton 2000). In addition, they are recognised as the primary wildlife reservoir of bovine tuberculosis (bTB) in the country. Possum management is consequently driven by both conservation and disease control motives. Concerted lethal control campaigns have resulted in the successful eradication of possums from a number of offshore islands (Brown and Sherley 2002). Possum eradication has the advantage of being permanent (barring further deliberate or accidental reintroductions) and the costs, which may be high, are nonetheless seen as a 'one-off' investment.

Mammal eradication programmes are much more likely to succeed on small and remote islands or where the target species has a restricted distribution. For eradication to succeed, removal rates must exceed birth rates, there must be no opportunities for immigration and all individuals must be available to be caught (Bomford and O'Brien 1995; Wittenberg and Cock 2001; Genovesi 2005). Complete eradication of a wild mammal population has not been reported for the specific aim of disease control. However, in Australia, over 7,000 feral Asian water buffalo (*Bubalus bubalis*) were culled from a 389 km^2 area of the Northern Territory between 1982 and 1984 as part of a successful bTB elimination campaign, reducing the estimated population in the region by about 99% (Ridpath and Waithman 1988). Complete depopulation has been proposed as a means of eliminating brucellosis and bTB from bison in Wood Buffalo National Park, Canada. Similarly, the eradication of feral pigs (*Sus scrofa*) has been proposed in the event of a Foot and Mouth Disease (FMD) outbreak in Australia. In the vast majority of cases however, culling has been used as a means of population reduction rather than eradication.

7.3 Culling for Disease Management

Culling is a well-established approach for the management of certain diseases in domestic animal populations. It may be employed when infected animals become an economic liability owing to reduced productivity, or in order to eradicate infectious diseases such as FMD. This can be effective in controlling disease in populations of domestic animals, which are clearly defined and tractable (Ferguson et al. 2001a). Culling wild mammal populations however is an entirely different proposition, as technical challenges and ecological complexities may profoundly influence the implementation and outcome of interventions (see Chapter 2). A wide range of mammal species have been subjected to culling in attempts to either eliminate disease or reduce transmission to a tolerable level. The generally accepted objective is to reduce the host population below some threshold density, K_T, required for the persistence of infection (Anderson 1991), although there is little empirical evidence for disease persistence thresholds in wildlife populations (Lloyd-Smith et al. 2005b) (see Chapter 2). Disease elimination is assumed to occur if the effective reproductive rate of the disease (R) falls below unity, such that on average, each infected animal gives rise to less than one new case (see Chapter 3). In reality, this means that we are not trying to reduce host density below a threshold, but to reduce contact rates below a critical threshold (see Fig. 4.2). Another way to reduce R is through vaccination (see Chapter 6). The choice of whether to cull or vaccinate depends on R_0, host density, and whether density dependence acts on host mortality or recruitment. Culling should be more effective than vaccination when R_0 is high or when host mortality is density dependent, assuming constant host behaviour at different population densities. In contrast vaccination may be more effective when R_0 is low and when density dependence acts on host recruitment, but particularly if transmission does not increase in a linear fashion relative to host density (Barlow 1996; Smith 2005). These differences can be partly explained by recalling that culling removes healthy and infected individuals, who can no longer make a reproductive contribution, and so population size and total productivity are reduced (although in the longer-term compensatory reproduction may occur, see Section 7.3.2). Vaccination on the other hand only 'removes' individuals from the pool of susceptibles (by making them immune), but they continue to contribute to total population size themselves, and to reproduce.

7.3.1 Practical Considerations of Culling

The two most important practical considerations of culling relate to feasibility and cost. When considering culling as a management option, managers must be clear that it is practical to implement in terms of the scale, efficiency and duration required to achieve the objective of disease control. The financial costs of a culling operation can be a significant constraint, as they tend to be expensive to perform and therefore should be examined for the economic return they generate (see Chapter 5).

7.3.1.1 Choosing the Right Method

Issues relating to practical implementation, target specificity, cost-effectiveness, sustainability, humaneness and public perception are of paramount importance when considering culling as an option for disease control. The four main methods used to cull mammals are hunting, trapping (including snaring), gassing and poisoning. There are advantages and disadvantages associated with each, and all have been used in attempts to manage disease in wildlife. The choice of method will be driven largely by the host species in question, although identifying the target may not be as simple as it sounds. Identifying the true reservoir of infection for multi-host pathogens is a persistent problem for wildlife managers (see Chapter 1) and the successful control of an infectious disease usually requires an understanding of the potentially complex reservoir dynamics.

Culling wild mammals by hunting typically involves shooting, either by dedicated teams or by recreational hunters. Hunting has been used for the management of disease in many carnivore and ungulate populations. It has the advantage of being highly species-specific and can sometimes be used to target gender or age classes, or even individuals. Consequently, non-target mortalities are generally low or non-existent. Hunting can also have the added benefit of providing data on the number of animals removed, thus allowing scale and cost-effectiveness to be monitored (see Chapter 5). Disadvantages are that it is labour intensive, requires a high level of skill and may risk dispersing surviving animals more widely. In addition, not all hosts may be accessible to hunters, although dogs *(Canis lupus familiaris)* have been used to track down individuals or find their den sites, restrain them until they could be despatched by hunters, and to catch and kill, where this is legal. For example, trained dogs were used to cull brushtail possums alongside shooting and trapping during their successful eradication from Kapiti Island, New Zealand (Brown and Sherley 2002). Dogs were regarded as essential during the final stages of the operation once the density of possums had been substantially reduced by other means, although this approach has not been widely adopted because of ethical and legal constraints.

Attracting target animals to fixed locations with the use of bait, scent or sound lures can increase hunting efficacy. For gregarious target species so called 'Judas' animals have been used to pinpoint herds, especially where culling has been carried out over vast expanses of wilderness or impenetrable terrain. The approach involves fitting a radio-transmitter to a captured animal and releasing it into an area targeted for control. Gregarious species will seek out con-specifics and hence by tracking the 'Judas' animal further individuals can be located and culled. This technique has been particularly useful in the eradication and control of introduced goats (*Capra hircus*) and pigs in parts of Australia, New Zealand and some oceanic islands (McIlroy and Gifford 1997; Campbell et al. 2004).

Recreational hunters have frequently been used to conduct or supplement culling operations. However, there may be a dichotomy between the interests of wildlife managers who generally wish to significantly suppress the target population, and sports hunters whose aim may be to maintain a healthy, viable population for future harvesting, through the removal of 'surplus' animals. Consequently, hunting pressure may not always be sufficient to reduce population densities to the levels required for disease control. This appears to have been the case in past attempts to control classical swine fever (CSF) in wild boar in Europe (see Box 7.1).

Trapping has typically been the main culling method employed for the management of disease in wild carnivores. The nocturnal or crepuscular habits of carnivores and their occurrence at relatively low densities, compared for example to wild ungulates, may decrease the cost-effectiveness of shooting as a control option. However, the most successful carnivore eradication campaigns have relied on a combination of trapping and shooting (Nogales et al. 2004). Traps essentially consist of devices to either kill or capture the target animal. Animals captured in live traps must be humanely dispatched, typically by shooting or lethal injection. In many countries there is a legal requirement to check traps and snares on a daily basis, regardless of whether they are designed to capture or kill. Consequently, all forms of trapping tend to be labour intensive and expensive. The use of cage traps followed by lethal dispatch is likely to be the most labour intensive method, but it does have the advantage of being highly specific as non-target species can usually be released unharmed.

Box 7.1 The role of hunting in the management of classical swine fever in wild boar

Classical swine fever (CSF) or hog cholera is a highly contagious disease of domestic pigs, which is causing increasing concern in parts of Europe where it is endemic in wild boar (*Sus scrofa*) populations (Artois et al. 2002; Kramer-Schadt et al. 2007). Wild boar may act as a reservoir of CSF, and both analysis of empirical data and mathematical models suggest that the persistence of infection is highly dependent on boar population size (see Fig. 8.2). Culling wild boar to suppress numbers below a critical threshold may therefore appear to be an appropriate tool to achieve disease elimination. Wild boar are already hunted extensively for sport across much of Europe, so an intuitive response to a CSF outbreak might be to consider increasing hunting pressure in the locality of the outbreak. In particular, it has been proposed that selectively targeting young wild boar would yield the greatest benefit as they are more susceptible and this would preserve the older, potentially immune, animals (Zanardi et al. 2003). However, attempts to increase hunting pressure and target piglets in response to CSF outbreaks have had unconvincing results in large wild boar populations (Rossi et al. 2005b). Several factors may help explain this failure. Firstly, hunting pressure was possibly not sufficient to reduce boar numbers below the disease persistence threshold. Indeed, the wild boar population has continued to increase in many European countries since the 1980s in spite of high levels of hunting (Acevedo et al. 2006). This may be related to insufficient targeting of reproductive females and a compensatory increase (see Section 7.3.2 and Box 7.5) in the recruitment of young sows in response to hunting (Rossi et al. 2005b). Secondly, hunting may favour the persistence of pathogens by enhancing the availability of susceptible individuals in seasonal pulses (Guberti et al. 1998; Choisy and Rohani 2006). Furthermore, hunting with dogs may cause wild boar to range further (Maillard and Fournier 1995) and so increase their probability of crossing physical barriers such as motorways (Vassant et al. 1993; Vignon et al. 2002) and hence promote the geographic spread of the virus. A better understanding of the effects of increasing hunting pressure on the demography of the host and the dynamics of disease transmission is therefore required to develop improved approaches to the management of CSF. Other options for disease control such as oral vaccination (Chapter 6) and contraception of wild boar (Section 7.4) are also worthy of further investigation.

Lethal trapping may result in considerable non-target mortality, and although this can be minimised through careful design and deployment of traps and other restraints, it is seldom (if ever) possible to eliminate it.

Capture rates vary widely between species but are often low relative to trapping effort, although this can be improved through the use of scents and lures (Roy et al.

2006). Another common problem is the behavioural avoidance of traps or baits. It is therefore advisable that trapping be carried out in conjunction with an independent means of population monitoring.

The use of snares is likely to be considerably cheaper than traps and in some cases may be more effective (Montague and Warburton 2000) although they arguably require a higher level of operator skill. The biggest drawback of this approach is that snares have been associated with causing significant suffering and so there are strong ethical arguments against their use. As a result, their use is banned or highly restricted in many countries and increasingly they are simply not considered as a realistic management option.

Culling programmes have on occasion taken advantage of the skills of professional and amateur hunters and trappers. One means of encouraging such involvement is by way of a bounty scheme, where individuals are rewarded when they supply an ear, or tail of the target species. Such schemes have been in existence for hundreds of years, and have targeted a variety of mammal species in the interests of 'pest' control, and in some instances for the purposes of disease management. However, this may encourage hunters to only remove 'surplus' individuals or even to import animals, thereby assuring a sustainable yield. Hunters may also tend to neglect populations that are least accessible. Consequently, bounty schemes have not been recorded as making a successful contribution to disease control (Debbie 1991).

Gassing generally involves flooding restricted spaces such as underground den sites or bat roosts with poisonous gas; hence its potential application is relatively restricted. It may be delivered by pumping into a confined space, or deploying a tablet, powder or cartridge, which produces lethal gas when exposed to moisture. A variety of poisonous gases have been used with apparent success in reducing numbers of red foxes (*Vulpes vulpes*) (Müller 1971), striped skunks (*Mephitis mephitis*) (Gunson et al. 1978) and vampire bats (*Desmodus rotundus*) (Fornes et al. 1974) for rabies control. Explosive gases have also been used in attempts to control prairie dogs (*Cynomys* spp.) and other burrowing rodents in the USA, although there are likely to be serious animal welfare concerns associated with such techniques. The use of gas can be target-specific if the restricted space is only occupied by the species of interest, although this is difficult to establish in many circumstances. Another substantial problem with this approach is the challenge of delivering a lethal concentration of gas throughout the enclosed space, particularly in complex burrow systems. In the UK during the 1970s hydrogen cyanide gas was pumped into Eurasian badger (*Meles meles*) burrows (setts) as part of a strategy to control the spread of bTB to cattle. This proved logistically difficult, and was ultimately curtailed on welfare grounds (Dunnet et al. 1986), as sub-optimal concentrations of gas in the further extremities of setts caused serious suffering in some animals.

Poisoning, with toxic baits, is an effective method of culling wild mammals over large areas. Toxic agents can be administered using fixed bait stations or by physically distributing bait in the environment. Baits may be distributed manually over small areas, or from aircraft over large areas and difficult terrain. Culling using poison baits is likely to be less labour intensive than either shooting or trapping unless it is necessary to monitor the fate of deployed baits. Aerial poisoning is the most cost-effective means of reducing brushtail possum numbers in New Zealand, where it has been the main method of control since 1956 (Morgan and Hickling 2000). Six toxicants have been registered for possum control in New Zealand but sodium monofluoroacetate (1080: ten-eighty) has been used most extensively, because of its ability to be degraded by microorganisms (King et al. 1994; Bowen et al. 1995; Eason et al. 1999). However, perhaps the most important challenge when considering using toxic bait on wild populations is to minimise the potential for any impact on non-target species. As 1080 is a broad-spectrum toxin it has the potential to cause secondary poisoning in a wide range of non-target species. Recent research on the development of carnivore-specific toxins (Marks et al. 2006) has been driven by the need to control populations of non-native predatory mammals, but may produce substances suitable for wildlife disease management. Impacts on non-target species can be minimised by tailoring the method of toxin deployment. This was the case when poison was used to control the spread of vampire bat-transmitted rabies in South America. Bats were captured in mist nets, painted with an anticoagulant and released to return to their colony where large numbers of other bats would be killed by ingesting poison during mutual grooming (Lord 1980). In this instance the delivery was highly species-specific, but it is the indiscriminate nature of most poisoning campaigns that is their main drawback. In 1952 an attempt to control the spread of rabies in Alberta, Canada, involved the distribution of nearly 500,000 strychnine baits which killed not only 50,000 foxes, the intended target, but also 35,000 coyotes (*Canis latrans*), 4,200 wolves (*Canis lupus*), 7,500 lynx (*Lynx* spp.) and 1,850 bears (*Ursus* spp.) (Ballantyne and O'Donoghue 1954). It is no surprise therefore that poisoning campaigns have met with most success when used to control introduced pest species on islands devoid of native terrestrial mammals. Examples include the control of black rats (*Rattus rattus*), brown rats (*R. norwegicus*) and Pacific rats (*R. exulans*) on island archipelagos (Howald et al. 2007). However, in some species, prolonged exposure to toxins can result in the development of resistance and aversion to poison baits through the consumption of sub-lethal doses (Leung and Clark 2005) and the selection of behavioural traits such as neophobia (see Section 7.3.2). Despite the proven effectiveness of poisons and continuing development of more environmentally benign formulations, the limited specificity of most existing toxins together with potential risks of environmental contamination and poor public perception, present substantial obstacles to their long-term use.

The concept of biological control involves the introduction of a natural enemy (predator or pathogen) to a population with the intention of reducing their numbers. Unfortunately, this approach has been associated with a catalogue of ecological disasters, which have typically involved the control agent establishing itself as a pest species, or the introduction of novel infections into non-target populations. Bio-control failures include the introduction of stoats (*Mustela erminea*) ferrets (*M. furo*) and weasels (*M. nivalis*) to New Zealand and the small Indian mongoose (*Herpestes auropunctatus*) to several tropical islands. However, target-specific pathogens such as myxomavirus and rabbit haemorrhagic disease (RHD) have been used to successfully control numbers of introduced rabbits (*Oryctolagus cuniculus*) in Australia (Fenner 2002). Such rare successes in using biological agents to control mammal populations are associated with highly species-specific pathogens. An extensive review of diseases in stoats identified potential viral agents for the lethal control of this introduced predator in New Zealand, but the authors advocated a cautionary approach to their development because of the risks posed to non-targets (McDonald and Larivière 2001). Recent advances in biological control include genetically modified organisms and immunotoxins (substances which may cause autoimmune disease), which can interfere with processes such as reproduction, and the use of parasites or viruses as vectors to deliver immunocontraceptives (see Section 7.4.2). However, there are no records of biological control being used to successfully manage a disease outbreak in wild mammals.

7.3.1.2 Selective Culling

So far we have considered culling as an indiscriminate tool for either eradicating, or reducing the size of a wild mammal population. Such approaches will result in the removal of individuals with no regard to their infection status, but opportunities may exist in some instances to target infected animals. This could enhance the efficiency of disease control substantially, particularly in situations where a relatively small number of infectious individuals make a disproportionately large contribution to the spread of disease in a population (see Section 2.2.4).

Selective culling has achieved some success in the control of chronic infections that spread slowly through the host population such as bTB and brucellosis, especially in ungulates that tend to form large aggregations (Tweddle and Livingstone 1994; Cross 2005). In practice this requires the capture and testing of large numbers of individuals to identify those that are infected. This requires considerable effort and, more importantly, is reliant on the availability of a diagnostic test that can be used to rapidly identify infected individuals in the field. A test and slaughter strategy was used to significantly reduce the prevalence of bTB in buffalo herds in Hluhluwe Umfolozi National Park, South Africa (Michel et al. 2006). Once a herd was located they were mustered by helicopters and vehicles and driven into a corral (boma). The construction of the boma allowed the animals to be segregated into smaller groups so that they could be anaesthetised and subjected to a diagnostic skin test. Infected individuals were subsequently shot and uninfected animals released.

A combined approach of culling adult bison and elk (*Cervus elaphus*) testing positive for brucellosis, whilst also vaccinating calves was successful in eradicating the disease from Elk Island National Park, Canada (Tessaro 1986). Nevertheless, test and slaughter is unlikely to be practical for most wild mammal species owing to the difficulty of capturing them in large numbers and the limitations of diagnostic tests. Chronic wasting disease (CWD) has recently emerged as a problem for wild deer in North America, causing considerable concern amongst wildlife managers, biologists and stakeholders (see Box 2.1). Mathematical models suggest that the test and slaughter of infected individuals may be more effective than non-selective culling if the majority of the population can be caught and tested, and if animals can be removed during the early stages of infection (Gross and Miller 2001; Wolfe et al. 2004). Unfortunately, this may not be feasible at the scale of the affected areas. Such problems mean that selective culling may have limited value for disease management in many wild mammal populations. Nevertheless, it may be a more attractive option for the elimination or control of disease in endangered species, where each non-infected individual is extremely valuable and populations are relatively small and potentially isolated (see Chapter 11).

Selective culling may also be directed towards specific age groups or sexes if they are known to be more susceptible to infection and can be identified (see Box 7.1). Conversely, some supposedly non-selective culling methods may inadvertently target some sections of a population over others as a result of inherent bias (Clutton-Brock and Lonergan 1994; Smith et al. 1995), and the subsequent skewing of age and sex ratios may have unpredictable and undesirable consequences for disease control.

7.3.1.3 The Spatio-Temporal Extent of Culling

Clearly, the smaller the geographic range and the shorter the duration of population control, the more feasible and less expensive culling becomes. Consequently, it has been most successful when used to prevent the establishment of an introduced pathogen, or to eliminate or control an existing disease within a restricted range. Both scenarios rely on early detection and diagnosis, and a rapid response, although this may be less important where there is limited scope for disease expansion such as on small, remote islands. The localised culling of carnivores and vampire bats has been successful in preventing epizootic waves of rabies spreading into disease free areas (Wobeser 2007). However, culling is less likely to be successful in the control of a disease that is already established over a wide area, particularly when it influences host behaviour and subsequent transmission rates (see Section 7.3.2 and Box 7.2).

There are few examples where the large-scale culling of a wild mammal population has demonstrated a measurable benefit in the control of disease. Indeed, extensive culling of foxes had no demonstrable effect on the control of rabies across Europe and may have made the situation worse in some areas (Aubert 1994), although it has prevented rabies spread when used in mountain valleys or on an isthmus. In Australia, the systematic culling of the introduced Asian water buffalo made a significant contribution to the eradication of brucellosis and the virtual elimination of bTB from Australian livestock (Cousins and Roberts 2001). The gregarious habits of the buffalo aided the culling approach, which was widely supported because of their exotic 'pest' status. However, much of the success of this campaign was also attributed to rigidly enforced cattle

Box 7.2 Positive and negative effects of culling for disease control: lessons from a large-scale field trial

Attempts to eradicate bovine tuberculosis (bTB) from British cattle have been hampered by the presence of persistent infection in badgers (*Meles meles*). For 25 years, cattle-based control measures were supplemented by various forms of badger culling on and around bTB-affected farms. These efforts were not, however, sufficient to prevent a nationwide increase in the incidence and geographical extent of bTB in cattle. An advisory committee of independent scientists therefore recommended that the utility of badger culling be tested in a large-scale randomised field trial (Krebs et al. 1997).

The UK Randomised Badger Culling Trial (RBCT) was conducted from 1998 to 2005 and may constitute the largest ecological experiment ever performed (Independent Scientific Group 2007). Ten 100 km^2 areas, located in areas of high cattle TB risk, were randomly allocated to receive 'proactive culling', that is, annual badger culls conducted on all accessible land. Ten similar areas were randomly assigned to receive 'reactive culling' which consisted of the localised culling of badgers associated with particular cattle TB outbreaks, while a further ten areas received no culling ('survey-only').

Proactive culling caused a substantial reduction in badger abundance inside culled areas (Woodroffe et al. 2008). Badger densities were also somewhat reduced on neighbouring unculled land, presumably as badgers immigrated into the culling areas. In all areas affected by badger culling, their normal territorial organisation was disrupted, with evidence of animals ranging more widely (Woodroffe et al. 2006a). These behavioural changes would have allowed greater contact amongst badgers and probably explain why the prevalence of infection with *Mycobacterium bovis* (the causative agent of bTB) rose among badgers taken on successive proactive and reactive culls (Woodroffe et al. 2006b; Independent Scientific Group 2007).

Proactive culling was associated with a modest (23%) reduction in the incidence of cattle TB inside culled areas (Donnelly et al. 2007). However, on neighbouring lands, and in reactive culling areas, the incidence of cattle TB appeared to show a short-term increase in response to culling, and infection also became less clustered within the cattle population (Donnelly et al. 2007; Jenkins et al. 2007; Jenkins et al. 2008). These changes are all consistent with a hypothesis of transmission from a badger population made more mobile, and more heavily infected, as a result of culling. The overall benefits of proactive culling, after accounting for detrimental effects on neighbouring land, were so small as to be trivial in disease control terms, and were greatly outweighed by the costs of conducting culling (Independent Scientific Group 2007). Localised culling, which was a more palatable policy option in terms of cost as well as numbers of badgers killed, had only detrimental effects (Donnelly et al. 2003). Hence, after extensive consideration of various possible forms of culling, the group overseeing the RBCT concluded that '*badger culling can make no meaningful contribution to cattle TB control in Britain*' (Independent Scientific Group 2007).

movement controls, full support from the farming community for the test and slaughter of domestic cattle, and the absence of other significant wildlife reservoirs. It took 27 years to achieve Australia's 'free from TB' status and a reputed cost of approximately AUS$840 million (Turner 2003). This status has been maintained through ongoing surveillance, followed by a rapid and aggressive response to the detection of bTB in cattle, although infection still occurs in some remaining buffalo herds.

Often, the extent to which a wild mammal population must be reduced in size in order to achieve the required level of disease control is unknown. Computer simulations are useful tools for generating estimates of the magnitude of population reduction required to eliminate or control disease in a wildlife population. For example, for many years the host density threshold required for disease persistence predicted by mathematical models was used to guide the control of brushtail possum populations in New Zealand for bTB reduction (Barlow 1995). Modelling was also used to investigate the importance of previously documented culling-induced changes in the behaviour of badgers in relation to the control of bTB transmission (Smith et al. 2007b). In the absence of replicated experiments or detailed population studies, models can provide valuable information to help inform decisions on the likely success and design of culling programmes. However, even the most complex model may not account for all the significant ecological and epidemiological processes (see Chapter 4), which can be particularly difficult to predict.

There will be a stochastic element to the removal of infected individuals during an indiscriminate culling operation, such that incomplete culling could result in removing none, few, most, or all of the infected animals purely by chance (see Box 4.1). The age, sex and infection status of the remaining animals may influence subsequent rates of disease transmission, immigration and population recovery. Furthermore, the density, demography and disease status of the population prior to culling may determine subsequent social and behavioural responses and their epidemiological consequences (see Section 7.3.2 and Chapter 2).

Simple statistical methods may be useful in determining the optimum duration of a culling operation in terms of removing the majority of the resident population (see Box 7.3), although in many cases it may be necessary to maintain a certain level of control for prolonged periods, or even indefinitely. Annual possum culling in a bTB infected area of central North Island, New Zealand achieved a 92% reduction in infection in the possum population and an 88% reduction in the incidence of disease in six cattle herds over a 10-year period (Caley 1997; Coleman and Livingstone 2000). The control of possums is thought to have reduced bTB infection in cattle and deer herds in New Zealand by over 50% between 1994 and 2000, and maintaining numbers at less than 20% of pre-control levels can lead to elimination of bTB from possum populations (de Lisle et al. 2001). However, such culling programmes are expensive and require sustained financial support. Without continual routine control of possums, bTB infection in cattle can recover to pre-control levels in 5–8 years (Coleman and Livingstone 2000).

Box 7.3 A stopping rule for culling

One potential consequence of culling a wild mammal population is an influx of animals from the surrounding unculled area. The magnitude of this effect will vary between different species and in relation to local densities. During a sustained cull the rate of immigration from outside is likely to increase, as numbers of residents decline (the vacuum effect). These immigrant animals may however not be the desired target of the cull. Hence, it would be useful to determine whether the animals being caught later in a culling campaign were likely to be immigrants from outside the area. Indeed, if the proportion of immigrants rises above some level, this could be used as a rule to determine when to stop culling. The approximate proportion of immigrants trapped on each day of a capture and cull campaign can be estimated using the following approach.

1. Delineate an inner and outer zone within the culling area. Ideally they should be approximately the same area, or initially be expected to contain approximately the same total number of animals. The outer zone should have sufficient width to ensure that any immigrants are likely to be caught before moving to the inner zone.
2. If the two zones are comparable then the capture rates would be similar in each. However, since we cannot be certain of this, it is more appropriate to use the capture rates on the first day(s) of trapping to determine the expected proportion that should be caught in the respective zones on subsequent days.
3. Using the numbers caught in each zone on days one and two we can compare the proportions in each zone. These results can be compared using a 2 × 2 chi-squared test. We can then test to see if the numbers caught in each zone differ over time by comparing day one with day three, and so forth. An increase in the proportion caught in the outer zone would imply that immigrants are arriving and being captured.
4. There are two reasons to combine data from different days. If the capture rate is relatively low and declines slowly in the first few days, it would be valid to combine the first two or three days (if they are not significantly different) when comparing with later captures, as the combined early sample would show less stochastic bias. It may also be necessary to combine later days to ensure that the expected captures in each zone are sufficient for a chi-square test to be valid (i.e. five or greater).

In a worked example, on each of ten days the following numbers of animals were caught in each zone:

Day	Inner	Outer
1	43	59
2	11	22

(continued)

Box 7.3 (continued)

3	11	10
4	5	10
5	5	8
6	4	4
7	5	7
8	4	7
9	2	8
10	1	12

On day two we would run a 2 × 2 chi-squared test comparing numbers caught on days one and two ($\chi^2 = 0.809$, $p > 0.100$). On day three we would compare captures on days one and three ($\chi^2 = 0.739$, $p > 0.100$), and so forth on a daily basis. On day six the expected values are less than five in both areas, so we would combine data from days five and six, and compare them with day one in a 2 × 2 test ($\chi^2 = 0.004$, $p > 0.100$). Similarly we would carry out similar comparisons with data from day one, on day seven with combined data from days six and seven, on day eight with data from days seven and eight, and on day nine with data from days eight and nine. In our example these comparisons yield no significant differences until day ten, when we combine data from days nine and ten and compare with those from day one ($\chi^2 = 6.840$, $p < 0.010$). Consequently, on this day we could be confident that we have more animals than expected in the outer zone, which we can assume are immigrants, and therefore cease culling.

7.3.2 Ecological Consequences of Culling

Wild mammal populations are part of ecological communities, with links to other species, some of which will be their prey, predators or parasites. The nature of these relationships may change as a result of culling, which may in turn affect other organisms in the ecosystem. Some of these effects are obvious and predictable but the complexity of ecological food webs means that many are not.

Culling predatory mammals could result in increases in the density of prey if this releases them from the pressure of predation. This in turn could have knock-on consequences depending on the ecology of the prey species. For example if the prey species is herbivorous, then an increase in their abundance may amplify grazing pressure. When large-scale culling of wild carnivores was employed in an attempt to stop the spread of rabies in Alberta, Canada, it resulted in an increase in deer and moose (*Alces alces*) populations, which gave rise to serious overgrazing and habitat damage (Macdonald 1980). When such chains of effects occur across multiple trophic levels they are known as 'trophic cascades'.

If the decline or disappearance of a top carnivore leads to increases in the abundance of smaller predators (a process known as meso-predator release) then this can have significant impacts on the prey of the smaller predators. For example, wolves are thought to limit coyote numbers in much of North America. It follows that wolf removal might release coyotes from this source of competition with knock-on effects for prey species like pronghorn antelopes (*Antilocapra americana*) (Berger et al. 2008). It is clear that the ecological consequences of culling can have significant economic and conservation implications. Consequently, these effects should be taken into account when considering population reduction as a disease management tool (see Box 7.4).

Box 7.4 Effects of badger culling on foxes

When badgers (*Meles meles*) were culled in the UK during an extensive field experiment to determine the effect on the incidence of bTB in cattle (see Box 7.2) a concomitant increase in red fox (*Vulpes vulpes*) numbers was observed (Trewby et al. 2008). The badger culling trial provided a rare opportunity to assess experimentally the consequences of culling one wild mammal species on populations of others. Foxes use badger setts (burrows) as breeding dens, share a similar diet and interact directly with badgers (Macdonald et al. 2004). Therefore, it was hypothesised that culling badgers, which are considered to be the dominant species, would result in mesopredator release and hence an increase in fox numbers. Changes in fox numbers may have significant economic, conservation and epidemiological consequences. Foxes kill and eat ground nesting birds (Reynolds and Tapper 1995a), hares (*Lepus europaeus*) (Reynolds and Tapper 1995b) and livestock (Moberly et al. 2003), and are likely to be the principal wildlife vector of rabies in the event of an outbreak in Britain (Smith and Wilkinson 2003).

As part of a project to assess the ecological consequences of badger culling, standardised surveys to estimate fox numbers were carried out prior to intervention in four of the areas where badgers were to be culled, matched with four experimental control areas where no culling was to take place. The surveys were subsequently repeated annually throughout the culling trial. After controlling for patterns of background variation in fox abundance, predicted mean fox densities in areas where badgers were culled were 1.6 to 2.3 foxes per km^2 higher than in the unculled areas (Fig. 7.1). Interestingly, in one area where badger culling was thought to have been less effective (only an estimated 40% removed) fox density did not change.

The results of this study clearly demonstrated that badger culling at the temporal and spatial scales applied in this trial is likely to result in markedly higher fox densities. This has potential implications for the costs of predation on livestock and game, the ecological impact of foxes in conservation terms as predators of ground nesting birds and hares, and risks to public health as potential vectors of rabies. This illustrates why it is necessary to take account of the broader potential ecological consequences when considering culling as an option for wildlife disease management.

(continued)

Box 7.4 (continued)

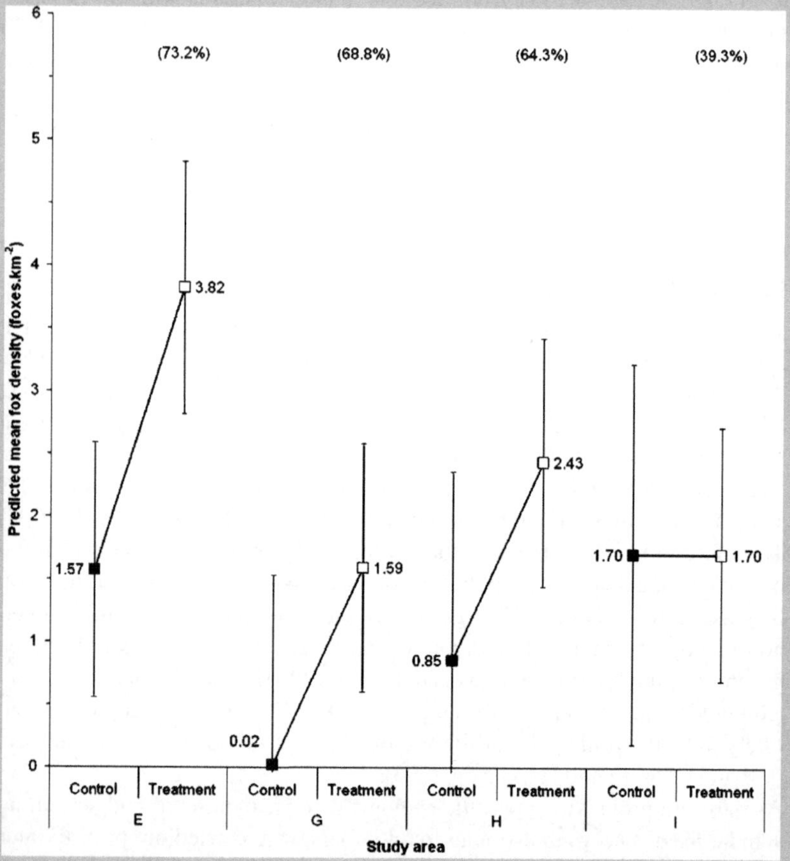

Fig. 7.1 The response of fox populations to badger culling. During the UK Randomised Badger Culling Trial (RBCT) fox density was estimated in culled (treatment) and unculled (control) areas in four locations (E, G, H and I). Predicted mean fox densities (with standard errors) were determined by distance sampling and adjusted for patterns of background temporal and spatial variation. The estimated efficacy of badger removal for the initial cull in each treatment area (Smith and Cheeseman 2007) is shown in brackets

Culling may also invoke behavioural changes in the target species. For example, sustained culling may select for behavioural traits such as neophobia, which results in higher levels of trap or bait shyness, and so reduces the efficacy of control. During a culling operation the population density of the target population declines and consequently the effort required to cull each individual will increase. This effect is likely to be exacerbated by the neophobic behaviour of remaining animals. Using a combination of methods can help to mitigate such effects.

Another behavioural response of wild mammals to culling is compensatory reproduction. Many populations exhibit some level of density-dependent reproduction, such that the production of offspring is greater at lower densities but is curtailed as numbers increase. Consequently, populations subjected to culling may respond by increasing productivity. This phenomenon has been observed in a range of mammal species, including brushtail possums (see Box 7.5) in which it was accompanied by enhanced juvenile survival rates at reduced densities. Enhanced productivity can increase the number (and proportion) of young susceptible individuals in the population, and can have a counter-productive effect on disease persistence.

Culling may also promote increased dispersal by surviving individuals and increased immigration into the culled area. The tendency for immigrants to move into culled areas is often referred to as the vacuum effect, but the extent of potential immigration will vary widely between species and with respect to local conditions. By definition, the distances travelled by dispersing individuals are greater than their typical movement patterns, and so may heighten opportunities for disease transmission. Dispersal is also likely to be a stressful process and so these individuals may be more susceptible to disease owing to poor physical condition or immuno-suppression.

Box 7.5 Possum control and compensatory reproduction in New Zealand

One of the factors that might hinder the desired outcome of culling is the compensatory response of host populations to density reduction. In New Zealand the introduced brushtail possum (*Trichosurus vulpecula*) is a principal source of bovine tuberculosis infection in cattle, and the target of widespread culling. However, brushtail possum populations that have been artificially depleted can recover rapidly because the species has a breeding potential well in excess of the requirements for immediate replacement. Female possums mature at one year and usually give birth to a single offspring each year. However, possums can potentially breed twice in a year with a main breeding peak occurring in autumn and a smaller peak in the spring. Both the proportion of yearling females that breed, and of females with second young were reported to be higher in colonising than in established populations (Green and Coleman 1984; Cowan 1993). Following the removal of 93% and 88% of the original possum populations from two 6 ha areas of native forest remnants in Coatesville and Huapai, North Island, New Zealand (Ji et al. 2004), a higher proportion of females bred, juvenile survival increased, and seasonal body condition fluctuated less in the colonising populations (Fig. 7.2). Two years later possum numbers in these two areas had recovered to 40% and 55% of their respective previous levels. Further evidence of the capacity for possum populations to recover can be found in the results of a 6-year possum eradication programme on Kapiti Island, New Zealand. In this instance, as density decreased, the proportions of young (≤3 years) and old (≥10 years) animals in the population increased. The observed changes in age structure are suggestive of density-dependent survival (Cowan 1993).

(continued)

Box 7.5 (continued)

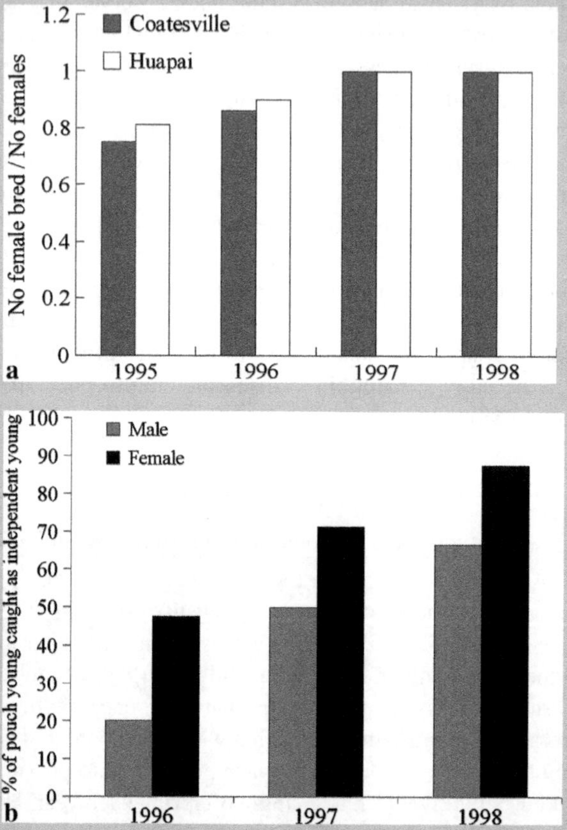

Fig. 7.2 Following depopulation of possums in two native forests in New Zealand, in late 1996, the proportion of females breeding (a) and the survival rate of young animals (≤3 years old) (b) both increased (Ji et al. 2004)

The capacity for disease management interventions to disrupt the social structure of wild mammal populations is increasingly being recognised. For example, the relatively stable social structure of undisturbed high-density badger populations has been shown to mitigate the spread of bTB (Vicente et al. 2007a). Badger culling can disturb this social stability leading to increased movement amongst the remaining and recolonising individuals (Cheeseman et al. 1993; Carter et al. 2007). Evidence from a long-term study showed that increased movement between badger social groups was correlated with increases in the incidence of infectious individuals (Rogers et al. 1998; Vicente et al. 2007a). Furthermore, a large-scale field experiment provided clear evidence that culling resulted in the social perturbation of badger populations and increased the prevalence of bTB in badgers and cattle (see Box 7.2). Evidence for culling-induced social perturbation has also been reported during the control of CSF in wild boar (see Box 7.1) and rabies in red foxes (Macdonald 1995). The combined impact of compensatory reproduction, enhanced immigration into culled areas and increased aggressive encounters (facilitating disease transmission) arising in culled fox populations, is believed to have negated any beneficial effects for rabies control (Holmala and Kauhala 2006).

7.3.3 *Public Perceptions of Culling*

Culling wild mammal populations often invokes strong reactions from conservationists, the public and other stakeholders. In 1923 public opposition prevented the slaughter of about 7,000 bison in Wainwright Buffalo Park (WBP), Alberta, Canada, and re-stocking with disease-free animals, for the purposes of bTB control. As a result, approximately 17,000 bison were culled between 1923 and 1940 as part of the alternative strategy of annual population reduction, in an unsuccessful attempt to remove bTB from the herd (Fuller 2002). In addition, between 1925 and 1928 nearly 7,000 supposedly "disease-free" bison were translocated from WBP to the newly inaugurated Wood Buffalo National Park (WBNP) on the border of Alberta and the Northwest territories, resulting in a new foci of infection. In 1990 an Environmental Assessment Panel recommended the slaughter and restocking of the entire herd of bison in WBNP, but this again met with public opposition. A 5-year Bison Research and Containment Programme aimed at preventing the spread of bTB and brucellosis to neighbouring uninfected populations, whilst facilitating research into bison disease ecology, was implemented in 1995 followed by the publication of interim measures to contain both diseases in the southwest of WBNP. However, no action was taken, pending the acquisition of stakeholder funding and the results of an extensive public consultation exercise (Nishi et al. 2006).

In the UK, culling badgers for the purpose of controlling bTB in cattle has been the subject of a particularly emotive and contentious debate for decades, resulting in highly polarised views between interested parties. Prior to a large-scale field experiment to assess the contribution of badger culling to the control of bTB in cattle, cage trapping, followed by humane despatch by shooting, was chosen over arguably more efficient methods such as gassing, or snaring and shooting. Prevailing public and political sensitivities associated with these methods, and the expectation that the field experiment would not be completed if either of these approaches were employed, played a large part in reaching this decision (Independent Scientific Group 2007).

Despite efforts to improve the humaneness, specificity and cost-effectiveness of culling, it is still regarded by many as ethically unacceptable, particularly in regard to native species. However, public opinion on culling varies considerably between countries, regions and sectors of society, and is often related to how the target species is perceived. For example, rats have a widespread reputation for disease spread and association with human squalor, and their lethal control is widely practiced and rarely the subject of controversy. In contrast, even when there is a clear case, attempts to control wild mammals with a more positive public image may provoke strong reactions. Hence, proposals to cull hedgehogs (*Erinaceus europaeus*) that had been recently introduced to some offshore islands in the UK sparked a national campaign of protest, even though their presence posed a serious threat to native bird populations (Jackson and Green 2000). Levels of public support for the culling of wild mammals for disease control are also likely to vary in relation to the perceived disease threat to human health, domestic animals or endangered species.

7.4 Fertility Control

As a result of public opposition to culling wild mammals and increasing restrictions on the use of such techniques, there is growing pressure to identify effective and sustainable alternatives. Fertility control has long-been considered by many to have the potential to replace or enhance culling for the purpose of wildlife management (Bomford 1990). However, one likely disadvantage of fertility control is that it will generally take longer to achieve equivalent population reductions simply because infertile animals remain in the population until they die. Nevertheless, in some circumstances fertility control may have some inherent advantages over culling, particularly with respect to disease control. The retention of treated (infertile) animals in the population may for example curtail population recovery through their contribution to density-dependent processes acting on recruitment and survival. Fertility control could therefore be particularly effective at maintaining a population at a lower density after initial reduction by culling (White et al. 1997; Merrill et al. 2003). This approach should be less prone to disruption of host social structures than continued culling, so reducing the potential risk of an associated increase in disease transmission (see above). It could also decrease vertical (i.e. from mother to offspring) transmission, which may be an important component of disease maintenance in a population. Other potential benefits of fertility control include removal of the physiological burden of reproduction and lactation, which may enhance the physical condition of females, and so potentially reduce their susceptibility to disease. Nevertheless, by inhibiting reproductive physiology, fertility control may induce behavioural changes, which could potentially either reduce (Ramsey et al. 2006; Ramsey 2007) or increase (Caley and Ramsey 2001) rates of contact between individuals, and the corresponding opportunities for disease transmission.

7.4.1 Fertility Control Tools

Despite a long-standing interest in the potential of fertility control in wildlife management, only in recent years have tools with real prospects for practical application begun to emerge. These include chemical and hormone agents, surgical sterilisation, intra-uterine devices (IUDs) and immunocontraceptives. The chemical agent with the greatest potential for controlling mammal populations is diazacon, which inhibits cholesterol production and blocks steroid hormone formation (Nash et al. 2007). The effects can last for several months after daily dosing for a 5 to 10 day period, so this approach holds most promise for species with a restricted breeding season. Silicone implants can be used to deliver fertility control agents by slow release, leading to infertility for the duration of the implant. Examples of such agents include progestins (synthetic forms of the hormone progesterone) like levonorgestrel (Sivin 1994), and Gonadotropin Releasing Hormone (GnRH) agonists (Bertschinger et al. 2001; Bertschinger et al. 2006). Implants can confer infertility for up to two years, although they usually only endure for less than 12 months. The main advantage of this approach is reversibility which is often an attractive option for captive wildlife, but seldom likely to be an issue for the management of free-living mammals. The use of such agents is not without risk and negative side-effects are possible (Munson, 2006).

Surgical sterilisation is used extensively around the world for rendering feral cats (*Felis catus*) and dogs infertile. It has also been used in experimental field trials to examine the population level effects of induced infertility, but is unlikely to be cost-effective for the management of most wild mammal populations. IUDs have been developed for use in feral horses (*Equus caballus*) (Daels and Hughes 1995; Killian et al. 2006) and white-tailed deer (*Odocoileus virginianus*) (Malcolm and Van Deelen 2007), offering long-term infertility with limited regulatory issues. However, the considerable effort needed for the capture and anaesthesia of individuals probably renders this method impractical for most wild mammals.

There have been significant recent advances in the development of immunocontraceptive vaccines, which induce the immune system to produce antibodies that interfere with a protein or hormone essential for reproduction. The possibility of using immunisation to control fertility in human and wildlife populations has been explored actively for many years with several sperm, egg or hormonal antigens considered as suitable targets for intervention (Delves et al. 2002). The most promising targets for mammalian vaccines, in terms of safety and effectiveness for a wide spectrum of species, are *zona pellucida* (ZP) protein and GnRH. The ZP vaccine affects female reproduction by blocking sperm penetration of the outer surface (*zona pellucida*) of an ovulated egg, thus preventing fertilisation (Miller and Fagerstone 2000; Dunbar et al. 2002). ZP proteins isolated from the ovaries of domestic pigs (porcine *zona pellucida*, PZP) are the most commonly used antigens and generate infertility in most species tested, although not in rodents (Miller et al. 1997). The GnRH vaccine prevents ovulation, the oestrous cycle, the production of oestrogen and progesterone in females, and the production of sperm and testosterone in males. It is known to be effective in many mammals (Fagerstone et al. 2006). An example of the application of this vaccine for disease management is the potential for reducing transmission of brucellosis in bison, which is primarily through contact with infected aborted foetuses, placentas and associated fluids, by rendering infected females infertile (Miller et al. 2004a).

Until recently, immunocontraceptives required an initial vaccination plus one or two booster doses to induce an immune response sufficient to render treated individuals infertile for one or two years. Consequently, these contraceptives have had limited practical utility for wildlife applications where the recapture of individuals to administer booster doses is often problematic. Recent technological advances have led to the development of so called 'single-shot' vaccines capable of inducing long-term infertility from a single dose (Miller et al. 2003; Miller et al. 2004b). This technological breakthrough makes the prospect of practical wildlife applications realistic and potentially cost-effective. The 'single-shot' PZP and GnRH vaccines have been shown to induce infertility for three years in a high percentage (>90%) of treated animals with the effect persisting longer in many, and permanently in some (Fraker et al. 2002; Killian et al. 2006). The adjuvant formulation is a key component of the longevity of the response to these single-dose vaccines (Miller et al. 2004b). An alternative delayed release approach to obtaining long-term infertility from a single injection offers two years of infertility with a PZP vaccine formulation (Turner et al. 2007).

7.4.2 Delivery Systems

Injection by hand has been used to deliver some fertility control implants and is currently the standard method of delivering immunocontraceptive vaccines. However, dart and biobullet technologies are improving prospects for remote delivery of vaccines and implants without the need to capture animals. Dart delivery has been used extensively to deliver PZP immunocontraceptive vaccines to feral horses (Kirkpatrick et al. 1990) and booster PZP vaccinations to African elephants (*Loxodonta africana*) (Delsink et al. 2007). The parallel development of automatic marking systems will allow the visual recognition of treated animals. Because immunocontraceptive vaccines can currently only be delivered by hand injection or dart, the potential range of applications for this emerging technology is restricted. The development of an orally effective immunocontraceptive vaccine would offer a far broader spectrum of potential applications. However, although oral delivery is possible for chemical agents of fertility control, such as diazacon, it does not offer the potential for inducing the long-term infertility afforded by immunocontraception.

An alternative approach to delivering an immunocontraceptive vaccine to a high proportion of a target population would be to engineer a biological dissemination agent. The development of such a system, known as Virally Vectored Immunocontraception (VVIC), was explored in Australia as a means to control introduced mammals. It requires the production of a recombinant virus containing an immunogen that renders infected animals infertile. This offers potentially high levels of specificity if the chosen virus only multiplies in the target host. Dissemination of the virus and its immunocontraceptive cargo away from the release site occurs through natural disease transmission processes thereby potentially inducing widespread infertility in the target population. However, such an approach is likely to be technically challenging, and to raise serious concerns over the release of a genetically modified organism (GMO) and the potential for adverse ecological consequences. For these reasons research into using VVIC for the management of wild mammals in Australia was curtailed. The technical feasibility of achieving fertility control in brushtail possums in New Zealand using, for instance, a genetically modified nematode parasite carrying a marsupial *zona pellucida* immunocontraceptive epitope has also been investigated, but this too would need to overcome many technical and regulatory challenges before becoming a viable management tool. Any attempts to develop such systems for the dissemination of fertility control agents will need to also take into account the heterogeneity in patterns of contact amongst individuals within host populations, as this is a major determinant of transmission dynamics (see Chapter 2).

7.4.3 Welfare Implications of Fertility Control

The effect of GnRH vaccines is essentially to revert animals to a sexually inactive state, which is a common feature of the life cycles of many wild mammals. GnRH immunocontraceptive vaccination has been employed in a range of mammal species with no significant welfare concerns recorded. However, there are contraindications for its use in males of species that develop vulnerable secondary sexual characteristics such as antlers in deer, which may fail to harden and remain persistently in velvet, with significant consequences for welfare (Killian et al. 2005).

PZP immunocontraceptive vaccines have a more specific physiological effect than GnRH vaccines and so might present a lower potential risk of inducing negative side effects. However, preventing egg fertilisation in species with a multiple cycling reproductive strategy (e.g. many deer species) has resulted in repeated cycling and an extended breeding season (Curtis et al. 2002). This may lead to a decline in condition, particularly amongst males who may attempt to defend access to repeatedly cycling females, and increase their movement rates with associated risks of enhanced vehicle collisions and disease transmission. Such effects might also occur in these species through the use of IUDs.

The only generic adverse reactions reported for both PZP and GnRH immunocontraceptive vaccines have been injection site granulomas, although these are generally of relatively mild severity (Dalin et al. 2002). Acute reactions have sometimes been reported but these may reflect an inappropriate injection site (Imboden et al. 2006). In assessing the significance of such welfare concerns it is important to make comparison with the negative effects associated with alternative management options.

7.4.4 Disease Management

The use of fertility control for the purposes of disease management in wild mammals raises several important questions. For example, what proportion of the target population must be rendered sterile to achieve a specified reduction in population size, and how long will this take? Also, for animals that have a lifespan exceeding the duration of the effect of the fertility control agent, how often should the agent be applied? As with culling, simulation models (see Chapter 4) have been usefully employed to explore these issues (Hone 1992; Hobbs et al. 2000; Cowan et al. 2006), in some cases with specific regard to disease management (Tuyttens and Macdonald 1998). With few exceptions the results of these modelling exercises generally suggest that fertility control would be likely to yield the most benefit to disease control programmes when employed alongside culling or vaccination, rather than in isolation. In the UK the introduced grey squirrel (*Sciurus carolinensis*) is implicated in the transmission of pox virus to the declining native red squirrel (*S. vulgaris*) population. Mathematical models suggest that when used in isolation, fertility control does not appear to offer any significant advantage over culling as a method of reducing grey squirrel populations, although it may have a role to play as part of an integrated management strategy (Barr et al. 2002). Similarly, it has been suggested that fertility control might be used alongside culling or vaccination, for the management of bTB in Eurasian badger populations in the UK. Fertility control has also been proposed as part of the optimum strategy, including culling and vaccination, to control a point-source rabies outbreak in red foxes. However, one important limitation of these simulation models is that they do not account for other factors such as behavioural changes in individuals subjected to fertility control treatments, which could influence disease transmission rates. Data from long-term field studies will be valuable in refining such models. Results emerging from studies of population-level responses to the injection of multiple and single-shot immunocontraceptive vaccines (Fayrer-Hosken et al. 2000; Turner et al. 2002; Gionfriddo et al. 2008) suggest the potential value of these approaches. There have now been a number of successful demonstrations of the use of immunocontraception alone to reduce populations of white-tailed deer (Rutberg et al. 2004); African elephants (Delsink et al. 2007) and feral horses (Ballou et al. 2008). Whilst the results of these long-term studies are encouraging they have also raised awareness of new issues specific to the fertility control approach that require consideration. Released from the costs of reproduction, infertile animals survive longer than their fertile counterparts, with consequent implications for the magnitude and rate of population reduction achieved. Enhanced longevity may also raise welfare issues associated with an increased likelihood of individuals reaching senescence. Future research will be required to address these new questions as the technology matures from individual based to population level studies of fertility control as a new wildlife management option with potential to contribute to disease control.

7.5 Conclusions

It is frequently assumed that reducing host population density will achieve a reduction in disease incidence. Although this has intuitive appeal and is supported by simple epidemiological models, a growing body of evidence is undermining the generality of this assumption. For example, there is scant empirical evidence for host population thresholds for disease persistence in wildlife (Lloyd-Smith et al. 2005b). In addition, the potential for culling to induce compensatory reproduction and immigration, to increase the proportion of susceptible or infected hosts and to alter host behaviour, means that the outcomes of such interventions can be unpredictable, and potentially counter-productive. Consequently, proposed strategies need to be carefully designed and adaptive in order to respond to any unexpected and undesirable consequences of intervention. It is also important that all the reservoirs of infection are correctly identified and their relative contribution to the maintenance of the disease evaluated.

Outbreaks of particularly virulent diseases in which wild mammals are implicated may represent such an extreme threat to human health or the national economy that there is considerable pressure for immediate action such as culling. In this regard it is useful to note that culling has been demonstrated to be far more successful in preventing the spread or establishment of diseases in wildlife, than as a means of controlling a disease that is already present (Wobeser 2002). Consequently, under certain circumstances culling may be useful as a short-term measure, particularly in response to a localised outbreak (see Chapter 9). However, the use of culling for sustained control or eradication of disease in a wild mammal population is likely to be difficult, protracted and expensive exercise, with unpredictable outcomes. Broad-brush, draconian approaches to disease management such as large-scale eradication and control campaigns are generally no longer regarded as being ethically acceptable or economically sustainable.

Given the potential problems associated with culling, it is likely to be increasingly considered as one component within a wider programme. Culling has been more successful in controlling disease in wild mammal populations when carried out together with other measures such as public education, improved husbandry of domestic livestock, habitat manipulation and vaccination (Rupprecht et al. 2006). In addition, developments in diagnostic testing and greater understanding of the dynamics of disease in wild mammals may enhance opportunities to use culling in a more efficient and targeted manner.

Fertility control offers an attractive alternative to culling. The underlying assumption of both approaches is that reducing host density can diminish the incidence of disease, but fertility control is unlikely to suffer from many of the disadvantages of culling, and is likely to be more publicly acceptable. However, a great deal more research is required before the use of fertility control becomes a practical reality for the management of disease in wild mammals.

There are clearly many factors that need to be addressed when contemplating the management of disease by targeting the wildlife host. To date, most attempts have been fairly heavy-handed, and have either been unsuccessful or impossible to evaluate. The importance of monitoring the success (or failure) of intervention cannot be overstated (see Chapter 10). It is imperative that the ultimate aim of the intervention is clearly defined and the probability of success critically evaluated. In particular, the cost of the intervention, including ongoing maintenance control if disease elimination is not the aim, should be evaluated against the longer-term economic benefit of success (see Chapter 5). Modelling the effect of intervention on subsequent disease prevalence in the targeted host and on any population requiring protection may be useful. However, it is vital that confounding ecological factors such as host ecology and behaviour, compensatory reproduction, social perturbation and the potential for intervention to have unpredictable outcomes are taken into account. Whilst culling may be a useful management option under certain circumstances, and fertility control holds some promise for the future, in general our approaches need to be refined quite considerably to realise their potential. An integrated approach combining conventional methods such as culling with fertility control, vaccination and biosecurity may need to play an increasingly important role in the management of wildlife diseases in order to achieve measurable, cost-effective benefits whilst also preserving biodiversity.

Chapter 8
Options for the Control of Disease 3: Targeting the Environment

Alastair I. Ward, Kurt C. VerCauteren, W. David Walter, Emmanuelle Gilot-Fromont, Sophie Rossi, Gareth Edwards-Jones, Mark S. Lambert, Michael R. Hutchings, and Richard J. Delahay

8.1 Introduction

Management of wildlife disease can be targeted at pathogens, hosts or vector populations, but may also focus on the environment. As constituent elements of any given environment, resident wildlife populations, and their pathogens, may be profoundly influenced by environmental change, in terms of their abundance, distribution and behaviour. Hence, it is reasonable to expect that incorporation of environmental manipulation into a programme to control wildlife diseases may potentially result in outcomes as effective as direct intervention aimed at hosts, pathogens and vectors.

Environments are not static, but are naturally dynamic, complex systems that exert strong influences on patterns of disease via their impact on hosts, pathogens, vectors and the interactions between them. Consequently, it can be difficult to identify which environmental variables are most important in influencing disease dynamics and hence which elements to target as part of a disease management programme. Nevertheless, environmental management has been used extensively to control

A.I. Ward, M.S. Lambert, and R.J. Delahay
Central Science Laboratory, Sand Hutton, York, YO41 1LZ, UK

K.C. VerCauteren and W.D. Walter
USDA, APHIS, Wildlife Services, National Wildlife Research Center, Fort Collins, CO, USA

E. Gilot-Fromont
Université de Lyon; Université Lyon 1; CNRS; UMR5558, Laboratoire de Biométrie et Biologie Evolutive, 43 Boulevard du 11 Novembre 1918, Villeurbanne F-69622, France

S. Rossi
Unité Sanitaire de la Faune, Micropolis, la Bérardie, Belle Aureille, 05000 GAP, France

G. Edwards-Jones
School of the Environment and Natural Resources, University of Wales, Bangor, Gwynedd, LL57 2UW, UK

M.R. Hutchings
SAC, West Mains Road, Edinburgh, EH9 3JG, UK

R.J. Delahay et al. (eds.), *Management of Disease in Wild Mammals*,
DOI:10.1007/978-4-431-77134-0_8, © Springer 2009

diseases in wildlife in many parts of the world, with some apparent success (Wobeser 2002). Anecdotal information arising from disease management projects and from studies of wildlife behavioural ecology and disease epidemiology suggests that environmental manipulation may offer potential opportunities for the long-term management of many diseases of wildlife. However, while more direct approaches to disease management, such as host population reduction (see Chapter 7) or vaccination (see Chapter 6), might have rapid effects, the benefits of environmental manipulation are likely to take much longer to accrue.

In this chapter we investigate relationships between wild mammals, their environment and disease dynamics. We then discuss the potential applications of environmental management as a tool for managing wildlife diseases, with reference to case studies.

8.1.1 The Environment – A Definition

The environment may be described in its widest sense as the conditions in which an organism lives, including the influences of all biotic and abiotic components. The topography of the physical environment is heavily influenced by the underlying geology, which influences the distribution of soils, vegetation and surface water. Superimposed onto this natural landscape are all the artefacts of human infrastructure. The vegetation communities that cover the land surface are a particularly important component of landscape structure in terms of mammal distribution. Their diversity provides a wide range of niches for mammals to inhabit. Even within a given vegetation community, structure varies, with canopy, sub-canopy and ground-level species contributing to the character of landscapes and influencing ecological processes. In this chapter we acknowledge this complexity and define the environment as the land, water bodies, natural and man-made structures, substrates and vegetation within which wildlife and their associated pathogens exist.

8.2 Environmental Management

Humans are prodigious engineers of their environments, pursuing management in the interests of agriculture, urbanisation and infrastructure development, and to enhance wildlife populations for food, leisure and (at our most enlightened) to conserve biodiversity. Environmental management has also been used historically to manage wildlife diseases. Such strategies have usually targeted host contact with pathogens, for example by using fencing to prevent wild mammals from gaining access to water holes infected with *Bacillus anthracis* (the causative agent of anthrax) (Hugh-Jones and de Vos 2002) and vector control, such as prescribed burning of forest vegetation to reduce tick populations (Allan 2001).

Few controlled experiments have been undertaken to determine the effects of environmental manipulation on wildlife disease dynamics or the distribution and abundance of pathogens of wild mammals. One exception was an experimental application of herbicides and vegetation burning to alter plant communities, which also affected the distribution, species richness, abundance and prevalence of helminths in their cotton rat (*Sigmoidon hispidus*) hosts (Boggs et al. 1991). It was suspected that vegetation management had altered local microclimates, thus affecting the survival of free-living stages of the helminth parasites. This study clearly illustrates the potential for environmental management to be used to target pathogens. An alternative approach would be to control pathogens by targeting environmental manipulations at their hosts or vectors, although reports of such experimental studies are rare. Nevertheless, countless ecological studies have described how wild mammal populations respond to environmental changes by altering their patterns of space use (see Box 8.1). For example, changing agricultural practices can lead to removal of food resources and cover for roe deer (*Capreolus capreolus*) causing them to shift their home ranges and alter their habitat use and spacing patterns (Cimino and Lovari 2003). Interpretation of these effects in the context of disease management suggests that alteration of habitat composition and structure could hold potential for manipulating local host densities and contact rates, with direct consequences for the transmission of infectious diseases.

Box 8.1 Habitat management and rat control

Rats are perhaps the most notorious of all mammalian disease vectors. Their historical association with the bubonic plague still endures, even though the ship rats (*Rattus rattus*) that carried plague (*Yersinia pestis*) throughout Medieval Europe have long since been replaced by the Norway rat (*Rattus norvegicus*) in most temperate regions. Norway rats rarely carry the Oriental rat flea (*Xenopsylla cheopis*), usually responsible for transmission of the plague bacteria from infected rodents to other animals, although they have been identified as reservoirs and vectors of many other zoonoses. Norway rats collected from UK farms were found to be carrying 13 zoonotic and 10 non-zoonotic parasites, including *Cryptosporidium*, *Pasteurella*, *Listeria*, *Yersinia*, *Coxiella* and *Hantavirus* (Webster and Macdonald 1995). Norway rats have also been suggested as potential vectors of foot and mouth disease in the UK (Capel-Edwards 1970), as they are highly mobile and could therefore carry infective material between farms.

Most disease transmission from Norway rats to livestock probably occurs indirectly, through contamination of food sources or incidental contact with rat urine and faeces. Rodent proofing of buildings can be an effective way of reducing direct and indirect contact between rats and livestock, but may not always be practical, especially on older buildings. Another option is to reduce rat populations using rodenticides. This can be effective in the short term, but rat populations have

(continued)

Box 8.1 (continued)

a considerable capacity for recovery through compensatory reproduction, and hence repeated applications of rodenticide sometimes become necessary. However, this incurs a serious risk that rodenticide resistance will develop (Cowan et al. 1995). The need for repeated lethal control could be reduced if attention were given to the reasons why rat populations become established, and if means could be identified of modifying the environment to make it less attractive to rats. The removal of scrub vegetation adjacent to Australian macadamia orchards helped control rat damage (White et al. 1998) and clearing refuse and overgrown areas reduced the size of rat populations in urban areas of the USA (Jackson 1998) and on UK farms (Lambert et al. 2008). Of course it is not possible to remove all areas of harbourage, so periodic and well-targeted rodenticide treatments may still be necessary. Reducing rat immigration from surrounding areas may decrease the need for rodenticides still further. Studies of radio-tagged rats suggest that they tend to avoid open areas, and probably move between farms using hedgerows and ditches as cover. The extent to which immigration contributes to the recovery of rat populations following rodenticide treatments is unclear, and in the UK it is unlikely that large-scale migrations across farmland occur. Even so, targeted trapping of rats along field margins and hedgerows might be useful in reducing the potential for disease transmission between farms.

8.2.1 *Effects of Environmental Management on Disease*

Naturally occurring host-parasite systems may evolve over time to reach a relatively stable equilibrium. However, dramatic changes, such as might be caused by human activities, can disrupt this endemic stability and result in disease outbreaks. The loss, degradation and fragmentation of wildlife habitats, largely through human encroachment, are not only responsible for substantial reductions in biodiversity but are also considered to be major causes of disease outbreaks in some mammals (McCallum and Dobson 2002).

Habitat fragmentation can result from expanding agriculture, silviculture or urbanisation and can lead to a reduction in available habitat for wildlife, thus altering space use and contact rates between wild and domestic animals and humans, with implications for the transmission of pathogens. African wild dog (*Lycaon pictus*) populations for example, have decreased in size in parallel with human population growth. While habitat loss and fragmentation, and increased persecution owing to human population expansion are considered to be the main causes of wild dog population declines, disease has been a significant source of mortality, particularly during episodic outbreaks (Woodroffe and Ginsberg 1999). Domestic dogs (*Canis lupus familiaris*) have probably been the predominant source of infection, and the likelihood of their contact with wild dogs has increased as human populations have expanded towards protected areas.

Human activities may degrade habitats in a variety of ways, including physical alteration, simplification of habitat structure and pollution. Some pollutants including heavy metals and polychlorinated biphenyls (PCBs), can directly compromise mammalian immune systems and thereby increase susceptibility to disease (Exon et al. 1985; Hilliam and Ozkan 1986).

Increased habitat fragmentation was predicted to result in the extinction of *Chlamydia psittaci* (a sexually transmitted infection) from wild koala (*Phascolarctos cinerus*) populations (Augustine 1998), which may, at face value, seem like a good thing. However, habitat fragmentation was also predicted to enhance the risk of extinction of koalas caused by infection with the parasite. In undisturbed environments koalas and *Chlamydia* co-exist within a natural, stable host-parasite relationship, and so it has been argued that loss of the parasite from this system would diminish native biodiversity (Augustine 1998).

Clearly, land management can have a considerable impact on diseases in wild mammal populations. The increasing global use of environmental impact assessments (EIAs) during development projects, offers a potential methodological framework in which to address and perhaps mitigate detrimental effects on disease dynamics. However, EIAs and risk assessments incorporating the effects on diseases of wildlife are far less common than those involving diseases of humans and livestock. An example of the latter is provided by an assessment of the impacts on human health of surface and sprinkler crop irrigation systems in Zimbabwe (Chimbari et al. 2004). The authors compared records of malaria and schistosomiasis from health centres serving areas with either type of irrigation scheme, and a location where no irrigation occurred. Their parallel risk assessment approach suggested that poor land management (e.g. inadequate drainage and accumulation of surface water) and poor maintenance of sprinkler equipment were most likely to be responsible for variations in disease incidence because they created suitable breeding habitat for mosquito vectors and snail hosts. Similar risk assessment methods could be used to assess the impacts of land development on diseases in wildlife. The limited use of this approach to date probably reflects our relatively poor understanding of the implications of changes in land management for wildlife disease dynamics.

8.3 The Importance of Landscape Structure

Landscape structure influences networks of host-pathogen contacts and thus the dynamics of diseases in wild populations. Models of disease in metapopulations (i.e. discrete but inter-connected patches of sub-populations of organisms) predict that spatial heterogeneity increases disease persistence (Post et al. 1983; Wood and Thomas 1996), drives epidemic cycles (Bolker and Grenfell 1995) and influences the evolution of parasite virulence through local adaptation (Lively 1989). These processes have yet to be demonstrated for wild mammals but the influence of spatial heterogeneity on pathogen transmission among invertebrates is well documented. For example, parasite transmission amongst barnacles (*Chthamalus dalli*) was enhanced by increasing both host density and the heterogeneity of their distribution (Blower and Roughgarden 1989).

The inclusion of landscape structure in disease management plans requires the availability of data on its influence on disease dynamics, ideally from experimental studies where cause and effect can be demonstrated. In practice however, correlative data may be all that are available and putative landscape effects may have to be cautiously inferred. Spatial modelling using geographical information systems (GIS) can be used to simulate the complexity of landscape structure and to investigate interactions with hosts and pathogens. Landscape data may be used to predict the environmental carrying capacity of a host population, contact patterns (diffusion) and the persistence of a pathogen in the environment. For vector-borne diseases such as malaria and West Nile virus, risks of disease spread may be predicted by mapping the distribution of habitat favourable to vectors. For example, remotely-sensed data within a GIS was used to monitor changes in artificial aquatic habitats in Wyoming, USA (Zou et al. 2006). This identified favourable sites for the development of larval mosquitoes, which may carry West Nile virus. Monitoring the location of such habitats could be used to predict vector distributions, and so help to more effectively target control efforts.

Landscape structure may also influence the efficacy of disease management measures where the terrain imposes limitations on the practical implementation of field operations. For example, aerial delivery of rabies vaccine baits to foxes (*Vulpes vulpes*) is less effective in hilly areas, because the density of baits per unit surface area is lower on slopes (Vuillaume et al. 1997), and aerial delivery is difficult in urban and suburban areas, which usually require delivery by hand (Müller et al. 2005).

Where wild mammals are organised into spatially distinct but inter-connected populations, the concept of metapopulation dynamics can be useful for predicting the likely impact of management interventions. Mathematical models to investigate optimal immunisation strategies, for example, suggest that for comparable levels of disease control, fewer individuals within a population are required to be vaccinated if they exist within metapopulations, than in a homogenous population of the same size (May and Anderson 1984). The local vaccination threshold necessary to eradicate a disease may be highest among high-density populations that are poorly connected, where individuals that are in contact with a given individual are not in contact with each other (Keeling 1999).

8.3.1 Habitat Quality and Seasonality

Landscapes can be dynamic structures, owing to seasonal changes in climate and vegetation growth. Food availability in particular may strongly influence intra- and inter-specific patterns of contact amongst mammals, with consequences for host-pathogen dynamics. For example, the seasonal availability of fruit may be associated with enhanced abundance and aggregation of mammals. This may help explain the seasonally increased incidence of Ebola haemorrhagic fever among Western gorillas (*Gorilla gorilla*) and common chimpanzees (*Pan troglodytes*), which congregate in areas of high fruit abundance (Pinzon et al. 2004). As seasonal changes in weather patterns are relatively predictable they may help improve the targeting of prophylactic campaigns or changes to management practices. For example, since uptake of vaccine baits by red foxes is higher during the summer, vaccination campaigns against rabies using oral baits are more successful when undertaken at this time of year (Hegglin et al. 2004). Such variations in bait uptake may relate to seasonal differences in the behaviour of the target species or the availability of alternative food sources.

8.3.2 *Habitat Corridors*

The preservation and creation of corridors of favourable habitat have been widely used by conservationists to provide connections between isolated habitat patches, and so promote the persistence of endangered species through increased genetic transfer between otherwise discrete populations. However, a downside to enhanced connectivity is that it may promote the persistence and spread of diseases between populations. Habitat corridors may allow disease to persist in metapopulations where it would have otherwise gone extinct by virtue of low host density. Occasional movements of infected individuals between metapopulations connected by corridors can result in the transportation of pathogens and potentially the re-seeding of infection. Indeed, persistence of classical swine fever (CSF) in wild boar (*Sus scrofa*), is more likely in populations comprised of a high number of connected metapopulations, and if these connections are defined by the presence of habitat corridors (see Box 8.2).

Box 8.2 The dynamics of classical swine fever in wild boar

Across much of Europe, wild boar (*Sus scrofa*) are considered to be a reservoir of classical swine fever (CSF), an infection that causes significant economic losses to the pig farming industry (Laddomada 2000; Moennig 2000; Artois et al. 2002). Animal health authorities are therefore interested in determining the factors that may play a role in the spread and persistence of CSF in wild boar populations (Artois et al. 2002). Environmental factors in particular, may influence the probability of contacts between social groups of boar. These include the continuity of forested habitat and the local density of wild boar, which is related to both food availability and hunting pressure (Rossi et al. 2005a).

CSF spreads as a continuous wave between contiguous administrative regions in Europe. This suggests that virus spread is more dependent on local contacts between boar than on long distance dispersal (Rossi et al. 2005b) and is consistent with their relatively sedentary habits. As wild boar movement patterns largely reflect the distribution of the forested habitat that provides them with food and shelter, so CSF transmission is determined by

(continued)

Box 8.2 (continued)

forest continuity. At the scale of an epizootic, in smaller, isolated forests the emergence of CSF is delayed and disease prevalence is lower compared to larger wooded areas. However, the relationship is more complex, because the effects of forest continuity (connectivity) and local boar density interact. Consequently, in small forested areas low wild boar density decreases the probability of CSF emergence and disease intensity (threshold effect), but within continuously forested areas (green corridors) CSF spreads regardless of boar density. In this environment, only significant barriers to boar movement, such as large rivers and fenced highways, may prevent disease spread (Laddomada 2000; Rossi et al. 2005b).

Environmental factors may also affect disease persistence after CSF has emerged and spread. CSF does not seem to persist locally, but it will remain in large forested areas where local epizootics are not in phase and cyclically recolonises uninfected patches (metapopulations) of wild boar (Rossi et al. 2005a; Rossi et al. 2005b). Within a large, connected landscape, virus persistence is not homogeneous, but occurs mainly in regions where wild boar density is high (Rossi et al. 2005a). This suggests that within a connected landscape, local areas of high boar density enhance the probability of a chain of transmission, and this is possibly related to high local food availability and birth rates (Fig. 8.1). The combined effects of forest continuity and local density result in a strong correlation between the persistence of CSF and wild boar population size (Fig. 8.2).

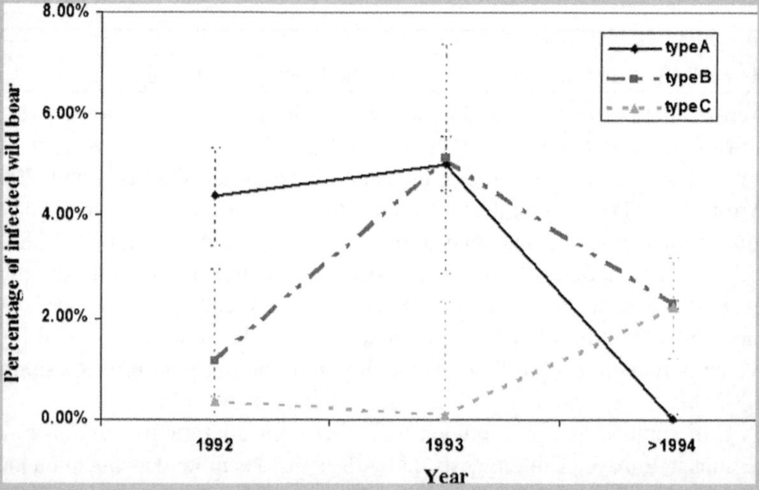

Fig. 8.1 From 1992 to 1997 in the Vosges mountains (France) the percentage of infected wild boar depended on the continuity of the forested habitat (green corridors): the peak of infection was delayed and lower in townships located in small and isolated forests (type C), compared to those located in large forests (types A and B) (Rossi et al. 2005a)

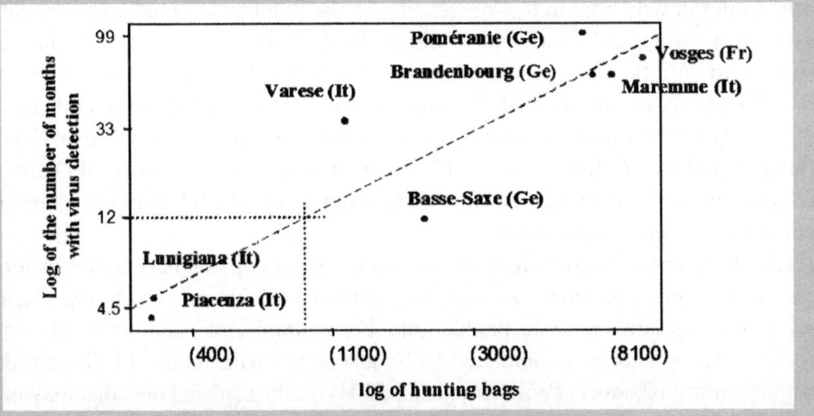

Fig. 8.2 In the 1990s and 2000s many CSF outbreaks were monitored in Germany (Ge), Italy (It) and France (Fr) (Laddomada 2000; Rossi et al. 2005a). The persistence of infection during these epizootics was highly correlated with the size of the wild boar population (as estimated from hunting bags)

The nature of connections between patches strongly influences disease spread and persistence in wild populations. For example, CSF among wild boar, and rabies in red foxes, spreads along forested corridors (Real and Childs 2005; Rossi et al. 2005a), whereas rabies in raccoons (*Procyon lotor*) is dispersed across unforested areas (Smith et al. 2002). The identification of such relationships can allow predictions to be made about the likely course of disease spread. However, corridors may not be as obvious as strips of woodland, particularly among more mobile species, and long-distance seasonal migrations may provide opportunities for the translocation of disease between distant regions along ill-defined corridors. Nevertheless, if data are available on migratory routes, then useful predictions of disease spread may be possible. Acquiring such information is likely to be much easier for long distance migrations of terrestrial rather than marine mammals.

8.3.3 Barriers

Managing disease at the local scale may influence overall transmission rates but might not necessarily lead to the desired level of disease control. Therefore, the area over which disease control is to be exerted must be clearly defined, and barriers, corridors and migratory routes must be taken into account. Ideally, this area should include all connected suitable habitat and population patches, but in reality these may be difficult to define, or too large to encompass (e.g. habitat patches at either end of a long-distance migration route). In France, during an outbreak of CSF originating in wild boar in the Vosges Forest, the putative infected area was defined using motorways and rivers that would probably limit disease spread by providing barriers to wild boar movement (Rossi et al. 2005b). The same approach was used to delineate areas within which Eurasian badgers (*Meles meles*) were culled as part of a study of the effects of wildlife management on bovine tuberculosis (bTB) in cattle in Ireland (Griffin et al. 2005). In such instances the choice of 'barrier' is critical, and must be based on a clear understanding of which features in a landscape will impede animal movements.

The presence of barriers (e.g. rivers, roads, lakes) is particularly relevant for disease management planning because they may slow down or prevent the spread of some diseases amongst wild populations. For example, reduced contiguity among social group territories is predicted to be associated with reduced bTB prevalence among Eurasian badgers (Wilkinson et al. 2004). Landscape features that may inhibit the spread of raccoon rabies in the USA have been identified by fitting observed data to mathematical models. Large rivers were associated with a seven-fold decrease in the local rate of transmission among habitat patches containing raccoons, and together with long-distance translocations were sufficient to explain the spatial pattern of rabies progression in Connecticut (Smith et al. 2002). This approach also successfully predicted the dynamics of rabies invasion in New York State (Russell et al. 2004).

For disease management purposes, it is important to note that while barriers may prevent disease spread between discrete populations, they may exacerbate the problem within the infected population along the barrier interface (Smith et al. 2002). Moreover, if used to aid disease prevention, by for example vaccination, then barriers must be sufficient to restrict emigration from the treated population. This is necessary because if host density increases in the vaccinated area due to the absence of disease, it could encourage dispersal of individuals (both vaccinated and unvaccinated) into the surrounding unvaccinated populations, thereby allowing disease to persist in the peripheral areas.

The configuration of suitable habitat patches and barriers may also affect the logistics and likely success of management efforts, because they influence the distribution and local density of hosts and the pattern of contacts between metapopulations. Mathematical modelling was used to predict the efficacy of culling brushtail possums (*Trichosurus vulpecula*) to control bTB under different scenarios of metapopulation patch arrangement (Fulford et al. 2002). The results showed that when patches of possum habitat were distributed as a chain (e.g. riparian habitat) or a loop (e.g. a woodland surrounding a lake), the model predicted that it was necessary to cull in several linked patches in order to counteract migration and thus eradicate the disease. The importance of curtailing immigration was further illustrated by the observation that when targeting control at a single patch surrounded by other patches to which it was connected, eradication was theoretically possible only if an exceptionally high culling rate was employed. Strategies to reduce the impact of immigration and so improve disease control were predicted to include culling in either the surrounding habitat patches only, across all patches, or in a single patch and a surrounding buffer zone designed to sever migration routes.

8.3.4 Scale and Clustering

Clustering of pathogens in the environment can lead to hotspots of disease at local, regional, national and international scales. Infection with *Mycobacterium avium* subspecies *paratuberculosis* (the causative agent of Johne's disease in cattle) clusters in some populations of European rabbits (*Oryctolagus cuniculus*). Infection is clustered locally in rabbits within regional hotspots in Scotland (Judge et al. 2005b). Rabbit distribution is also clustered at national, regional and local scales, being influenced by availability of suitable habitat patches and the structure and quality of corridors between them (Wilson et al. 2002; Carvallo and Gomes 2003). Such clustering of disease may allow effective targeting of management efforts at the host species if hotspots are geographically stable, although this approach may not be without its problems (see Chapter 7) and its success relies crucially on the accurate identification of the hotspots. This requires the collation of suitable data on disease incidence or prevalence in the target host, or a proxy for this such as levels of infection in sentinel species. In order to optimise disease control efforts, it may also be necessary to determine the distribution of infection within the hotspots themselves.

The scale at which disease is studied can have a considerable effect on the subsequent impression of its spatial and temporal distribution. Taking a 'snapshot' at a particular spatial or temporal scale can lead to serious misrepresentation of the disease status of an area, thus risking misinforming any management programme. If hotspots are not stable in space and time then subsequent targeting of hosts within discrete patches may, at best, be ineffective. In this case it may be more profitable to target corridors through which pathogens (and/or their hosts) may spread, in order to break the transmission chain.

8.4 Targeting Pathogens and Vectors

The most obvious direct method of targeting pathogens in the environment is by disinfection. Chemical disinfection of drinking water has been widely practiced to control anthrax in wild game mammals in southern Africa, but is not appropriate in many circumstances, such as in large water bodies (Berry 1993). This method is only likely to be successful where localised foci of pathogens can be identified, since wider scale disinfection of the environment is likely to be uneconomical, and potentially environmentally damaging.

The carcasses of infected animals may represent highly localised foci of infection. *Trichinella spiralis* (the causative agent of trichinosis), for example, is transmitted during scavenging. Also, the investigation of infected wildlife carcasses by brushtail possums, red deer (*Cervus elaphus*) and domestic cattle, particularly after they have been opened up by scavengers, is considered to be the main route of inter- and intra-specific transmission of *M. bovis* among wild mammals in New Zealand (Nugent 2005). Carcasses also play an important role in the transmission of anthrax in parts of Africa where the removal and burial or burning of wildlife carcasses has been central to efforts to control the disease in wild mammals. Although it is unlikely that all carcasses can be located, even following intensive searches, reducing the overall availability of such sources of infection by disposing of what can be found, may be expected to provide some benefits. Nevertheless, the effectiveness of this approach is not clear, as when employed during disease outbreaks in wild birds, it does not appear to have reduced avian mortality (Wobeser 2007).

Vectors, and the free-living stages of parasites, can be indirectly targeted by manipulating the environment to make it unfavourable for their persistence. For example, removal of vegetation from Acacia savannah in sub-Saharan Africa rendered the environment inhospitable to tsetse flies (*Glossina* spp., the insect vector of *Trypanosoma* spp.), thus controlling trypanosomiasis and Chaga's disease in resident wild mammals, livestock and humans (Molyneux 1982). However, such action may not be without collateral ecological costs, and in this case the resulting habitat was also rendered unsuitable for wild mammal populations that had traditionally foraged there (Molyneux 1982).

Where pathogens persist in the environment in the faeces of infected hosts they may pose a risk of infection. *M. bovis* bacilli for example, may survive in the faeces of infected Eurasian badgers, particularly in dark, moist environments, but are vulnerable to desiccation and ultraviolet light. Badger faeces are often concentrated at latrine sites, which may represent a potential source of bTB infection for cattle. It has been suggested that introducing cattle to pasture in the afternoon would maximise the exposure of bacilli present in badger latrines to the weather, and hence reduce their infectivity to grazing livestock (Phillips et al. 2003).

Direct targeting of insect vectors with insecticides has been widely practiced in the past, but has fallen out of favour owing to the problems of insecticide resistance and health risks to humans and livestock. In recent years interest has focused on integrated approaches to vector control, which include environmental management, chemical, biological and mechanical control (Lacey and Lacey 1990). Many species of anopheline and culicine mosquitoes carry pathogens causing a variety of diseases such as malaria, Japanese encephalitis, West Nile virus and Rift Valley fever. Intermittent irrigation, flushing fields and changing the timing of crop plantings have been used to discourage mosquito breeding in rice producing areas, in order to reduce disease risks for humans and livestock (Lacey and Lacey 1990). Similar approaches might be applicable for the control of pathogen vectors for wild mammals.

8.5 Targeting Hosts

Direct targeting of wildlife hosts for disease management has in the past often involved the reduction of population density by culling (see Chapter 7). Environmental manipulations may provide an alternative means of reducing intra and inter-specific contact rates, through their effects on mammal distribution and local density. However, since mammals are typically highly mobile and make complex decisions regarding space use and movement patterns, the outcomes of environmental manipulations targeting hosts may be less easily predicted than those directed at pathogens or vectors.

A reduction in the availability of crucial resources will result in a concomitant reduction in the abundance or distribution of a population. If environmental carrying capacity is pushed sufficiently low so as to reduce the population below the density threshold at which a pathogen can persist (i.e. where $R < 1$; see Chapter 3), then infection should disappear from the population.

8.5.1 Manipulating Host Density and Behaviour

Optimal foraging theory predicts that animals will distribute themselves according to the availability and abundance of resources. Hence, higher densities of individuals are expected in resource rich patches, with lower densities in sub-optimal areas. Consequently, local density may be suppressed by reducing the availability of critical resources, such as food or shelter, or distributing them more evenly across a landscape. However, such approaches are not without their potential problems. Reductions in the availability of resources could in the short-term result in malnutrition and hence increased susceptibility to disease. Also, the dispersal of animals seeking alternative food sources could potentially spread disease if infected individuals ranged further and made contact with susceptible hosts elsewhere. Finally, the use of environmental manipulation to reduce food resources may cause significant suffering (starvation), particularly among more sedentary species, and therefore raises concerns over whether such an approach is ethically acceptable.

In each situation the resource requirements and likely behavioural responses of wild populations need to be understood in some detail before environmental manipulation can be seriously considered as a disease management tool. Responses of host populations may be complex and can defy simplistic assumptions. For example, the population density of red foxes in temperate Eurasia and North America influences the spread and incidence of rabies. As fox distribution and density are dependent on the availability of food and shelter, it seems reasonable to expect that fox density could be influenced by manipulating the distribution and abundance of these critical resources. In practice however, because foxes are highly adaptable and can exploit a diversity of food items and environments, attempts to control rabies outbreaks through environmental manipulation (Steck 1982) have met with far less success than culling (Müller 1971) and vaccination (Holmala and Kauhala 2006). This is likely to be the case for other adaptable, generalist species with broad diets and habitat requirements.

The local density of wild mammals has been profoundly altered by changing agricultural practices (Cimino and Lovari 2003), burning (Van Dyke and Darragh 2007), and planting unpalatable foods (Conover 1991). Attempts to alter the density of wild mammal populations by manipulating resources, whether for the purposes of pest control, game production or conservation may also have consequences for disease dynamics. For example, diversionary feeding strategies have been employed in order to discourage wildlife from congregating in sensitive areas where they were considered to cause damage or nuisance, and supplementary feeding has been widely employed for game production. In the context of disease control however, supplementary feeding areas can themselves pose a risk of enhanced transmission by encouraging aggregations of individuals. Large numbers of white-tailed deer (*Odocoileus virginianus*) congregated at supplementary feeding stations in Michigan, USA, and the local increases in deer density were implicated in an increased prevalence of bTB amongst wild deer and domestic cattle herds (Miller et al. 2003). Deer culling was successfully employed to reduce local deer densities below the threshold at which bTB could persist. However, restrictions on the supplementary feeding of deer also made a major contribution to the reduced prevalence of bTB in both deer and cattle (Miller et al. 2003). The dispersed planting of attractive food sources across the landscape may provide an alternative means of reducing local densities of herbivores.

Predator control is usually implemented with the intention of protecting prey populations that are of economic or conservation value. But the actions of predators may influence levels of disease in prey populations, by for example removing heavily infected individuals and reducing prey density. For some density-dependent diseases, predator removal has the potential to increase disease incidence within the prey population by allowing their local density to increase. The converse may also be true, such that an improvement in resources for predators may increase their abundance or predation success rate, and thereby disperse or reduce the density of their prey, and so potentially impede disease spread. However, unless predators are maintained at artificially high levels it is likely that the density-dependent feedback of a reduced or dispersed prey population will lead to a reduction in predator abundance in time, thus providing only short-term disease control until an equilibrium is reached between predators and their prey. An alternative scenario is that a high density of predators may promote high local abundance of pathogens that may be transmissible to other animals sharing the same environment. These hypotheses have yet to be tested empirically, and other outcomes are possible, so we are at a early stage in understanding how the manipulation of predator pressure could be used as a tool to control disease in prey populations. Nevertheless, the potential role of predator populations should be considered when developing any plan to manage disease in a wild mammal population.

8.5.2 Disease Spread

It is possible that the rate at which disease progresses within a population may influence the extent to which it can be controlled through environmental manipulation. The differing potential effects of habitat heterogeneity on disease spread were identified in a model simulating a chronic (i.e. bTB) and an acute (i.e. rabies) infection in Eurasian badgers. The model outputs suggested that increasing habitat heterogeneity would lead to a gradual decrease in bTB prevalence. However, a threshold effect was detected for rabies transmission, such that low levels of habitat heterogeneity had no effect on transmission, but high levels limited its spread (Smith and Wilkinson 2002; Wilkinson et al. 2004). These effects probably arose as a result of the different ways in which chronic and acute diseases persist and spread across landscapes. A chronic disease, such as bTB, does not require a high frequency of host contacts in order to persist since infected individuals can survive over longer timescales. Hence, increasing habitat heterogeneity should be expected to maintain chronic diseases in localised foci, which should fade with time in the absence of host contacts. In contrast, an acute disease, such as rabies, requires a higher frequency of host contacts in order to persist and so also requires a minimum level of habitat connectivity to ensure sufficient host interactions. The implication is that enhancing habitat heterogeneity may in some cases be used to manage disease spread in wild mammal populations by controlling contact rates, and the benefits may accrue quickly, but in the case of a rapidly progressive disease this is only possible after a contact rate threshold has been reached. For a slower progressing disease, the benefits may not accrue so quickly. At the moment these are only theoretical possibilities as no empirical evidence has yet been generated experimentally.

8.5.3 Reducing Susceptibility to Disease

Nutrition influences immune system functioning and hence susceptibility to disease (Lochmillar and Deerenberg 2000; Wobeser 2006). The availability of essential nutrients, protein and energy are directly associated with habitat quality and can be influenced by numerous factors. Density-dependent competition may decrease the ability of some individuals to acquire sufficient food resources, reducing their overall protein and energy intake. The competition between conspecifics that may arise as population density increases is also likely to cause stress, which can impact adversely on the performance of the immune system. It follows that reductions in population density, below the level at which inter-specific competition for resources is detrimental, could potentially improve the physical condition and resilience of individuals to disease. However, accurately predicting when this point has been reached is a considerable challenge. In addition, the demographic and behavioural consequences of reducing host population density may be counter-productive for disease control for other reasons (see Chapter 7).

The absence of adequate shelter for the purposes of thermoregulation, predator avoidance and rearing young is likely to be another potentially important cause of enhanced stress. Therefore, management of the environment in ways that maximise the availability of suitable cover may help to decrease stress and disease susceptibility among some mammals, although of course this may also increase host density.

As disease susceptibility can vary between conspecifics of differing sex and age classes (see Chapter 2), the effects of habitat quality on disease occurrence may exhibit similar variation. Such potential differences will need to be considered when planning disease management through environmental manipulation.

8.5.4 Reducing Transmission Between Wild Mammals and Livestock

Transmission of pathogens at the wildlife-livestock interface can occur in both directions and may therefore pose a threat to either agriculture or conservation. Foot and mouth disease (FMD) in domestic cattle serves as a case in point, because although they are the most important source of infection for wild mammals on many continents, in parts of Africa they are themselves susceptible to transmission from a reservoir of infection in wild buffalo (*Syncerus caffer*) (Bengis et al. 2002).

The most obvious means to prevent contact between wild and domestic mammals is the use of fencing. Numerous fence designs have been successfully employed to this end (Vercauteren et al. 2006) but the cost and practicality of fencing extensive areas may limit the range of potential applications. Moreover, fences may be ineffective if not deployed at a sufficiently large scale or if positioned far from the disease front. For example, inadequately positioned fences failed to prevent transmission of brucellosis between bison (*Bison bison*) and cattle in the USA (Cheville et al. 1998). Numerous national parks have constructed high fences either to contain wild mammal populations or to prevent access from those outside (Kassilly 2002; Whitehouse and Kerley 2002; Sievers 2004; Walter et al. 2005), and they routinely deploy significant resources for their periodic inspection and repair. Typical problems include damage from water run-off, bad weather, fallen trees and vandalism. Electric fences have been designed specifically for the purposes of restricting the movements of wild mammals and have been deployed in South Africa and Zimbabwe in order to protect cattle from bTB and FMD transmission from wild mammals (Taylor and Martin 1987). More recently 'invisible fences' have been tested to assess their efficacy at reducing contact between livestock and wild mammals (see Box 8.3).

Box 8.3 Livestock protection dogs for deterring deer

In parts of Michigan, USA, white-tailed deer (*Odocoileus virginianus*) are the primary wild maintenance host of bovine tuberculosis (bTB) and are considered to be the main reservoir of infection for local cattle (O'Brien et al. 2002; O'Brien et al. 2006). The principal route of transmission is thought to occur when deer contaminate cattle feed put out on pasture, (Palmer et al. 2001; O'Brien et al. 2006) although direct contact cannot be discounted. Methods to reduce deer activity near cattle may offer options to control direct and indirect disease transmission, but while field-scale exclusion may be effective, deer-proof fencing may not always be appropriate.

For thousands of years, livestock producers have used domestic dogs to deter predators from pastures and paddocks. It follows that dogs could potentially be used to reduce direct and indirect contact between white-tailed deer and cattle and hence contribute to disease control. In a field trial of this approach, dogs were kept alongside cattle within discrete areas of pasture on a deer farm where they could be surrounded by an artificially high density of deer. Dogs were kept within the enclosures by an Invisible Fence® (IFCO Enterprises, Malvern, Pennsylvania, USA) and cattle were confined using a traditional electric fence. The Invisible Fence system involved each dog wearing a collar carrying an electronic device that responded to a signal from a wire encircling the enclosure. The collars emitted an auditory cue when a dog approached to within 1 m of the wire and an electric shock if they failed to move away. The dogs were quickly conditioned to the Invisible Fence, and treated it as if it were a physical boundary. The results showed that dogs were effective at substantially reducing deer incursions onto pasture and almost entirely prevented contact between deer and either cattle or their feed, even at high deer densities.

Dogs specifically trained to remain with grazing cattle may therefore offer a practical tool to minimise contact between deer and cattle, and thereby limit opportunities for transmission of bTB and potentially other infectious diseases. Even in larger pastures, dogs may effectively exclude deer from using spatially concentrated sources of cattle feed, which probably present the greatest risks of transmission from deer to cattle. Livestock protection dogs may therefore provide a valuable biosecurity tool, particularly for small cattle operations and the use of modern invisible barrier systems may facilitate their efficient deployment where traditional physical barriers are not appropriate.

The effect of physical fencing on the behaviour of non-target species should be considered prior to installation. Fencing along waterways and highways may have delayed wolf (*Canis lupus*) population expansion in Spain for nearly two decades because they obstructed dispersal routes (Blanco et al. 2005). In sub-Saharan Africa fences have been used to segregate wild mammals from livestock for disease control (Molyneux 1982) but in Kruger National Park, South Africa they also severed a wildebeest (*Connochaetes taurinus*) migration route (Whyte and Joubert 1988). Fenced motorways may prevent CSF spread between wild boar populations but they also constrain lynx (*Lynx lynx*) dispersal (Rossi et al. 2005b; Klar et al. 2006). Restricting dispersal may also have an undesirable impact on disease management if the density of hosts inside fenced areas increases and so enhances transmission rates.

Various types of deterrent that have been employed to protect crops and other resources from wild mammals could potentially also be used to influence contact rates with domestic stock and hence disease transmission risks. An example would be the use of domesticated animals (usually dogs) as guardians of livestock or farm facilities (see Box 8.3). Devices employing visual (e.g. scarecrows and predator-mimicking devices) and auditory (e.g. exploders and distress calls) stimuli have been used as area deterrents, although these approaches may result in eventual habituation (Vercauteren et al. 2005). In general, such devices are more effective if they are animated (e.g. by using automated motion sensors), and if the stimuli are unpredictable and associated with a strong negative experience. The use of deterrents is likely to be most appropriate when the aim is to deter wild mammals from a specific area, such as farm buildings or a field of livestock, where risks of disease transmission are deemed to be high.

A variety of changes to domestic animal husbandry practices may help to reduce the risks of transmission of infection from wild mammals. Livestock that are housed in facilities to which wild mammals can gain access may be exposed to direct contact or environmental contamination from infectious hosts (Dolan 1993; Flanagan 1993; Hutchings and Harris 1997; Meerburg et al. 2006; Ward et al. 2008a). Where it is practicable, exclusion of wild mammals from such locations is likely to be a worthwhile livestock biosecurity measure. However, potentially infectious excretions may also be distributed across open pastoral landscapes, where the prevention of exposure to domestic stock may be more difficult.

8.6 Turning Information into Policy

Increasingly, policy development in many countries is required to be evidence-based, and this provides scientists, conservationists and land or wildlife managers with opportunities to influence the opinions of policy makers. Information collected with scientific rigor can provide a robust and defensible evidence base, but the length of time it can take to collect may frustrate policy makers. Hence, it is not uncommon to find policy underpinned by observation and anecdote as a substitute for scientific evidence. However, there are considerable risks associated with sources of evidence that are not robust, and are subject to selective personal interpretation. In circumstances where environmental manipulation is being considered for disease control purposes, few empirical data may be available, but it is nevertheless important that whatever information can be obtained is assessed in a systematic and objective manner. Qualitative risk assessment (see Chapter 9) may provide a useful framework for this purpose.

An excellent example of a strategy considering the potential impacts of a wildlife disease management plan, is the environmental impact statement on the control of chronic wasting disease (CWD) in white-tailed deer populations produced by the Wisconsin Department of Natural Resources, USA (Bartelt et al. 2003). The authors reviewed what was known about the pathology, transmission and detection of CWD, deer ecology and behaviour, and how they might affect the spread of infection, how other states managed the disease and contemporary control methods. They explored options for controlling wildlife diseases (including doing nothing) and the potential consequences for a variety of stakeholders including state agencies, hunters, landowners, farmers, wildlife enthusiasts, local businesses and native American Indian communities, and potential impacts on vegetation and animal communities. The comprehensive report served to inform both decision makers and the public of the likely consequences of options to control CWD.

8.7 Changing Attitudes and Behaviour

Manipulation of the environment may offer opportunities to manage disease in wild mammals without resorting to potentially controversial lethal control or costly vaccine development and deployment, and so may be an attractive option for policy makers. However, environmental management is likely to require the co-operation of several key stakeholders (e.g. farmers and other land managers) and this raises a major challenge for policy makers. These parties may be reluctant to alter their long-established management practices, especially when the benefits may be uncertain or take a long time to accrue. For example, whilst the potential risks of disease transmission from wild mammals via contamination of livestock feed had been clearly demonstrated (Hutchings and Harris 1997; Garnett et al. 2002; Daniels et al. 2003a), few UK farmers appeared willing to invest in the necessary protective husbandry measures (Bennett and Cooke 2005). Moreover, wild mammal populations transcend land ownership boundaries, and disease management strategies may therefore require co-ordinated action amongst many parties. Achieving consensus on a disease management strategy may however be difficult, particularly where neighbouring landowners have different values and opinions. The same will be true for all other sectors of society who may have an interest in the issue, including stakeholder groups, the general public, government policy makers and politicians.

Understanding the prevailing attitudes of stakeholders and how to change them in the face of scientific evidence is a substantial challenge for the development of sustainable approaches to wildlife disease management. Hence, the discussions that follow are of generic importance, although they are particularly relevant to environmental management programmes because these often require co-ordination across landscapes and land-ownership boundaries, and are therefore hostage to the values, attitudes and opinions of multiple stakeholders.

8.7.1 Understanding Attitudes

One way to enhance adoption of innovation is to understand how people make decisions. Once this process is better understood, it will become easier to influence it in order to encourage people to adopt practices related to disease management. Many farmers, for example, are unusual in that their business interests, lifestyle and culture are all closely related. As a result, their decision-making processes are influenced not only by financial considerations, but also by a range of social factors, such as the age and structure of the family, sources of off-farm income and their connection to the local community (Potter and Gasson 1988). These socio-demographic issues can easily affect farmers' attitudes to risk, willingness to invest large sums of money and their likelihood to change long-standing practices (Edwards-Jones 2006). Decisions are also likely to be influenced by people's fundamental personality, attitudes and objectives (Edwards-Jones et al. 1998; Willock et al. 1999). Early adoption of environmental schemes and improvements to animal health and welfare, is often linked to a farmer having a personality and set of attitudes that are open to new ideas (Austin et al. 2005; Dutton et al. 2008). Typically, only a minority will adopt new ideas quickly, a larger number will consistently resist change, while most may adopt change over time as their social and financial situation permits.

8.7.2 *How to Influence Attitudes and Behaviour?*

Although there is no single blueprint for bringing about behavioural change, the key elements of a successful campaign typically include:

- Communicating a convincing message
- Gaining trust with the stakeholder community
- Embracing stakeholder participation
- Developing practical demonstrations
- Developing credible champions for the message
- Minimising administrative burdens
- Removing perverse incentives
- Supporting the campaign with wide scale communication
- Helping stakeholders feel good about what they have achieved

8.7.2.1 Communicating the Message for Change

It is vital that the basic message about why change is necessary is credible and makes inherent sense to stakeholders. It is likely to be necessary to demonstrate that a management approach can deliver net benefits to the stakeholder, before they can be expected to implement or accept such measures themselves.

While benefits may be demonstrated to scientists and policy makers through experimental investigations, land managers may be more readily convinced by practical demonstration in a realistic setting, such as a working farm. Preferably such a farm would be managed by someone who is trusted and respected (i.e. a champion). It is clearly important to have a good understanding of the financial costs and benefits of any environmental manipulation and these may be presented in the form of a series of investment appraisals if net benefits resulting from behavioural change are expected to accrue to a business. If most benefits are expected to be external to the business, such as an improvement in the health of wild animals, then it may be more difficult to make the case for change financially appealing to business stakeholders.

In order to consider the wide-scale benefits that may accrue to society from changed behaviour, economists tend to undertake cost–benefit analyses (CBA; see Chapter 5). CBA requires the identification and valuation of all elements of a system that will be impacted by some intervention. Benefits may be relatively straightforward, such as increased profit for local businesses, but they may well also include beneficial changes in so-called ‘non-market goods’ such as landscape, biodiversity and animal welfare. Although these benefits do not typically have market prices associated with them, economists use a range of techniques to estimate their monetary value (see Chapter 5). Through considering all relevant costs and benefits in this way, the viability of a project can be determined in quantifiable monetary terms. Although CBA is a powerful and widely used technique it tends to be better suited to informing major business and policy decisions, than to persuading individual farmers to adopt certain practices. This is because in essence the CBA is suggesting that if the farmer undertakes certain actions (which may cost him time and money) other people in society may reap some of the benefits (i.e. through improved wildlife health). This almost always raises the inevitable response from farmers that if society is getting all this benefit then why are they not paid more for delivering it? For this reason, in many cases, it may be more productive in the long run to appeal to the farmers’ better nature, rather than involve them in discussions of CBAs.

8.7.2.2 Regulation, Incentives and Administration

In many countries agricultural policy and the regulatory framework are complex. In addition, a variety of different organisations are typically responsible for the various components of the system. For example, within the UK, separate agencies are responsible for payment of agricultural subsidies, agri-environment schemes, animal health, waste disposal, food processing standards, farm worker safety and planning. However, many of the activities regulated by these different agencies interact at the farm level. This type of organisational structure is not confined to the UK, and is characterised by the typical observation that changes in one activity may relate to regulations that originate from more than one agency. This can create a frustrating and complex administrative burden, which means that changes to management practices are hindered or even prevented.

8.7.2.3 Peer Support and the ‘Feel Good’ Factor

A successful campaign may persuade stakeholders to change their behaviour. However, if this situation is to persist, then stakeholders require support from their peers. It is difficult for any individual to maintain a behaviour when their peers disapprove of their actions. So when planning a campaign to alter stakeholder behaviour, it is important to use the media and other sources to communicate the message to the wider community. In this way the stakeholders will find themselves living and working in a supportive community, rather than one that is unsympathetic to their activities. Finally, nothing sustains desired behaviour like positive feedback. Communicating positive messages about stakeholder activities to other stakeholders and the wider community can be a powerful tool for encouraging sustained effort (Ward et al. 2008b).

8.8 Conclusions

Environmental management has been used historically to control many diseases in wild mammals. While experimental studies demonstrating efficacy are rare, some predictions can be made on the basis of what is known about the relationships between environmental structure, mammal hosts and their pathogens. From the evidence presented here it is clear that while environmental management may be a useful tool for the control of disease in wild mammals, its success rests on a sound understanding of the ecology of the host–pathogen system. Of key importance is an understanding of how pathogens persist and spread in space and time within and between populations and environments. In this respect field studies and experiments are fundamentally important in providing robust empirical data, although this process can be frustratingly protracted. Developments in geographical and mathematical modelling tools can help by providing platforms on which to construct predictive models of disease spread and control, although their value is directly related to the quality of input data and their post hoc validation using independent data (see Chapter 4).

It is important to consider both target and non-target impacts of proposed management plans since environmental manipulations are likely to impact on other components of ecological communities, including other human activities. EIA may provide a useful framework for the review and assessment of the potential impact of such approaches to disease management. However, this may be a considerable challenge given that the benefits of environmental manipulations are less certain than for other disease control methods, may not accrue directly to stakeholders expected to undertake the manipulations and may take some time to materialise. This makes it all the more important to understand stakeholders' attitudes and values in order to develop and implement sustainable policies.

Chapter 9
Risk Assessment and Contingency Planning for Exotic Disease Introductions

Vicky S. Jackson, Selene Huntley, Alex Tomlinson, Graham C. Smith, Mike A. Taylor, and Richard J. Delahay

9.1 Introduction

Globalisation has greatly enhanced opportunities for the spread of infectious diseases throughout the world, giving rise to serious threats to human and animal health. This is illustrated by the recent introduction and subsequent spread of West Nile virus in the USA, and outbreaks of Severe Acute Respiratory Syndrome (SARS) in South-East Asia. It is therefore becoming increasingly important that national (and potentially regional) governments should not only have robust systems in place to reduce the risk of disease introductions, but that they need to also consider how to identify and deal with outbreaks of pathogens in wild and domestic animals. In this chapter we will discuss the roles of risk assessment and contingency planning in the management of exotic disease risks involving wild mammals.

The principal purpose of contingency planning is to ensure that a state of preparedness exists in the event of a disease introduction. This requires that the most likely risks of pathogen introduction are identified, that there are adequate means of detecting the pathogen's presence, and that a set of instructions exists describing the best available methods for its rapid and cost-effective containment and control. Contingency planning will involve some of the approaches to disease surveillance (Chapter 10) and management (Chapters 6–8) discussed in other chapters, and so will entail many of the associated challenges, costs and benefits. However, as the aim of a contingency plan is likely to be the rapid containment and subsequent elimination of a pathogen (that is either exotic or endemic but emergent) within a restricted area, the methods of management should reflect this urgency. This may mean that it is appropriate to deploy more severe or costly measures over a short period than would be considered for the sustained control of an endemic pathogen.

It would be impractical to attempt to develop contingency plans for every pathogen of wildlife that could theoretically be introduced. Instead, risk assessment approaches should be used to identify those pathogens where the risks of importation and subsequent establishment are high, and the potential effects, usually in terms of human

V.S. Jackson, S. Huntley, A. Tomlinson, G.C. Smith, M.A. Taylor, and R.J. Delahay
Central Science Laboratory, Sand Hutton, York, YO41 1LZ, UK

R.J. Delahay et al. (eds.), *Management of Disease in Wild Mammals*,
DOI:10.1007/978-4-431-77134-0_9, © Springer 2009

health or economic impact, are sufficiently serious. Some diseases of wildlife, such as rabies, are of such public health concern that almost all disease-free countries have some form of contingency plan to deal with them. A cursory internet search will reveal a number of local and national rabies contingency plans for the UK, USA and Canada. Similar national contingency plans exist for several serious diseases of livestock. In the UK, there are plans covering a range of diseases in livestock, including foot and mouth disease (FMD), avian influenza, Newcastle disease, classical swine fever, African swine fever, swine vesicular disease, rabies and bluetongue. Many of the most important global pathogens of wild mammals are listed by the World Organisation for Animal Health (OIE, Office International des Epizooties). However, the risk from other diseases may also be worth assessing, owing to their geographical proximity, the probability of entry through trade routes or changes in their global status. Some pathogens such as rinderpest, are close to being eradicated (2010 is the target date set by the Global Rinderpest Eradication Programme), while others such as West Nile and Nipah virus are considered to be emerging diseases, and some such as FMD appear in a variety of strains, each of which may need to be considered separately. For those diseases where the risk and potential impact of an outbreak is deemed to be sufficiently great, it would seem prudent to plan an appropriate response.

An exotic disease could be imported in infected domestic animals, wildlife or animal products, or through natural movements of infected hosts or disease vectors. The first line of defence against introduced pathogens is prevention through systems such as quarantine, screening, animal movement tracing and import controls. Risk assessments can inform the development of measures to prevent disease introduction, and contingency planning will provide a level of preparedness should this fail.

Clearly, historical surveillance data on wildlife diseases could provide a background against which to identify novel or emerging pathogens. However, surveillance systems for diseases in wildlife are almost certainly likely to be more poorly developed than systems for monitoring disease in livestock. Surveillance of disease in wild hosts is unlikely to produce strong evidence that a particular pathogen was previously absent (see Chapter 10 for more details). In practice therefore, records of novel pathogens in domestic animals may provide the most reliable predictors of exotic disease introductions that require intervention.

Contingency plans should have a clear overall objective. The choice of acceptable outcomes might include elimination of the disease, or containment within defined limits in terms of geography, prevalence or economic impact. The chosen outcome will determine the scale and characteristics of the response. The level of resources required to achieve this outcome (both equipment and trained personnel) must be available. The response should be described in a series of instructions that should also indicate which organisations are responsible for the implementation of each part of the plan. As the precise conditions of the outbreak cannot be accurately predicted beforehand, the plan should describe the action required under a variety of circumstances. For example the appropriate response may differ depending on whether the disease was initially detected in domestic animals or wildlife, or in relation to the extent of its geographic spread. Indeed, the overall aim of the intervention may be influenced by the conditions (e.g. extent of disease spread, time of year) at the time when the problem is initially identified. It is important that the plan describes how the outcome of intervention is monitored, and ultimately the conditions required for action to be terminated (e.g. no disease detected for x weeks or months). It is clear from this brief overview that any effective contingency plan needs to consist of instructions that are adaptable to a range of circumstances, which may change in the course of a disease outbreak.

Mathematical modelling (see Chapter 4) can be a valuable tool for the development of contingency plans and to guide intervention during their implementation. Models provide a means of simulating a range of disease outbreak scenarios and of estimating the level of effort and resources required to eliminate or contain infection. The incorporation of an economic dimension in the modelling will provide information on the relative cost-effectiveness of different approaches (see Chapter 5).

Contingency planning for the control of some diseases of domestic animals is likely to require careful consideration of the potential role of wild hosts. The development of effective contingency plans to control disease in both wild and domesticated species will have many features in common. However, management of disease in wildlife raises particular additional challenges related to determining their abundance, distribution and disease status, the practicalities of capture and handling, and the potential for complex behavioural and ecological responses to intervention (see Chapter 2). Inevitably, disease contingency plans involving wildlife are substantially more challenging than those designed for domesticated animals alone, and require dealing with higher levels of uncertainty in terms of both available data and the predictability of outcomes.

There has been a recent trend in modern government in some parts of the world towards cost sharing for animal disease control in livestock. The increasing freedom of information and an inclination towards greater government transparency has led to many contingency plans being publicly available, and even subjected to consultation while in draft form. Both these trends have tended to foster ever-greater stakeholder participation, which is now generally welcomed and encouraged by many governments. This allows all organisations with any interest in the disease outbreak or the methods of control to be involved at an early stage of contingency plan development, and should in theory achieve maximum 'buy-in' from organisations prior to the implementation of any plan.

9.2 Risk Assessment

Since it is impractical to produce a contingency plan for all potential exotic pathogens of wildlife, we need to prioritise by identifying those that present the greatest risk. There are two broad approaches to assessing such risk. A qualitative risk assessment uses subjective categories (e.g. low, medium, high) whereas quantitative assessments use numerical data to quantify parameters (and their uncertainty). The use of real data is clearly preferable but may be difficult to obtain for some parameters. Risk assessments that are an amalgamation of the two approaches are sometimes referred to as semi-quantitative.

Qualitative approaches are useful for rapid assessments of disease risks and in particular for those pathogens and hosts for which little or no quantitative data exists. A qualitative risk assessment often takes the form of a series of questions, each of which may be assigned a rank or score, or alternatively all questions may be considered equally important. One or more experts may then be asked to answer each question with a categorical response. Wherever possible, the uncertainty in their response, and their level of expertise in each area should be estimated. This allows the depth of knowledge of different assessors to be taken into account and areas of uncertainty and data shortfall to be identified. The UK government has published a generic qualitative risk assessment scheme for non-native organisms (Baker et al. 2008), which considers the introduction of exotic animals, plants and pathogens. The advantage in using a generic risk assessment is that the risks posed by a range of organisms that may be important in the epidemiology of a disease (including both hosts and vectors) can be assessed and compared. This provides a framework for the rapid assessment of the relative risks of different pathogens. The scheme can be used to assess the potential for entry, establishment and impact of an exotic pathogen in the UK. The magnitude of the potential consequence of pathogen introduction was included in the risk assessment as a weighting. Scores relating to the likelihood of entry and establishment, and the magnitude of impact, were then given a numerical value as an aid to interpretation. Such an approach allows direct comparisons to be made between risk scores associated with different pathogens and hosts.

A quantitative risk assessment uses numerical data (probabilities, rates, etc.) and is thus regarded as more objective. This approach is useful for determining the risk of disease importation, particularly through trade, where importation rates, and routes are quantifiable. While quantitative risk assessments could be regarded as more accurate, they are not necessarily more useful. Generic risk assessments are much more difficult to produce, as all the necessary quantitative data may not be available; hence comparative quantitative risk assessments are difficult to accomplish. Comparative assessment of the likelihood of importation of different diseases through trade routes may be a relatively straightforward exercise, but it may be problematic to attempt to compare these with risks of disease introduction by other means (e.g. via migrating birds or wind-blown invertebrate vectors). A quantitative risk assessment requires that all routes of disease introduction are quantified, which can be very difficult where illegal importation occurs (e.g. bush-meat). A UK government report on the potential introduction of terrestrial rabies used a quantitative risk assessment approach, and determined that the likelihood of introduction into the country was approximately once in every 36 years (range 21–87 years) under a regime of six months quarantine for all imported domestic cats (*Felis catus*) and dogs (*Canis lupus familiaris*) (Kennedy et al. 1998). This report was used as the basis for the introduction of a pet travel scheme involving vaccination, identification and blood testing of companion animals as a replacement for quarantine, since this led to only an estimated 2% increase in risk: much less than the uncertainty of the assessment for quarantine. Subsequently, as the movement of pets increases and the scheme is expanded to a wider selection of countries, so the changing risk can be reassessed. Such re-assessments cannot be performed well using a qualitative approach as the final measure of risk is only described in relation to categorical levels (e.g. low, intermediate, high). Quantitative risk assessments are therefore far superior when the aim is to assess changes in risk following the adoption of new policies or procedures.

Risk assessments can be extended spatially, to provide measures of local risks across a broad geographical area. One such approach involved assessing regions at risk of West Nile virus by matching local temperatures to the competency of arthropod vectors to incubate the virus effectively (Zou et al. 2007). A comparative risk assessment of the relative roles of Eurasian badgers (*Meles meles*) and wild deer as sources of *Mycobacterium bovis* infection for cattle, also included a spatial dimension (Ward et al. submitted). The density of badgers and deer across England and Wales were estimated in 10 × 10 km squares. Information on host body weight and the mean prevalence of *M. bovis* infection (from surveillance studies) was then used to estimate the potential relative levels of environmental contamination with bacilli from each host species. When data on cattle stocking density in each 10 km square was overlaid this created a spatial map of potential TB risks to livestock. Thus risk assessments can be performed to predict the effects of policy changes, or when linked to GIS, to determine the geographical risks in different areas with the potential to inform local decisions on disease management for exotic or endemic diseases.

The risk of disease establishment, and the rate of spread, will depend on the availability of suitable host species, and their combined density and behavioural characteristics (e.g. dispersal rates and distances). This is difficult to determine because of the uncertainty in the suitability of many wild mammals to be competent hosts, and the lack of information on their local density and dispersal behaviour.

9.3 Detection/Surveillance

The challenges and approaches to disease surveillance in wild mammals are discussed in detail in Chapter 10. Here we describe how the detection of disease relates to the implementation of contingency plans, and the value of ongoing surveillance during the course of dealing with an outbreak.

There are a variety of approaches to the diagnosis of infection in the host, including serological tests, identification of gross or histo-pathology and isolation of the pathogen itself. All such tests have their limitations, which are usually expressed in terms of sensitivity and specificity. For many diseases, the appropriate tests will not have been validated in wild hosts, as it is frequently the case that they were initially developed for use in livestock or humans (see Chapter 10). Also, by the time infection has been identified in a given individual, there will almost certainly be further cases present. Hence contingency plans need to take account of the probability of case detection and the likely rate of disease spread prior to detection. Some system of surveillance should be initiated (or any existing system should be intensified) as soon as possible once the first case has been detected. The initial aim of this surveillance will be to determine the spatial spread of disease and to gain some information on the likely time since its introduction. When such disease outbreaks are detected in livestock the first response is often to stop movements and perform contact tracing. However, this course of action is rarely possible for wildlife, although restrictions on the movement of captive wildlife and susceptible livestock or domestic animals may help contain the spread of disease.

Within the European Union (EU), the standard alert system for confirmation of disease status after a suspect report relies on referral to the relevant EU reference laboratory (EU 2008). For diseases with high mortality or morbidity the submission of suspect individuals is the most efficient method of detecting disease. The detection of a novel disease will usually increase the submission rate of suspect animals, thus increasing the absolute number of infected cases detected. As a result, this sample is not sufficiently representative to provide an estimate of disease prevalence, and should rather be used to indicate the detection rate during a disease outbreak.

Determining disease prevalence is best performed by active stratified sampling so as to minimise detection bias. Where relevant, hunting bags may be a secondary choice, although the processes of capture and submission of carcasses by hunters can be prone to inherent bias. It is important that such bias is minimised (or at least quantified or constant) in samples obtained for disease surveillance during control measures, or this data will be inadequate for monitoring progress during implementation. During the latter stages of disease elimination, unbiased sampling can estimate the likely maximum level of undetected disease. For example, if we assume we apply a sufficiently sensitive test to an effectively infinite population, then a sample consisting of 300 negative cases would demonstrate (with 95% confidence) that the disease, if present, was at a prevalence of less than 1%. Consequently, it would require a sample of 3,000 negative cases to demonstrate a maximum prevalence of less than 0.1%. This demonstrates how sampling effort becomes increasingly critical for diseases that occur at low prevalence, and importantly, for detecting initial cases of an introduced disease. Hence, the results of disease surveillance can be particularly influenced by sample size and sources of bias during the closing phase of any contingency plan, when they should inform the exit strategy (e.g. the time which must elapse after the last case, before "freedom from disease" can be established).

Diseases which are deemed to be important, but which do not cause mass morbidity or mortality, can only be reliably detected by continuous surveillance. The effort (or sample size) required should be determined by the cost of sample collection, the ability to respond to the disease if detected, and the cost of an outbreak. There is little point collecting data for the presence of a disease, if it is not very costly and cannot be managed or eliminated once detected. Economic analysis could be used in these situations, but it should be borne in mind that once a disease detection strategy is in place, it is likely to cost relatively less to investigate each additional pathogen.

9.4 Contingency Plans

9.4.1 Design

Wherever possible the principal aim of any contingency plan for an exotic pathogen should be elimination. However, this may be unachievable in a practical sense, or the available approaches may be too costly to implement, or have undesirable consequences. Where infection cannot be eliminated from the wild host, then the aim may be to reduce it to such levels that spillover into domestic animals or people is acceptably infrequent. This could be achieved by reducing disease prevalence below some level, and containing the disease within specified geographic boundaries.

It is critical that any contingency plan clearly identifies those hosts that will be the subject of management action. Definitive lists of pathogens are rarely available, and are non-existent for most wild mammals, so the susceptibility of different species often has to be inferred from knowledge of the disease in other hosts. Another area where data is frequently limited relates to the distribution and population density of wild mammals. This data is unlikely to be available at a suitable resolution to provide information relevant to local disease outbreaks. In contrast, relatively good quality information is often available on the abundance and distribution of people and domestic livestock. Developing contingency plans for the containment of disease in wild mammal populations is therefore likely to be far more challenging than planning for livestock disease management. Improved abundance and distribution estimates for wild hosts, and information on their behaviour and ecology, will greatly enhance our ability to predict the likely spread of disease and hence the timing and area over which control should be applied. In addition, the ongoing monitoring of disease outbreaks and the impact of interventions will be greatly improved by the application of practical methods to rapidly assess wildlife presence and abundance in targeted areas. The future development of such methods should be considered as a priority amongst those organisations with responsibility for the control of disease in wildlife.

As not all hosts will necessarily be important in the maintenance and spread of an infectious disease, it follows that control measures do not need to focus on all affected species. However, the identification and targeting of true reservoir hosts (see Chapter 1) will be instrumental to effective disease management. Both reservoir and spillover hosts may be involved in perpetuating disease in wildlife or livestock, but effective action targeted at maintenance hosts will also reduce infection in spillover hosts. In some circumstances host status may vary in space and time, such that a particular species only constitutes a reservoir of infection when population density is sufficiently high, for example. This is the case for *M. bovis* infection in feral ferrets (*Mustela furo*) in New Zealand; where in places they occur at densities above 2.9 km^{-2}, and can be considered as maintenance hosts (Caley and Hone 2005). Conversely, some species may not only be reservoir hosts but may also be carriers. Such hosts do not exhibit clinical signs of disease, but are able to transmit the pathogen. Furthermore, where infection is indirectly transmitted (by an arthropod vector for example) then control of vector species may also be required. This will require information on the biology and distribution of vector species, which in practice, however may be incomplete or simply unavailable.

In its simplest form a contingency plan will comprise of three steps: (1) a trigger for implementation (e.g. detection of disease), (2) a set of procedures to adopt (e.g. wildlife vaccination) and (3) exit strategies to decide when to cease action. However, the last of these three steps is missing from many contingency plans. A hypothetical contingency plan for a disease outbreak in wildlife is given in Fig. 9.1. The plan is triggered

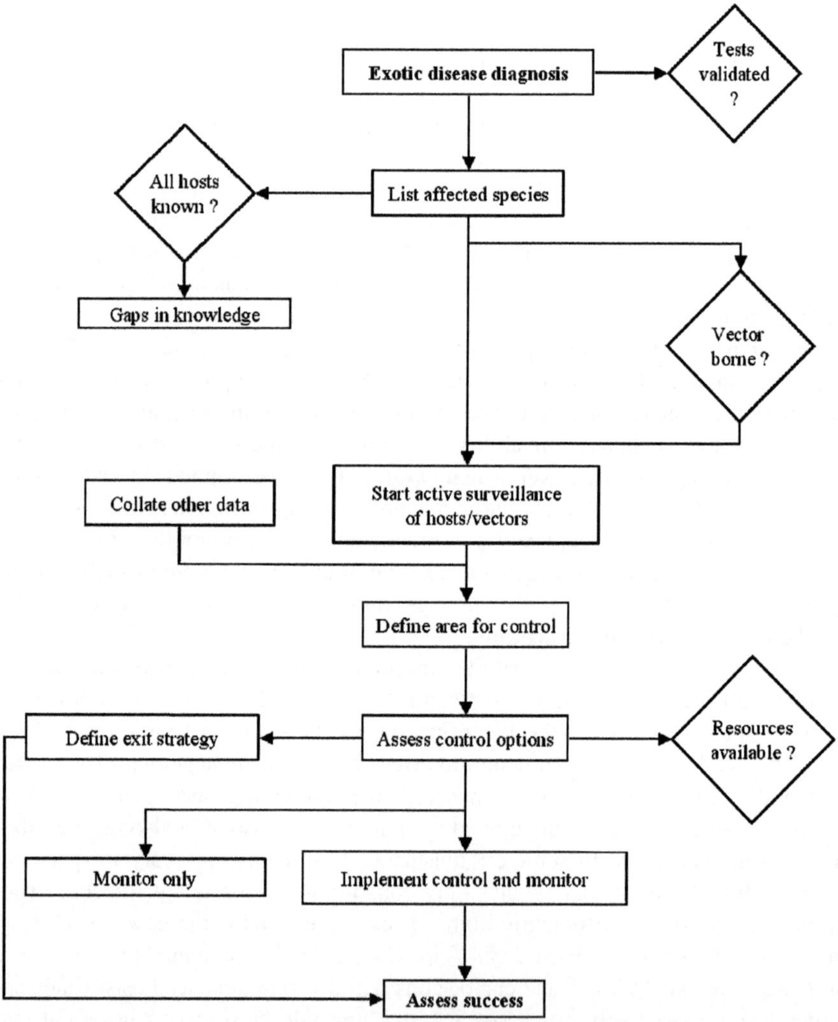

Fig. 9.1 A generic contingency plan for the implementation of control measures to deal with an exotic disease outbreak in wildlife

by the detection of a disease or pathogen, although the validity of the tests in the species of concern will be instrumental in determining the probability of a false positive result. It is important that a list of species that can be affected and are regarded as competent hosts is formulated, as this will identify where there are important gaps in our epidemiological knowledge. Another early consideration is whether the disease is vector borne, and if so, the relevant species and their distribution should be determined. Once this information has been gathered then an active surveillance strategy can be designed and implemented to determine the area currently affected. This should include data from all sources and feed into risk assessments for livestock and humans, so that suitable containment and prevention strategies can be adopted as required. Knowing the current area of spread, we can then assess the options for disease control. If management or elimination appears to be achievable, then the control plan is implemented, and monitored until a pre-determined point at which the exit strategy is triggered. This point may be reached once the disease has been eliminated, whereupon control activities will cease (although heightened surveillance may continue for some time). Alternatively, failure to contain and control the disease may require a switch from the current plan to adoption of measures for the management of endemic disease (i.e. having to live with it). In practice, disease control will be implemented before the geographical area of spread can be determined for two reasons: (1) successful disease control is more likely if control starts as early as possible and (2) it is politically difficult to do nothing while further data are obtained.

The principal options for disease control and management in wild mammals have been discussed in detail in previous chapters, and many of the same considerations apply when these approaches are employed in contingency plans. Hence, the optimal approach may be to combine methods, perhaps applying diverse approaches in different zones or periods of time. However, as contingency plans are often concerned with the rapid containment or eradication of an exotic pathogen in a limited area, the appropriate responses will frequently differ markedly from those required for the sustainable management of an endemic infection. Hence it may be appropriate to employ more severe measures over a restricted area or time frame. For example, culling of wildlife is considered more acceptable when used to control an epidemic outbreak than to manage endemic disease. This approach has been successful in halting rabies incursions over an isthmus (Westergaard 1982) and along alpine valleys (Irsara et al. 1982).

The UK Rabies Contingency Plan (see Box 9.1) includes the possibility of focal population reduction by culling, surrounded by vaccination (often referred to as ring vaccination). The objective is to vaccinate those individual animals that would disperse into the culled area during and after culling, and to reduce the number of susceptibles in the area into which infected foxes may disperse. Modelling studies suggest that this is the optimal strategy for control of a focal outbreak of rabies in a high-density area of foxes (Smith 1995). The effectiveness of this approach for rabies control in racoons has been demonstrated in the field (Rosatte et al. 2001). The clear aim of such a plan is to rapidly contain and eradicate this exotic zoonotic pathogen. In contrast, where rabies is endemic in wild mammal populations, the deployment of vaccine in baits has been demonstrated to be the most effective

Box 9.1 The UK rabies contingency plan

In Great Britain, domestic dog (*Canis lupus familiaris*) (urban) rabies has been recorded historically from the 11th to the 19th century. This includes two recorded human fatalities resulting from being bitten by pet foxes, and two outbreaks in park deer (King et al. 2004). There are no definitive descriptions of rabies spread in wildlife, so we must assume that these historical records are of dog rabies. By 1902, dog rabies had been eliminated following legislation enforcing quarantine, muzzling and the rounding up of strays. This approach has also been successful in eliminating dog rabies in many other countries. Terrestrial rabies was subsequently absent from Britain until a brief period (1918–1922) after the First World War when returning servicemen brought infected dogs back with them. Following two cases of dog rabies outside of quarantine in 1969 and 1970, a contingency plan was established with statutory powers set out in the Rabies Control Order 1974. These included powers to leash and muzzle domestic animals, seize strays and prohibit gatherings of animals and hunting. The Order also permitted the establishment of an infected area, within which wild mammals could be destroyed and the deliberate feeding of wildlife, and their access to waste food, could be controlled. Since then the UK Rabies Contingency Plan has gradually evolved in line with our improved understanding of the range and biology of hosts, and the differentiation of species-specific viral strains. The current plan describes the basic approach for dealing with a rabies outbreak, including implementation, zoning and the logistics of control. In essence, following a confirmed positive case, a decision is made (based on the viral strain and the case history) as to the likely risk to wildlife. 'Minimal risk' would result from a case in domestic animals or livestock where the infected individual has had no opportunity for contact with wildlife, and so no wildlife response would be deemed necessary. The risk would be considered 'possible' if an infected animal had been in an environment where some contact with wildlife might have occurred, but where the rabies strain was unlikely to spread in British wild mammals (e.g. a dog strain of rabies). The appropriate action in this case would be monitoring of wildlife (i.e. enhanced passive and potentially some active surveillance, depending on the case history) for a period of time not exceeding two years. Risk would be considered 'likely' where contact between wildlife and a compatible rabies strain were considered probable. In this instance the control of infection in wildlife would be initiated. This would also be the course of action where the first confirmed case of infection was identified in a wild species. The area and species targeted for control are determined from data on case history, host ecology and the outcome of simulation modelling. In the UK, mathematical models are used to estimate the rate of spatial spread of disease, and to determine the optimal control method (vaccine or culling), area and cost-effectiveness (see Box 4.3 and Box 5.1).

The default control strategy for wildlife rabies in the UK is vaccination following the EU protocol of twice-yearly bait distribution up to a radius of between 20–50 km from the outbreak (European Commission 2002). However, at high fox densities, or where other host species such as the Eurasian badger (*Meles meles*) may be involved, the optimal control strategy may include the focal deployment of poison baits and ring vaccination (Smith and Wilkinson 2003; Smith 2006). An additional reason for the adoption of focal culling is the apparent lack of immunity in the badger following vaccine bait distribution (Smith 2002). It is likely that in some areas of the UK, the badger may be at a sufficiently high density to be a reservoir host, and so should be included in the contingency plan. Recent work has focused on predicting the costs of different control strategies, and the expected time to achieve rabies elimination, so that their relative cost-effectiveness can be determined (see Box 5.1).

approach to its control (see Chapter 6). In practice however, vaccination of wild hosts may seldom be an option, as vaccines are only available for a limited number of diseases of wild mammals, and even these may not work in all host species. Nevertheless, appropriate vaccines may be available for those livestock, domestic animals and humans that are most at risk during an outbreak of disease in wildlife.

One means of reducing the risks of spillover from infected wildlife is to manage the opportunities for contact with people or domestic animals (see Chapter 8). In effect, this amounts to improving systems of biosecurity (see below), but has no effect on the circulation of disease in the wildlife host species. In the event that the disease cannot be eliminated, then this may remain the only option.

For some diseases of wild mammals, control measures may need to be targeted at vector species in addition to hosts (see Box 9.2). However, this may be logistically difficult, particularly when highly mobile arthropod vectors are involved. If a pathogen is only transmissible via a vector, or an intermediate host is necessary for the completion of its life cycle, then control of the vector or intermediate host is often more effective than controlling the definitive host species. However, in many instances vector control is extremely difficult, if not practically impossible, in which case minimising infection of the host species may be the best policy. Bluetongue virus is transmitted by midge vectors and a recent outbreak in the UK (serotype BTV8) is thought to have occurred due to infected midges being blown from Northern Europe, following its introduction there from sub-Saharan Africa. The virus is consequently now likely to be circulating in midge species that were not previously exposed to the virus. This demonstrates how it may be necessary to consider the potential for involvement of novel vector species in the spread of an exotic pathogen, following its initial introduction. Control of bluetongue in the UK through vaccination of domestic ruminants is considered a more achievable strategy than attempting to control midge populations. Wild ruminants are also known to be susceptible to bluetongue but their role, if any, in the UK outbreak is unknown. Targeting all vectors may also be impractical when there are many species or where the full array of potential vectors is unknown. Control is often most challenging when winged arthropods are involved in the epidemiology of

Box 9.2 Management of plague in wild hosts and vectors

Plague is caused by the bacterium *Yersinia pestis*. The disease is relatively stable in enzootic cycles in maintenance hosts but can cause mass die-offs in amplification hosts when epizootic cycles arise. Both maintenance and amplification hosts are usually rodent species, although the bacterium is able to cause disease in a wide range of mammals. Although a susceptible animal may acquire infection by inhalation or ingestion, inoculation via bites from infected fleas is regarded as playing the most important role in disease transmission between individuals, with some flea species acting as more competent vectors than others. Control of spillover into wild or domestic mammal populations should focus on pulicide (an insecticide that kills fleas) treatment either simultaneously or prior to rodent control, to prevent infected fleas leaving dead hosts and disseminating disease (Perry and Fetherston 1997). Care must be taken in selecting appropriate insecticides and rodenticides as resistance has been observed in some flea and rodent populations. Pulicides are formulated for application either in the environment or on individual animals (although the latter is impractical for controlling disease in wildlife hosts). Both environmental and individual animal treatment must be practised for the control of fleas on pets and it is important to carry out prophylactic treatment of households and pets in enzootic areas or where an outbreak occurs. This is both to prevent disease in pets and to minimise transmission to humans, which is particularly associated with inhalation of aerosol during close facial contact with domestic cats (*Felis catus*).

Plague has proven difficult to eliminate from some parts of the world. In such enzootic areas control methods are usually employed in an attempt to quell or prevent an epidemic rather than to achieve local elimination. For example, since the introduction of the disease to areas of the western USA, foci of infection have persisted in ground dwelling rodents such as the black-tailed prairie dog (*Cynomys ludovicianus*) and the California ground squirrel (*Spermophilus beecheyi*). Attempts to eliminate the disease have proven unsuccessful. However, a programme of public education and the existence of state emergency plans, with rapid reporting of cases and efficient risk communication, have ensured that the number of human cases per year in the USA has remained relatively low (approximately 3.7 cases per million people in the period 1992–1999) (Change et al. 2003).

Still more information is required on the epidemiology of plague, in particular its ability to persist in wild mammal populations despite control efforts. It has been postulated that the bacterium may survive for long periods in burrows outside the mammal host, possibly in dormant fleas. Certainly the bacillus does not appear to have long-term environmental stability and is thought to die quickly outside the vector or mammalian hosts, particularly in dry conditions with high temperatures and on exposure to sunlight. However, viable bacilli have been recovered under natural conditions after at least 24 days in soil contaminated with infected blood. The reason for this extended survival time is unknown but may be attributed to the blood serving as an enrichment medium (Eisen et al. 2008).

The epidemiology of plague is complex and epizootics appear to arise in relation to sudden increases in populations of mammal hosts and competent flea vectors. Flea populations are dependent on host availability, which is influenced by environmental conditions. In addition, flea species differ in their competence as vectors and in the extent to which they are host specific. Hence the occurrence of an epizootic, or the maintenance of the disease as an enzootic, is the result of a complex interplay between hosts, vectors and environmental conditions. The development of transmission models using existing knowledge will further our understanding of the factors that may influence the persistence of plague in host and vector populations, and so help to identify the most appropriate control strategies.

the disease, as there is the potential for transmission of infection over long distances, particularly in windy conditions. Mass application of insecticides can reduce the likelihood of exposure to arthropod vectors but potential benefits will need to be weighed against the risks of environmental contamination and adverse impacts on human health and non-target species. Following the introduction of West Nile virus to New York City in 1999 attempts at controlling human exposure to the mosquito vector included the widespread distribution of insect repellent and both terrestrial and aerial deployment of insecticide. Although such measures may reduce exposure to arthropod vectors, changes in weather conditions can ultimately play the most decisive role in determining their abundance.

Control of livestock in the event of a disease outbreak is made easier by the availability of areas on farms where animals can be concentrated and if necessary isolated for treatment or dispatch. Wild mammals, however, are considerably less tractable and if disturbed may disperse over a wide area. The likely behavioural responses of wild mammals to interventions, and their potential to exacerbate disease spread (see Chapter 2), are crucial considerations when devising contingency plans. Sufficient understanding and assessment of these possibilities will require input from experienced wildlife managers.

9.4.2 Modelling of Control Options

The utility of mathematical models for improving our understanding of disease transmission, and the likely impact of control options, is generally accepted. Models permit a low-cost assessment of the potential outcome of various management interventions, which can be compared with a non-intervention scenario. However, for the risk of success (or failure) to be determined, model outputs must be stochastic so that they can demonstrate the range of potential responses to intervention. A full discussion of how models should be used is given in Chapter 4. In relation to contingency plans, modelling can be performed in advance of any outbreak to help determine policy (e.g. quarantine, importation restrictions), or to help devise a control strategy. Models incorporating spatial aspects of disease control and heterogeneity in host density (Smith and Harris 1991 and Chapter 4) were used to inform the design of the UK rabies contingency plan (see Box 4.3). Modelling can also be used to simulate conditions during an outbreak, and so assess the likely outcome of interventions as the epidemiological situation changes. This can provide valuable guidance to operations on the ground during the course of the outbreak, particularly when integrated with Geographical Information Systems (GIS). In recent years there has been an increasing reliance on GIS to provide 'real-time' monitoring of the spatial distribution of cases to aid rapid interpretation of the changing situation during disease outbreaks (Kroschewski et al. 2006).

In 2001, three different models were used to predict disease dynamics and inform control options during a foot and mouth disease (FMD) outbreak in UK livestock. A critical appraisal of these models indicated that each had its strengths (Kao 2002), but we should be mindful that all models need to be adaptable to changing conditions during an outbreak. Of overriding importance is that such models are transparent, can be easily communicated to non-specialists, and the methods and levels of control are achievable in real life. During interpretation of model outputs, care should be taken to consider the underlying assumptions and limitations and the 'real world' practicalities of management. Model predictions should therefore be used to guide policy decisions, rather than to make them.

9.4.3 Economic Analysis of Control

A contingency plan can only be considered as a realistic proposition if the proposed actions are cost-effective. Hence, the development of a potentially effective contingency plan will require careful consideration of the economic implications of intervention, and indeed the costs of non-intervention. We can determine whether any intervention is economically worthwhile by means of a cost–benefit analysis (see Chapter 5). Assuming that the economic case is made for intervention, then an assessment of the cost-effectiveness of different management options will be necessary. For example, either culling or vaccination may be the most effective means of controlling disease under certain circumstances, but the most effective approach may also be the most costly. If either vaccination or culling were performed through the distribution of baits, then the costs may be similar. However, poison baiting may be much more expensive if the baits need to be recovered. As a general rule, culling is usually more expensive than vaccination (if the costs of vaccine development are not included), since the major component of wildlife control is personnel time.

Just like any other components of a contingency plan, costs are likely to change over time as the epidemiological situation develops and methods are adapted.

This will alter the cost:benefit ratio, making it necessary to periodically review the economics of control. This is likely to become particularly critical during the later stages of disease control when the detection of cases becomes infrequent and the cost of control can appear to be increasingly disproportionate. At this point there is likely to be pressure for the early curtailment of control measures on financial grounds, but the risks of disease resurgence are unknown if control ceases before the time specified by the exit strategy.

9.4.4 Implementation

It may be simple enough to design a hypothetical plan of action to eliminate, contain or manage a disease in wild mammals based on our understanding of disease dynamics and host ecology. However, a wide range of practical considerations will determine whether such a plan can be implemented. Even a relatively simple exercise such as a spatially restricted cull of a wild host population during a focal outbreak of exotic disease, might require the availability of a range of resources such as adequately trained staff (including both veterinary and wildlife management expertise), vehicles, specialist equipment, stocks of consumables (such as poison baits, protective clothing and disinfectants), transport, laboratory and carcass disposal facilities, GIS skills for mapping, administration and the establishment of local offices. Hence, systems need to be in place to allow many of these essential resources to be swiftly released from storage, purchased or seconded in the event that the contingency plan is invoked. But resources are no use whatsoever unless they can be deployed effectively, and this will require the existence of an appropriate organisational structure with clearly defined roles, responsibilities and lines of communication. There may be one or more areas where resources are strictly limited: for example, laboratory diagnosis, specialist equipment or suitable numbers of highly trained staff. Such limitations need to be considered when producing the overall plan. The availability and time required to access these resources will become important if it is necessary to expand the contingency plan from the scale of a local outbreak to that of regional control.

Following an outbreak of FMD in UK livestock in 2001 the government carried out extensive re-evaluation of its animal disease contingency plans, in the light of the experience gained. Although this was primarily concerned with the containment and elimination of infectious diseases in livestock and poultry, the basic principles are also relevant to any civil contingency response. They suggested a four-stage alert system to define the status of a disease outbreak: (1) disease not present; (2) disease risk higher than normal (because it is present in a nearby country for example); (3) suspicion of disease on clinical grounds, and (4) disease presence confirmed. During the last two stages, suspect animals may be slaughtered as a preventative measure, and samples taken for diagnosis. When the final stage is reached the Chief Veterinary Officer is obliged to set out the objectives for disease control and must establish a National Disease Control Centre (NDCC) and Local Disease Control

Centres (LDCC). The NDCC is responsible for policy and operations at a national level and advises Government Ministers. The LDCCs on the other hand, are responsible for the local co-ordination of disease control, including tasks such as the implementation of biosecurity, cleansing and disinfection of farm premises, disposal of carcasses, handling samples, GIS mapping, licensing animal movements, record keeping, surveillance, contact tracing, and health and safety. This complementary approach allows policy and strategic decisions to be made nationally, and tactical implementation to be performed locally. During an outbreak all relevant data is collected and checked locally, and communicated daily to the national centre. Both bodies can have a prescribed daily timetable that will include communications meetings, media briefings, daily report compilation and 'birdtable' meetings (where defined participants contribute in the same order at each meeting to communicate between all operational partners, provide situation reports, identify emerging issues and a structure for dealing with action points).

The roles and responsibilities of all those involved in the implementation of a contingency plan should be clearly defined. The range of expertise required to co-ordinate action should not only include those with veterinary and wildlife management experience, but also those versed in statistics, modelling, GIS, economics, management and finance.

In the case of diseases of wildlife that affect livestock, additional biosecurity measures may be required. These may include the restriction of livestock movements to stop further spread, and reducing opportunities for contact with wildlife and vector species. The latter may require that steps be taken to control wild mammal incursions onto farm premises. In areas where livestock are culled, stringent measures may be needed to ensure that wild mammals (carnivores and rodents in particular) cannot gain access to infected carcasses.

In order to reveal the full extent of logistic considerations it may be advisable to carry out trial exercises to simulate real-time outbreaks. This will have the added benefit of familiarising key staff with the necessary procedures.

9.5 Conclusions

Contingency plans should play a vital role in disease management, but their value depends on the accuracy and level of information underpinning the decision processes. As has been seen in previous chapters, there are examples where intervention has not always been successful in terms of disease reduction, and may have even exacerbated the problem. However, a detailed contingency plan based on risk assessments can provide practical advice for rapid implementation once disease in wildlife is suspected.

In summary, horizon scanning should identify which diseases may be imported by natural or anthropogenic means and risk assessments should be performed to identify those diseases which merit intervention. Expert opinion and the availability of vaccines will inform on the design of control strategies that can then be modelled and economically evaluated to produce an overall contingency plan. This plan will also depend upon the availability of resources and suitably trained personnel, and should be publicly discussed with all appropriate stakeholders in order to maximise concensus on the control strategies.

It is likely that there will be shortfalls in the data required for qualitative and quantitative risk assessments, although it is important that some attempt is made to formulate initial contingency plans for those diseases of most concern. These should not only include description of the most appropriate methods of control, but should also indicate the personnel, organisational framework and resources that will be necessary. In particular, the plan should define the exit strategy, to determine when control should cease (i.e. how long after the last recorded case) or change. Contingency plans should also be subject to regular review as risks change, new data becomes available and novel management techniques are developed.

Chapter 10
Wildlife Disease Surveillance and Monitoring

Marc Artois*, Roy Bengis*, Richard J. Delahay, Marie-José Duchêne, J. Paul Duff, Ezio Ferroglio, Christian Gortazar, Michael R. Hutchings, Richard A. Kock, Frederick A. Leighton, Torsten Mörner*, and Graham C. Smith

Authors are given in alphabetical order.

M. Artois
Université J. Fourier, Laboratoire TIMC-IMAG, Unité Environnement et Prévision de la Santé des Populations F-38000 Grenoble; Ecole Nationale Vétérinaire de Lyon, France

R. Bengis
Veterinary Investigation Centre. P.O. Box 12, Skukuza 1350. South Africa

R.J. Delahay and G.C. Smith
Central Science Laboratory, Sand Hutton, York, YO41 1LZ, UK

M.J. Duchêne
AFSSA LERRPAS, Technopôle Agricole et Vétérinaire, B.P. 40009, 54220 Malzéville, France

J.P. Duff
Veterinary Laboratories Agency Diseases of Wildlife Scheme (VLADoWs), VLA Penrith, Penrith, Cumbria CA19RR, UK

E. Ferroglio
Dipartimento Produzioni Animali, Epidemiologia ed Ecologia Via Leonardo da Vinci, 44-10095 Grugliasco (TO), Italy

C. Gortazar
Instituto de Investigación en Recursos Cinegéticos IREC (CSIC-UCLM-JCCM), Ronda de Toledo s.n. E-13071, Ciudad Real, Spain

M.R. Hutchings
SAC, West Mains Road, Edinburgh, EH9 3JG, UK

R.A. Kock
Conservation Programmes, Zoological Society of London, Regents Park, London, NW1 4RY, UK

F.A. Leighton
Canadian Cooperative Wildlife Health Centre, Department of Veterinary Pathology, University of Saskatchewan, Saskatoon, Saskatchewan S7N 5B4, Canada

T. Mörner
National Veterinary Institute, 751 89 Uppsala SWEDEN

*Members of the OIE Working Group on Wildlife Diseases together with Drs C. Bunn, J. Fisher and M. Woodford

DOI:10.1007/978-4-431-77134-0_10, © Springer 2009

10.1 Introduction

Emerging diseases of human or veterinary importance are a major challenge to human society. As previously discussed, infectious diseases of wild mammal populations can have significant economic impact, may threaten human and livestock health (Artois et al. 2001), and can affect the welfare and conservation of game (Gortazar et al. 2006) and species of high conservation value (Cleaveland et al. 2002). Wild mammals are also implicated as sources of emerging diseases (Daszak et al. 2000a; Cleaveland 2003; Cunningham 2005). Comprehensive epidemiological investigations and disease surveillance of wild mammal populations will enhance our capacity to detect and control infectious diseases that may emerge in the future in human and domestic animal populations. Given that the majority of diseases that have emerged in the last couple of decades had a wildlife origin (see Chapter 1), surveillance for wildlife diseases may be seen as an essential tool for the protection of human health.

For these reasons, the development of effective programmes for the surveillance of disease in wildlife populations is becoming increasingly important. Epidemiological investigations in wildlife are similar in many respects in terms of their objectives, concepts and methodology to those undertaken for domestic animal health surveillance and monitoring. However, there are also substantial differences, owing to the zoological, behavioural and ecological characteristics of wildlife populations. Consequently, definitions, methods and procedures must often be adapted to suit the unique conditions of wildlife disease surveillance.

10.1.1 Definitions

Several terms can be used to describe an investigation of disease in a population (see Table 10.1), but as they may refer to distinctly different concepts, or time frames, it is important to clarify their respective definitions. The main difference between surveillance or monitoring on the one hand and surveys on the other, is their duration. Surveillance and monitoring usually refer to an ongoing process, whereas surveys are more often limited in duration (i.e. a 'snapshot' in time). The term surveillance is commonly used to refer to the monitoring of behaviour or events from a distance. In an epidemiological sense however surveillance (sometimes called epidemiosurveillance) should be restricted to the ongoing recording of diseases in wildlife populations with a view to disease management (OIE 2006). It has been traditional to separate surveillance into scanning (or passive) surveillance (recording cases as they occur) or targeted (or active) surveillance (targeting individuals to detect the disease). An epidemiological survey on wildlife should not be considered as disease surveillance unless the survey is continuous and specifically designed to analyse and manage any associated health risks. In contrast, surveillance data are used to identify the areas to be targeted for control, and to anticipate spatial and temporal resurgences so that pre-emptive management interventions can be used to reduce disease risks.

Table 10.1 A comparison of definitions for terms referring to disease studies

Source	Investigation	Monitoring	Surveillance	Survey
Oxford Dictionary (OUP 2008)	"An inquiry into an incident or allegation so as to establish the truth"	"Keep under observation, especially so as to regulate, record, or control"	"Close observation, especially of a suspected spy or criminal"	"A general view, examination, or description"
Centre for Disease Control, (CDC)			"Systematic collection, of data to control and prevent disease"	
World Organisation for Animal Health, (OIE)			"Ongoing collection, of data to inform on the control and prevention of disease"	"Systematic collection of information on a sample within a defined time period"
Saunders Dictionary (Blood and Studdert 1999)		"Continuous measurement of a variable"	"Watching over a population with the aim of early detection …"	"Comprehensive examination of an area or population for a particular purpose"
Thrusfield (1995)		"The routine collection of information on disease, productivity and other characteristics possibly related to them in a population"	"An intensive form of monitoring, designed so that action can be taken to improve the health status of a population, and therefore frequently used in disease control campaigns. Appropriate action to control disease thus follows surveillance"	"An investigation involving the collection of information and in which a causal hypothesis usually is not tested …. It may suggest aspects worthy of study"
World Health Organisation, (WHO)			"Systematic ongoing collection, of data. So that action can be taken"	"Comprehensive compilation of baseline information on the health of populations"
Our definitions	Searching for the origins of disease events (in particular, outbreaks of infectious disease in humans and domestic animals which can originate in wildlife)	The systematic recording of epidemiological data, with no other specific purpose than detecting temporal trends. Ideally this should include or integrate with data on host abundance and distribution	A system for continuously collecting and analysing information on the health of wild species and associated risk factors, in order to meet the objectives of controlling or potentially eradicating disease in a population or community of wild animals	Collection of data on diseases or species, over a specific time frame (e.g. to analyse factors affecting disease distribution or to assess disease prevalence in a given population)

10.1.2 Importance of Monitoring and Surveillance

This chapter focuses largely on epidemiosurveillance and monitoring of disease in wildlife populations, and less on investigations and survey studies. Epidemiosurveillance and monitoring are important tools in public health, agricultural disease management and wildlife conservation. Surveillance and monitoring are both important for understanding and documenting emerging epidemiological situations and should be used not only in response to disease threats and outbreaks but also in association with high risk activities such as the translocation of wild animals from one geographic location to another.

Table 10.2 New pathogens identified in wild mammals in Italy (from 1995 to 2005) that were linked with previous wildlife translocation or other sources

Pathogen	Affected species	Suspected source of infection	Zoonosis	Source
Thelazia callipaeda (nematode)	Fox	Unknown	Yes	Rossi et al. (2002)
Physaloptera sibirica (nematode)	Fox, Badger	Unknown	No	Ferroglio and Ragagli (2008)
Setaria tundra (nematode)	Roe deer	Translocated wildlife	No	Favia et al. (2003)
Camelostrongylus mentulatus (nematode)	Roe deer	Camel from a circus	No	Rossi and Ferroglio (2001)
Brucella abortus (bacteria)	Chamois	Cattle	Yes	Ferroglio et al. (2003)
Brucella melitensis (bacteria)	Alpine ibex	Sheep	Yes	Ferroglio et al. (1998)
Hypoderma diana (diptera)	Roe deer	Translocated wildlife	No	Rambozzi et al. (2002)
Brucella suis (bacteria)	Wild boar	Translocated hares	Yes	Grattarola et al. (2006)
Ashworthius spp. *(nematode)*	Red deer	Translocated wildlife	No	Rossi unpub. data
Mycobacterium paratuberculosis (bacteria)	Red deer, roe deer, Alpine ibex	Unknown	Yes	Ferroglio et al. (2000); Nebbia et al. (2000)
Neospora caninum (protozoa)	Red deer, roe deer, chamois, Alpine ibex, European brown hare, field mouse	Unknown	No	Ferroglio et al. (2001); Ferroglio and Rossi (2001); Ferroglio and Trisciuoglio (2003); Ferroglio et al. (2007)
Mycobacterium bovis (bacteria)	Wild boar	Cattle	Yes	Bollo et al. (2000)
Mycobacterium bovis (bacteria)	Red deer	Translocated wildlife	Yes	Ferroglio unpub. data

Translocation is a commonly employed tool in wildlife management, with substantial health risks (Woodford and Kock 1991; Griffith et al. 1993; Viggers et al. 1993; Woodford and Rossiter 1993; Cunningham 1996; Daszak et al. 2000a). By way of illustration, Table 10.2 lists those pathogens which have probably spread as a result of wildlife translocations in northwest Italy during a ten-year period. The health risks associated with wildlife translocations, and other wildlife management practices, can be reduced by incorporating robust qualitative risk assessments into all levels of planning and implementation. These should ensure compliance with legislation covering these activities, and the relevant guidance from the World Organisation for Animal Health (OIE 2007). Such risk assessments require sufficient reliable information on the pathogens and host species present in both the source and destination ecosystems, so as to identify those to target for screening or treatment.

One fundamental but demanding aspect of wildlife disease surveillance is the early detection of outbreaks. In terms of public health (Hashimoto et al. 2000) and veterinary science (Doherr and Audigé 2001) 'early warning' can only be provided through adequate monitoring and surveillance (i.e. to find it you must first look for it).

10.2 Surveillance Targets and Cases

In this chapter we define wild mammal species as non-domesticated and free living. Any species legally exploited for recreational hunting can be termed 'game'; and may be divided into large (mostly ungulates) and small (mostly lagomorphs) game. The differing levels of management and husbandry to which game populations are subjected, categorise them into three broad groups: (1) unrestrained and self-sustaining, hunted populations, (2) fenced or managed game and (3) farm-reared game. In natural ecosystems, the practical and logistic aspects of disease and health monitoring of wildlife are challenging and require the development and implementation of novel techniques.

10.2.1 Targets

The most familiar method of recording the frequency of occurrence of a disease in a population is to record the number of individual cases, often expressed as a percentage of the total population size (see Section 10.3). This is usually sufficient to monitor a disease that is frequently encountered and easy to detect. However, wild mammals may inhabit remote areas and are often difficult to approach and examine. In addition, when an infection is acute, clinical expression in individuals may be brief, and hence the probability of detecting a diseased (or infected) animal is reduced. One option for dealing with this problem is to increase the size of the unit of sampling. For instance rather than targeting individuals, a group (e.g. herd, pack or social group) or a specific area (e.g. a forest, or pond) may become the sampling unit. To be considered as affected a herd or area would therefore need to contain at least one infected individual. The main advantage of this approach is that it allows epidemiologically useful information to be derived from relatively poor data. An example is the definition of rabies-affected areas for treatment with vaccine baits, which could be made on the basis of only a handful of rabid foxes.

By definition, a pathogen imposes adverse effects on the health of susceptible individuals. Some pathogens have been intensively studied because they cause detectable harm to humans or domestic animals (and often have an economic impact). However many parasites may be harboured by wild mammals in the absence of any visible signs of clinical disease. Modern microbiological and immunological techniques may however allow epidemiologists to detect the presence of such organisms, or previous exposure of the host without the need to rely on clinical signs.

A syndrome is a collection of clinical signs, frequently observed in association and putatively linked with some aetiology or disease risk factors. Syndromes are of most value in helping us to recognise diseases that are incompletely defined. An example was rabbit haemorrhagic disease (RHD), now known to be a calicivirus infection of rabbits (see Box 10.1). In contrast to traditional surveillance, a syndromic approach (Henning 2004) does not attempt to detect known etiologic agents or diseases, rather

Box 10.1 Monitoring in practice – rabbit haemorrhagic disease

Rabbit haemorrhagic disease (RHD) is an emerging viral disease of domestic and wild rabbits (*Oryctolagus cuniculus*), which rapidly spread around the world following its initial recognition in 1984. In farmed rabbits it caused high mortality and was not similar to any other disease previously reported in the species. Liver changes at the microscopic level were characteristic. As is usually the case, it was more difficult to be precise about the situation in free-living rabbits, although outbreaks resulting in high mortality were frequently observed in wild colonies and the clinical signs were again unlike those of any previously reported diseases. For example, nothing resembling the epidemic RHD outbreaks reported in wild rabbits in Britain in the mid and late 1990s had ever been reported before. In addition, the spatio-temporal distribution of outbreaks around the world, following the initial case in China in 1984, was typical of radiating disease, spreading first in Asia, followed by Europe, and subsequently to areas around ports throughout the world. Outbreaks of disease in wild rabbits were usually reported in these countries after disease in farmed animals. In Australia RHD was initially introduced by accidental escape from a field trial site in 1995, but subsequent deliberate releases occurred both there and in New Zealand.

We can be relatively confident that this was a new clinical disease spreading to farmed and wild rabbits around the world, primarily because it was readily observed and had not been recorded previously. The severe mortality observed in rabbit populations, and an initial lack of information on the causative agent, gave rise to concern over the potential risks to the health of humans, livestock and other wild species, which made a compelling case for monitoring and risk assessment. As the causative agent was unknown, monitoring and surveillance was based on the characteristics of the syndrome, which allowed the detection of typical cases. RHD remained a syndrome for several years due to the length of time it took to definitively identify the pathogen (a novel calicivirus). Risk factor monitoring involved identifying areas with high populations of wild rabbits and monitoring for new patterns of mortality. Targeted surveillance for the pathogen itself proved difficult because of the limitations of the diagnostic tests. Furthermore, no single characteristic alone defines an RHD case. The early case definition was important because it allowed temporal and spatial tracking of disease incidents.

Rabbits affected with RHD died within 24 hours, usually underground, or were removed by predators, and therefore RHD morbidity monitoring was of limited practical use. Case mortality monitoring was important, however counting dead animals was problematic. Wherever spatial and longitudinal analyses of scanning surveillance data demonstrated mass mortality incidents in wild rabbits, then RHD was considered a possible cause together with other differential diagnoses including myxomatosis, juvenile coccidiosis and poisoning. As RHD was almost invariably fatal, carrier disease status occurred infrequently and so was not an important consideration for disease monitoring. In addition, other wild species were not identified as susceptible to RHD infection and were therefore thought unlikely to be virus carriers.

Until 1994, the diagnostic test of choice for detecting the agent was electron microscopy of the liver. This is a direct test in which the RHD virus is observed, however it required technical expertise, which made monitoring expensive and limited to specialised diagnostic laboratories. It was several years before RHD ELISA and PCR tests were developed. Detecting exposure by identifying antibodies was of little practical use in recognising new outbreaks of RHD in wild rabbits because the majority of animals died in a matter of days, before they had time to produce antibody.

In conclusion, monitoring for RHD is influenced by the characteristics of the pathogen, the host and the ecosystem they inhabit. In the case of RHD it was initially difficult to detect the disease agent. In some respects, the presence of the disease in nearby domesticated rabbits provided sentinel surveillance for RHD in wild populations. Syndromic surveillance was important for detection, where obtaining fresh carcasses was not possible. The spatial and temporal occurrence of RHD in wild rabbit populations is now seen as sporadic and difficult to model. This relates to the complicated epidemiological picture caused by the recognition of several different serotypes of varying pathogenicity (White et al. 2004). It is noteworthy that in Europe, monitoring in recent years was largely confined to Spain where the disease has caused prolonged depression of rabbit populations with significant consequences for biodiversity, in particular for threatened predators such as the Iberian lynx (*Lynx pardinus*) and the Spanish Imperial Eagle (*Aquila adalberti*).

it seeks to use the clinical or epidemiological characteristics of disease occurrences to provide evidence to establish whether they are likely to be linked.

Disease risk is the probability of an occurrence (OIE 2006) and the use of the term denotes an intention to deal with the associated potentially negative impacts (e.g. threats to human health, economic losses). Risk surveillance often focuses on areas where the probability of occurrence or the seriousness of the consequence for target populations is high. Hence, it seeks to bias the collection of data in favour of species, areas, seasons or circumstances where risks are expected to be greatest.

10.2.2 Cases

A case is a unit for quantifying a health risk under epidemiological investigation. The science of epidemiology is largely concerned with quantifying and describing trends in data related to health events and so the definition of such events is at the root of any epidemiological study. As many of the pathogens of wild mammals are not routinely studied, accurate definition of a case is a fundamental challenge for wildlife disease surveillance. A positive case needs to be defined on the basis of the presence of a specific disease agent, a clearly described response to a diagnostic test, or in the case of a syndrome on a detailed description of lesions or clinical signs. In addition, it is important to accurately identify the host whenever possible, as this will help determine the epidemiological status of different species (e.g. are they reservoir or spillover hosts: see Box 3.4). This has often been a problem in the past when for instance on several occasions, European Bat Lyssavirus has been recorded as EBLV1a in 'a bat', and avian influenza cases as HPAI H5N1 in 'a duck'! The criteria that define positive cases need themselves to also be clearly defined, so that they can be routinely referred to as standards, compared, and challenged in the face of new data.

As mentioned above, a case may refer to an individual with a given disease, affected by a precisely described syndrome or carrying a specific pathogen. A case also may refer to a spatial or social unit (e.g. herd or region), when it may be described as an 'outbreak'; this term generally implies that several animals are affected (Thrusfield 2007). It is important that the units are clearly defined, in terms of geographical delineation (e.g. of an area or region) or composition (e.g. single cases or social groups of mammals).

10.2.2.1 Morbidity

Morbidity refers to the state of being diseased; from the Latin *morbidus*. Diseases causing macroscopic (visible) lesions such as infectious kerato-conjunctivitis in Alpine chamois (*Rupicapra rupicapra*) (Hars and Gauthier 1984) or obvious mortality like RHD (Villafuerte et al. 1994) may be relatively easily detected and monitored since the public (including hunters and gamekeepers) may provide useful epidemiological information. However, early stages of such disease are likely to be under-reported. In reality, the expression of clinical signs in wild mammals may be difficult to observe, and quantify, particularly when no comparative information is available on the infection in humans or domestic animals. Furthermore, even when such data is available, it may not always be useful because of the potential for wide inter-specific variation in the nature of the host-pathogen interaction. Clinical diagnosis has only been useful in a limited number of disease outbreaks where groups of free-ranging wild animals were subject to continuous monitoring by trained personnel. In such instances the observer must ensure that quantified clinical data on any sample of animals is reliable and representative. This may only be possible when dealing with health disorders affecting visible parts of the body or those that profoundly modify the behaviour of mammals which are habituated to the presence of humans.

10.2.2.2 Mortality

Accurate identification of a mortality event requires that a pathologist with particular expertise in examining wildlife carry out a detailed necropsy. This should be performed in accordance with a standardised procedure, regardless of the size and state of preservation of the carcass (Woodford et al. 2000). For the purposes of opportunistic surveillance, the carcasses of animals that have died from traumatic injury (e.g. road traffic casualties) may be used to screen for pathogens, even where they present no macroscopically visible signs. The spatial and longitudinal analysis of wildlife mortality statistics and the results of the associated systematic screening provides a useful resource for investigating health risks to, and emanating from, wildlife (provided the sampling is adequate). Again we must stress the importance of accurately recording the species, and where possible the sex, age and condition of hosts.

10.2.2.3 Pathogen Carriers

Clinical manifestations or lesions caused by many zoonotic or economically important pathogens that occur in wildlife can be difficult to observe. Hosts may for example be apparently healthy carriers. Therefore, disease surveillance for these pathogens must not be based on the collection of clinical data (i.e. mortality or morbidity). Below we describe approaches to detecting such pathogens, although there is little published information available to help investigators in the design of surveys for such conditions in wildlife populations (Kaandorp 2004).

10.2.2.4 Test Sensitivity and Specificity

Sensitivity and specificity are qualities of diagnostic tests that seek to distinguish individuals that are infected or have been previously exposed (see Section 10.2.2.6) to a pathogen from those which have not. When an animal is known to be affected, the sensitivity of a test is its ability to give a positive response. When an animal is known to be unaffected, specificity is the ability of the test to give a negative response. The evaluation and interpretation of diagnostic tests is a complex issue. For the sake of simplicity, it is common practice to divide responses into positive or negative results. This often requires the identification of a cut-off value for test results. However it is important to understand that there is an inverse relationship between sensitivity and specificity, such that one characteristic is achieved at the expense of the other (Thrusfield 2007). Diagnostic test results should therefore always be interpreted with these limitations in mind.

Many diagnostic tests designed to screen for infectious diseases in domestic mammals do not have the same levels of sensitivity and specificity when used in wild mammals. However, as a general rule, tests aimed at directly detecting the pathogen tend to give similar results in both domestic and wild species. The same cannot be said for indirect tests, which are often based on detecting the immune response of the host to the pathogen, and so depend on recognising specific proteins associated with that response. Variations in host response amongst species means that indirect tests such as antibody ELISA tests or skin tests, may not be accurate indicators of exposure to the pathogen. For example, other pathogens may elicit antibodies that cross-react with the test, causing a false positive result. Validation of existing diagnostic tests in wild hosts can be difficult owing to the practical challenges of acquiring sufficient numbers of known positive and negative controls. Test sensitivity and specificity are also difficult to determine where there is no 'gold standard' test, for example when pathogen identification is difficult as can be the case in sub-clinical cases of bovine tuberculosis. Nevertheless, there may be opportunities to usefully employ insensitive tests to detect exposure at the group level (e.g. the herd). The OIE (Office International des Epizooties) Working Group on Wildlife Diseases maintains an updated list of recommended diagnostic tests for screening wildlife (OIE 2008a).

10.2.2.5 Detecting the Agent

Infectious agents can be directly detected using a wide variety of techniques including cultivation in laboratory animals, or preferably on cell culture or other media, identification of phenotypical characteristics (as identified by staining techniques for example), or genetic tests such as genomic amplification, PCR or RT-PCR and sequencing. Frequently, evidence of contact with the disease agent requires laboratory analyses based on agent isolation, PCR testing or serology.

For macroparasites (such as helminths and most arthropods), disease monitoring should ideally also include isolation of the relevant life stage of the parasite, such as larvae, nymphs etc. In many situations however, the mere presence of a parasite may be of less consequence if it is generally benign. Certain new technologies (e.g. PCR) are so sensitive that they can detect extremely small amounts of genetic material, such as the remnants of the pathogen, and so the results of these tests need to be interpreted with caution. Likewise, when a test fails to detect a pathogen this does not exclude the possibility that it is in fact present, because all tests have their limitations. Understanding and quantifying these limitations is essential and consideration of their influence should be central to the interpretation of epidemiological data.

10.2.2.6 Detecting Exposure

Many techniques are now available for detecting prior exposure of an individual to a specific pathogen. One approach is to use physiological or biochemical changes, such as the level of chemical compounds in the blood or tissues which act as markers for previous exposure. Exposure to most infectious agents can provoke the appearance of antibodies in blood, excreta or secretions. These antibodies are not necessarily linked with immune resistance, but can be used to evaluate what fraction of a population has been exposed. However, antibody responses can wane with time after exposure, thus decreasing the sensitivity of detection. This can vary between individuals, so the amount of antibodies present does not tell you how long ago the animal was exposed to the agent. It also does not tell you if the animal had been diseased, or infectious, only that it was exposed to the agent. Many studies use blood serum samples to detect antibodies, and their results are often referred to as seroprevalence. In a UK study of European Bat Lyssavirus 2 in the Daubenton's bat (*Myotis daubentonii*) the observed seroprevalence was approximately 5%, but the virus itself was not identified in a single case, and hence disease prevalence was zero (Harris et al. 2006).

10.2.2.7 Non-Invasive Tools

Animal welfare concerns and the need to limit manipulation of highly endangered species have prompted the development of non-invasive disease monitoring techniques. Available tools include faecal sampling for parasitological or bacteriological surveys, and feather and hair sampling for genetic and toxicological analysis. Non-invasive approaches are currently rarely adequate substitutes for traditional sampling techniques. Nevertheless, this is an area of much recent research activity which may yield valuable surveillance tools for wildlife diseases in the future.

10.3 Indicators and Statistics

The most useful parameters to quantify disease presence and describe patterns in space and time are prevalence and incidence (Thrusfield 1995). In practice, however it is difficult to accurately determine the number of cases and the size of the target wild mammal population. This difficulty may be compounded by the influences of the spatial and social structure of mammal populations on the distribution of cases (see Chapter 2) and the probability of their detection.

10.3.1 Prevalence and Incidence

Prevalence is the total number of cases (expressed as a proportion or percentage) in an exposed population over a given sampling period. Incidence is the number of new cases (expressed as a proportion or percentage) that arise in a population per unit of time. Both are usually given as proportions of the total sub-population sampled, and this is often assumed to be an unbiased estimator of the true population prevalence or incidence.

In practice, it is unlikely that the absolute size of a population of free-living wild mammals is known. The size and social organisation of wild mammal populations can often only be crudely estimated, and the development of improved methods for estimating animal abundance is a fundamental challenge for wildlife disease management. Hence, the proportion of cases in a sample of wild animals can only be considered as an indication of the probability of infection or exposure to the pathogen. However, the more representative the sample is of the wider population, the more accurate the final estimate is likely to be.

10.3.2 Issues of Host Abundance

The denominator for prevalence and incidence estimates is the size of the 'local' population from which the sample was derived, rather than the national population. Since disease is often aggregated, and most populations are continuous, defining the extent of this sample population is difficult. Diseases are often expected (not always correctly) to increase in prevalence as host density increases, so an estimate of population density would also be useful in many circumstances. Estimates of mammal population size can be performed by capture-mark-recapture studies, but these are expensive and time consuming, since they involve the repeated capture of animals, and ideally estimates of population turnover and emigration. A population census (i.e. a complete count) can be performed in limited circumstances, where the species is large and distinct. Alternatively, population size can be estimated from survey data using methods that correct for the probability of detection, as have been developed for rabbits (Poole et al. 2003), and badgers (Hounsome et al. 2005) in the UK. For many mammalian species, field signs such as footprints and droppings can be used as crude estimators of abundance but such methods often have serious limitations (Wilson and Delahay 2001). Genetic methods, such as the non-invasive: sampling of faeces or hair, are becoming more reliable (e.g. Wilson et al. 2003). Quantitative comparisons of the various techniques for estimating abundance are urgently required for many species, as different approaches all have their advantages and disadvantages (Wilson and Delahay 2001; Acevedo et al. 2008).

10.3.3 Spatial and Temporal Trends

Recording cases of morbidity and mortality in a given area can provide information on spatial and temporal trends of infection in wildlife. However, the distribution of hosts in space and time will influence the temporal and spatial distribution of morbidity.

It is important to be able to describe 'background noise' in morbidity and mortality rates, in order that any significant deviations indicative of emerging disease events or new diseases can be identified. A variety of statistical techniques have been developed for the explicit purpose of identifying clusters of cases that cannot be explained by chance occurrence (Lawson and Kleinman 2005).

Pathogens can survive and propagate in populations in different spatial and temporal patterns (Begon 1995), for example the invasion of pathogens into susceptible areas can lead to spectacular waves of new cases. Mathematical modelling allows epidemiologists to describe the most significant factors that are likely to contribute to the spatio-temporal trends observed. These trends are often categorised into a few basic types (forms) that are used to describe disease events (see Thrusfield 1995; Toma et al. 2001). Morbidity or mortality events that oscillate above and below an average over time are indicative of an endemic situation (the term enzootic is used to specify that the population is composed of animals). An outbreak suddenly appearing in a place where it was previously unrecorded is called an epidemic (or epizootic for animal populations). Morbidity events, which occur in an unpredictable manner in time and space, are called sporadic.

Another important concept in epidemiological investigation is whether a morbidity event is propagating from individual to individual (direct or indirect transmission), or if the event is clustered around focal point sources (e.g. a water-borne source in an arid environment). At an early stage of the event, it can be difficult to distinguish which disease pattern one is dealing with, but analysis of the distribution of cases in time and space will give some indication of the potential transmission dynamics.

10.3.4 Detection of New Diseases

Detection of new diseases is a challenging task. The definition of 'new disease' should include the occurrence of known disease agents in novel host species, in addition to completely new agents. Detection probability will depend on disease prevalence, patterns of transmission and disease-induced mortality. Sampling effort will therefore be crucial and the resources available for this are likely to be greater for disease agents that could spill over to humans or have a potentially substantial economic impact.

For new diseases to be confidently identified, a sound baseline knowledge of the pre-existing disease status of a range of hosts in a given area is required. This is not always available for wild hosts, but at the very least the detection of new pathogens will require systematic investigation of those clinical cases where the aetiology is unclear or potentially novel. This can be achieved through careful scanning (or passive) surveillance focused on specific syndromes or areas perceived to be at greater risk (see Box 10.2). This flexible capability should be possible as part of any existing programme for the surveillance of disease in wildlife.

Box 10.2 Early disease detection in wildlife: European brown hare syndrome

European brown hare syndrome (EBHS) is caused by a calicivirus that is related to, but distinct from, the rabbit haemorrhagic disease (RHD) calicivirus. The detection of EBHS in the UK illustrates several principles and problems in the early detection of new diseases in wildlife.

Unexplained mass mortality incidents in brown hares (*Lepus europaeus*) had been observed in England for many years by the *ad-hoc* and non-systematic surveillance schemes employed at the time. A toxicological aetiology was suspected but assay results were consistently negative. Tissues from some of these mortality incidents were archived by freezing. A syndromic description was not produced for EBHS at the time, and in retrospect this significantly delayed the detection of the disease. Instead, the description of 'large numbers of dead hares found at one location' was sufficient to alert workers to the possibility of a new disease, and to archive incident reports and tissues, but was too vague to provide any indication of aetiology.

Identification of the first case of EBHS in England in 1989 occurred when a live but non-responsive hare, exhibiting no fear of humans, was submitted for veterinary examination. This focused on central nervous disease and the brain was examined. However, the investigator had read a surveillance report on hare deaths in Germany where liver disease was suspected, and as a consequence electron microscopy revealed many calicivirus particles in the liver. This first case of EBHS exhibited hepatic encephalopathy in which impaired brain function occurred secondarily to severe liver dysfunction. Retrospective examinations of the archived hare livers and their associated reports showed that the disease had been present in England since at least 1982. Archived reports hinted at suspicious incidents from the mid-1970s and archived sera showed a high seroprevalence to EBHS in hares sampled as far back as 1963 (Duff et al. 1996).

This example illustrates how difficult it can be to detect a novel disease in wildlife, even when the condition is an acute infectious disease such as EBHS. In retrospect, we can identify several reasons why this syndrome was not identified earlier. Firstly, at this time in England there was no systematic scanning surveillance scheme for diseases in wildlife, which would have detected unusual hare mortality incidents and then targeted carcass submissions. Also, the gross pathology of EBHS is usually unremarkable, there was no systematic approach to laboratory investigation for wild mammals, and there was no routine microscopic examination of tissues (histopathology). Finally, EBHS incidents frequently lasted only a few days and by the time investigators received negative laboratory results they rarely had access to more dead animals. Critically, a clearer and more detailed syndromic description at the time of the outbreaks would undoubtedly have allowed earlier detection of the condition.

10.3.5 *Precision, Bias and Accuracy*

Precision, bias and accuracy are characteristics of any sampling design, and it is important to understand their respective meanings and the way they may influence results.

10.3.5.1 Precision

The term precision refers to the repeatability of a result. Confidence intervals provide a measure of the precision associated with a prevalence estimate (*p*) and can be calculated from the sample size (*n*). A frequently used formula for estimating the confidence interval associated with an estimate of microparasite infection prevalence is

$$\text{S.E.95\%C.I.} = 1.96[p(1\text{-}p)/n]^{1/2} \quad \text{(Martin et al. 1987).}$$

Prevalence and similarly proportions can be compared using a variety of statistical tests (Siegel and Castellan 1988). Macroparasites usually exist in aggregated distributions, whereby a relatively small number of hosts carry many parasites, but more hosts carry fewer or even none. This left-biased frequency distribution is best described by the negative binomial distribution, and specific approaches have been developed for the calculation of prevalence estimates (Rózsa et al. 2000; Rózsa 2005). As a general rule for both micro and macro-parasites, increasing the sample size will increase the precision of any prevalence estimate.

10.3.5.2 Bias

The bias of an estimator is a reflection of the extent to which it (e.g. observed prevalence in the sample) differs from the true value of the parameter being estimated (e.g. actual prevalence in the exposed population). Wildlife sampling using carcasses from hunting bags, road casualties or cetacean strandings for example, may include bias that could lead to either over or under-estimation of disease prevalence. If bias is likely, then sampling should be random or preferably stratified (e.g. split into sub-samples) relative to those factors of concern, such as habitat, region, date, age and gender. This will allow comparison between different sub-samples, although small sample sizes can become an issue. Also, there may be additional logistic or economic reasons why it is not possible to adopt such a systematic approach.

The design of any sampling strategy should generally seek to minimise potential sources of bias. In most situations, stratified random sampling is the most advisable design for investigating wildlife populations. This is likely to require some basic knowledge of host population structure and distribution. Larger sample sizes however, will not necessarily reduce or remove the influence of bias. For example, increasing the size of a survey based on the collection of trapped animals will not reduce bias resulting from diseased hosts being more or less likely to be captured. Furthermore, the trappability of some categories of individuals may change over time and hence could modify the perception of the temporal trend in cases (Courchamp et al. 2000). Unlike precision, bias cannot generally be quantitatively estimated.

Bias can nevertheless also be beneficial, for example when trying to detect a novel disease, or where the aim is to establish that a disease is absent. The submission of suspect carcases for rabies surveillance is for example highly biased because it concentrates on those animals displaying aberrant behaviour, and is consequently more effective at detecting cases than random sampling would be. However, this approach does require that the direction of bias is known.

10.3.5.3 Accuracy

The term accuracy relates to how close a given result is to the true value. Hence a prevalence estimate, based on a given sample size, is more accurate the closer it reflects the true prevalence in the whole population. This can only be determined by sampling a sufficiently large and representative proportion of the total population. It is however, often difficult in studies of wild mammals, to achieve adequate sample sizes. Nevertheless, even if the entire population were able to be sampled (i.e. a census) then the observed prevalence would still be subject to the limitations of test sensitivity and specificity.

10.3.6 Disease Absence and Limits of Detection

The likelihood of being able to detect the presence of a pathogen increases with prevalence and sampling effort. Hence it is relatively easy to obtain prevalence estimates for diseases that affect a large proportion of the population such as tuberculosis lesion prevalence in wild boar in Spain (Vicente et al. 2006). But this becomes increasingly difficult when prevalence is below 1%, such as for transmissible spongiform encephalopathies in European cervids (Schettler et al. 2006). It follows that confirmation of the absence of a disease is a difficult task. For example, in order to be 99% confident that disease is absent or below 1%, a sample size of 448 undiseased individuals would be required from an estimated total population size of 10,000. This calculation is derived from the formula:

$$n = [1-(1-a)^{1/D}][N-(D-1)/2],$$

where n is the required sample size, a is the probability of observing at least one diseased animal in a sample when the disease affects at least D/N, and D is the number of diseased animals in a population of size N (see Martin et al. 1987). In fact, practical constraints mean that most wildlife disease surveys can only provide information on sample prevalence, with difficulties in extrapolating accurately to population prevalence.

10.4 Data Collection, Storage and Interpretation

Surveillance and monitoring may be carried out by the 'passive' collection of samples or alternatively by an 'active' process of collecting material for diagnostic testing. When animals are routinely submitted for investigation on an *ad hoc* basis, for example as a result of road casualties, 'pest' control, abnormal individuals in a game bag or mortality during rehabilitation, and this information is collated, then this constitutes scanning (or passive) surveillance. Alternatively, we use the term targeted (or active) surveillance when animals are proactively sampled (either dead or alive), by various means (e.g. by dedicated capture or sub-sampling of game bags) specifically for the purpose of examining and testing them for evidence of exposure to pathogens. Such studies can provide data in the form of the number of cases or outbreaks observed during a given time period. These data can then be centralised and (where necessary) cases may be notified to local, regional or national authorities. However, notification may also be based on continuous (real-time) reporting of results as they arise (as part of mandatory activities involving results of laboratory diagnoses or examination of game and game meat at inspection points), in which case it is referred to as *ad hoc* or routine sampling.

Scanning surveillance based on official notification, is not sensitive and is inevitably biased towards species and diseases of priority interest. Nevertheless, this provides a non-representative indicator of events and trends, which may be of interest for public health, veterinary and wildlife management purposes, and may be particularly useful for the initial detection of exotic diseases.

10.4.1 Recording and Storage of Data

Before data are recorded, they must be coded in order to standardise case definitions and to allow comparisons in time and space. Such standards are rarely used in surveillance of disease in wild mammals. There is currently no internationally agreed standard, although in 2002 the American Veterinary Medical Association approved support of a Systematized Nomenclature of Medicine (SNOMED) as a standard for veterinary data recording and management (Anon 2002). Across Europe, harmonised standards are either absent or are inadequately implemented (Klein 2002).

10.4.2 Effects of Management on Disease Prevalence and Distribution

The most obvious effects of successful wildlife disease management are reductions in disease prevalence (in either the wildlife, domestic or human population) and in the spatial and temporal range of infection. The monitoring of disease prevalence in a given area allows one to distinguish between endemic situations (e.g. rabbit myxomatosis) and emerging or epidemic situations (e.g. the arrival of rabbit haemorrhagic disease). This distinction is important for establishing the appropriate management actions (if any), and the optimal design of surveillance to detect new cases. When calculating disease prevalence, the relevant confidence interval, or level of uncertainty associated with the result should also be known. Where trends are being examined it is important to remember that a change in the prevalence estimate for a given sample does not necessarily equate to a measurable change in the population prevalence, particularly if the sample size is small. When looking at local disease prevalence, sub-dividing the total sample quickly reduces sample sizes and levels of confidence in the results.

10.4.3 Effects on Disease Intensity and Transmission Risks

In some cases, wildlife disease management can target disease intensity and transmission risks rather than disease prevalence. In the case of most parasitic diseases for example, hosts in good body condition may have lower parasite burdens than undernourished or stressed individuals. The body condition of red deer (*Cervus elaphus*) was improved by supplementary feeding, at the cost of increased host contact rates. Deer in good condition carried lower nematode burdens possibly related to the nutritional costs of improved immune function. However, supplementary feeding encouraged the aggregation of individuals and enhanced the potential risks of bovine tuberculosis (bTB) transmission (Vicente et al. 2007b). In such cases risk surveillance (focused on clinically affected animals, or intensively managed populations) would be advisable for management purposes.

10.4.4 Effects on Other Species that Share Disease

Disease control in an abundant wild host may reduce risks to less abundant and more valuable wildlife species. For example, in Spain the endangered Iberian lynx (*Lynx pardinus*), is threatened with spillover of viral infections from feral cats (*Felis catus*) and bTB from their wild ungulate prey (Delibes et al. 2000). Disease surveillance is almost certain to be more straightforward if focused on the more abundant feral cats and ungulate prey, which are likely to be the subject of management actions. A different situation exists where wild boar are implicated as potential sources of several notifiable diseases in domestic pigs. In this instance, surveillance data on disease incidence in domestic pigs can be used to monitor the success of the management actions that target the wildlife species (e.g. control of CSF by vaccination, see Box 6.3). Both examples illustrate how disease surveillance carried out on hosts that are not necessarily either the species of most concern or the direct target of management efforts can be useful in assessing the impact of interventions.

10.5 Existing Monitoring and Surveillance Systems (MoSS)

The Internet and the World-Wide Web have introduced major changes in the way we observe and record events, and share data. New sites and information networks are constantly appearing and provide opportunities to continuously update information in 'real-time'. Perhaps the oldest global surveillance network for wildlife health and diseases is that organised by the OIE, which has been collecting data since 1993, particularly on diseases of importance to international trade and agriculture. These pathogens are described as 'listed diseases'. Initially, data was only collected on listed diseases in domestic species, but following the creation of a Working Group on Wildlife Diseases in 1993, the surveillance system began to expand to include wild animals (OIE 2008b). Data are collated from notifications submitted by each member country's designated wildlife disease reporter (or 'focal point').

Among the earliest general surveillance programmes for wildlife diseases in Europe were those established in the 1930s in Scandinavian countries (Mörner et al. 2002). Another comprehensive wildlife disease surveillance programme is the SAGIR network in France, which started in 1986 (Terrier et al. 2006). The World Health Organization (WHO) created a rabies-specific centre for surveillance and research which has published a quarterly bulletin since 1977 (WHO 2008b). At the scale of the European Union, although there is informal coordination of organisations conducting disease surveillance in wildlife populations, this is not yet formalised. Most EU countries have appointed a focal point to notify the OIE annually of significant wildlife disease events, and this informal network is coordinated under the auspices of the European Section of the Wildlife Disease Association (EWDA 2008). In addition, EU funding allows groups to formalise surveillance for important or notifiable diseases (e.g. EDEN 2008; MedVetNet 2008).

In European countries, the organisation of these systems for surveillance and monitoring follows one of two basic models. In the first, one or more laboratories with relevant skills and facilities gathers samples from all over the country (or region, province etc.), conducts analyses, processes data and disseminates the results. This approach operates in Austria, Scandinavian countries and Switzerland (and, at a regional scale, in Italy, Germany and Spain). In the second system, one organisation, or in some instances, a person appointed for this duty, collects results from various laboratories or sources and publishes a synthesis. This system has been employed for many years in France, the UK, Italy and the Netherlands.

Various other types of surveillance systems have been implemented elsewhere in the world. In Canada, for example, a multi-centre organisation deals with scanning and targeted surveillance (see Box 10.3). In South Africa, most surveillance is based in conservation areas such as Kruger National Park, where scanning surveillance is ongoing and coupled with campaigns of active detection of specific diseases (see Box 10.4). The variety of surveillance systems is therefore broad, from active to scanning, from general to targeted. The future challenge will be to find effective ways to share and exchange data on a global scale so as to improve our capacity to identify new health risks in wildlife populations and enhance our capability to manage them when necessary.

Box 10.3 Wildlife disease surveillance in Canada

In Canada, a national programme of monitoring and surveillance of pathogens and diseases in wild animals has been carried out since 1992 by the Canadian Cooperative Wildlife Health Centre (CCWHC 2008), a partnership among Canada's five veterinary colleges and federal, provincial and territorial government agencies (Leighton et al. 1997). The central pillar of this programme is scanning disease surveillance based on *post mortem* examination of wild animals found dead or diseased. Data and knowledge developed by this core program have given rise to numerous additional projects and programmes in targeted disease surveillance and other research. This national programme now plays a key role in developing, testing and improving Canada's overall capacity for disease detection and responses, and management of animal and human health.

The primary objectives of the surveillance programme are to develop a complete national inventory of pathogens, their vertebrate hosts and their geographic ranges, to assess changes in these over time, to detect diseases of socio-economic and zoonotic importance as early as possible, and to inform decisions by government agencies responsible for public health, domestic animal health and wildlife conservation and management. Secondary objectives are to use the material generated by the programme to educate the wildlife health personnel who will be needed by Canada in the future, and to identify priorities for research related to wild animal health and disease.

The core disease surveillance programme of the CCWHC integrates four separate activities: (1) detection of dead or diseased wildlife, (2) identification (diagnosis) of pathogens and disease processes in those specimens, (3) management of the information derived from these two activities through a national wildlife disease database, and (4) communication of relevant information to government decision makers and the public.

The CCWHC model has proven highly effective and cost efficient. The CCWHC provides wildlife health services to the nation and, thereby, generates knowledge, specimens and infrastructure for scientific research and education in the wildlife health field. The veterinary colleges provide the CCWHC with much of its professional expertise and all of its physical space, laboratory and information facilities. Government investment in the operation of the CCWHC assures access to expert wildlife health services for government agency programmes, and the education and training of a much-needed pool of potential future employees. As an organisation outside of government, the CCWHC is particularly well-positioned to coordinate complex national disease surveillance and management programmes among a wide range of government agencies at all levels, and with non-government agency partners.

Box 10.4 Disease surveillance and monitoring in free-ranging African wildlife

Disease surveillance in wild mammals is generally weakly structured and usually passive in approach, because free-ranging wildlife are not visited and observed on a regular basis, frequently do not have owners, and are not easily manipulated for 'hands on' examination or specimen collection. For these reasons, surveillance techniques for wildlife should be structured so as to maximise the information gained from the limited availability of captured animals and carcasses. Opportunities for investigations into causes of morbidity and mortality are infrequent because carcasses are either not found or have been scavenged. Hence, one must make full use of every opportunity to monitor animal and environmental health indicators in extensive free-range ecosystems

Here we describe the surveillance and monitoring techniques currently in use for four common infectious diseases and one possibly eradicated disease of African wild mammals.

Anthrax

In sub-Saharan Africa, anthrax outbreaks are generally driven by dry climatic conditions with hydrological stagnation, coupled to relative or absolute overabundance of preferred hosts. Outbreaks are generally short lived and are dramatically terminated by the onset of the rainy season. Anthrax is an acute multi-species disease caused by a bacterium (*Bacillus anthracis*), and its preferred hosts vary amongst habitats and ecosystems. Scanning surveillance for anthrax is mainly executed by trained field staff, including rangers, game guards, biologists and veterinary technicians. Suspect carcasses of most mammal species that die of anthrax, are usually in good body condition, and frequently have no signs of predation, when found soon after death.

In the Kruger National Park (KNP) field personnel are issued with blood smear collection kits, which include two glass slides wrapped in a small data sheet and a waterproof pouch. Blood smears are taken from all suspect carcasses, data sheets are completed and the samples are dispatched for staining and microscopic examination, or for culture when necessary. Once an outbreak has been detected, surveillance and monitoring moves into a targeted mode, involving moderate-scale deployment of staff, vehicles, a mobile laboratory and a helicopter. A central command centre is established at the nearest rest camp, and collected data is collated, stored and mapped on a daily basis to identify spatio-temporal trends (De Vos and Bryden 1996). Circling and descending vultures are one of the most important indicators for pinpointing carcass locations. GPS co-ordinates are collected for every carcass. The use of GPS technology facilitates data management and mapping using GIS imaging and layering.

(continued)

Box 10.4 (continued)

Foot and mouth disease

In sub-Saharan Africa, the endemic cycle of foot and mouth disease (FMD) is maintained in buffalo (*Syncerus caffer*) herds with virus cycling between adult carriers and the annual cohort of calves (Thomson et al. 1992). Most buffalo calves are born in the rainy summer season, and receive colostral antibodies against FMD from their dams. As this passive immunity wanes at between 5 and 9 months of age, most juvenile buffalo become susceptible to infection during the dry season of mid-winter and early spring, a time when many species are congregating around the remaining permanent sources of surface water. During primary infection, buffalo calves shed large amounts of virus, and the infection (usually sub-clinical) rapidly spreads to the other buffalo calves in the herd, and may spill over into other sympatric cloven-hoofed species, resulting in an epidemic cycle.

In the KNP, impala (*Aepyceros melampus*) are the most abundant wild cloven-hoofed ungulates, are highly susceptible to FMD and develop clinical disease when infected. Hence, to detect FMD epidemic outbreaks, impala are targeted through surveillance of herds (Bengis et al. 1994). Clinical signs of FMD in impala include pilo-erection (febrile response), "walking on eggs" (weight shifting from one limb to another), overt lameness, lagging behind the herd and lying down. Animals with clinical signs are sampled (non-lethally or lethally) to obtain blood and tissue samples for virus isolation and serology. During epidemic outbreaks, clinical disease may also be diagnosed in kudu (*Tragelaphus strepsiceros*), and less frequently in giraffe (*Giraffa camelopardalis)*, bushbuck (*Tragelaphus scriptus*), nyala (*Tragelaphus angasii*) and warthog (*Phacochoerus aethiopicus*). More recently, active sero-surveillance for FMD in impala has been employed, whereby 30–40 animals are randomly selected, chemically immobilized, examined and blood sampled on a monthly basis. This sampling is applied to three geographically distinct populations of impala in the Kruger National Park, on a three monthly rotation cycle.

Bovine tuberculosis

This bacterial disease has a wide host spectrum, and has entered several free-ranging buffalo populations (Guilbride et al. 1963; Woodford 1982; Bengis et al. 1996; De Vos et al. 2001), as well as kudu (Thorburn and Thomas 1940; Keet et al. 2001a) and lechwe (*Kobus leche*) (Gallagher et al. 1972). These species all appear to be efficient maintenance hosts, with aerosol transmission predominating. Infection spills over into predators and scavengers that ingest infected material, and frequently involves the mesenteric lymph nodes, with secondary haemotogenous spread to distal sites, including lungs, bones, joints, spleen, kidneys and serosal surfaces (Keet et al. 1996; Keet et al. 2001b). Aerosol and percutaneous infection are also important transmission modes in lions (*Panthero leo*).

Bovine tuberculosis (bTB) is a slow progressive disease with a long subclinical phase, lasting months to years. In buffalo, lechwe, baboons (*Papio spp.*) and warthogs, only animals with disseminated or advanced disease show any clinical signs, which may include coughing, emaciation, staring hair-coat, non-healing skin lesions, depression and lameness. Therefore scanning surveillance generally only detects the tip of the iceberg. Kudu, however, frequently develop overt swellings of one or more of the lymph nodes of the head, at a relatively early stage. The parotid lymph nodes, in particular, tend to enlarge massively due to abscess formation, and sinus tracts draining muco-purulent material are commonly seen below the ears.

Lions frequently present with emaciation, swellings of bones and synovial structures, and non-healing bite wounds with underlying granulomatous infection of the subcutaneous and muscular tissue.

In most species necropsies, or non-lethal sampling with *ante mortem* testing using the intradermal tuberculin or blood-based tests (if validated for the species), are necessary for bTB detection and monitoring. There are unfortunately no sensitive or specific *ante mortem* diagnostic tests currently available for pachyderms.

Rabies

On the African continent, rabies has been diagnosed in 33 carnivorous and 23 herbivorous species, with regional variation in the dominant role-players (Swanepoel 1994). Scanning surveillance for rabies involves the sampling of individuals of any species that display abnormal behaviour such as extreme aggression, dumbness, tameness, aimless wandering, paralysis, hypersexuality and excessive vocalisation. Salivation and an inability to drink or swallow may also be seen. For diagnostic reasons, suspect animals should not be shot in the brain.

Rinderpest

Rinderpest was possibly the most serious infection ever to affect mammals on the African continent, causing the devastation of populations of susceptible species from the late 19th to the 21st Century. Over this period, several strains affected a wide range of artiodactyls with particular virulence expressed in buffalo, tragelaphine antelope, giraffe and warthog. The disease significantly reduced populations, with mortalities anecdotally quoted at 90% during the early pandemic and confirmed at 60% amongst buffalo in Kenya in the mid 1990s (Kock et al. 1999a). Such was its impact that it may have influenced the current distribution of some species (Rossiter 2003). Intermittent *ad hoc* surveillance of wildlife populations was undertaken, but the apparent persistence of the virus in wild mammals at the end of the Pan African Rinderpest campaign, resulted in the launch of a major epidemiosurveillance initiative. This has involved over 30 African countries, and employed passive and active

(continued)

Box 10.4 (continued)

methods of surveillance, focusing on buffalo but also including a wide range of other species (Kock et al. 2006; Kock 2008). Epidemiological evidence from buffalo and some other species showed that wildlife were not able to act as a long-term reservoir but that they could prolong epizootics, and that the current strategy for eradication remains valid. There has been no confirmed infection in wildlife since 2001 (in Meru National Park, Kenya), and it is hoped that the disease is now eradicated. During the campaign it was also possible to monitor for peste des petite ruminants (PPR) as part of differential diagnosis and evidence for circulation was established in some species (particularly buffalo). The invasion of old rinderpest strongholds with PPR is perhaps not surprising given that vaccination is cross protective between these two diseases.

Summary

These examples illustrate the multi-faceted approaches that are required to establish a meaningful disease surveillance system for free-ranging populations of wild African mammals. The sampling opportunities presented during any wildlife management activities need to be maximised and followed by intensive diagnostic screening and detailed necropsies, where appropriate. Efficient data collection, storage and management are essential, and the value of serum and tissue storage banks for retrospective studies and analysis cannot be over stated.

10.6 Conclusions

Traditional wildlife epidemiosurveillance based on passive and active ongoing reporting should be expanded to all countries and areas where sufficient resources are available. The results need to be collected by international organisations such as the OIE and then shared at a global scale. Once such a system is available, efficient early warning of emerging risks will require further development of approaches in three fields in particular.

10.6.1 Sentinel Surveillance

Wild animals can be used for the detection of emerging infectious diseases (EID) or pathogens, because they are often more at risk of infection than humans or domestic animals. The use of wildlife sentinels may be a particularly valuable approach to surveillance for emerging zoonotic infections, many of which have their origins in wild hosts. For example, wild lagomorphs (Lane et al. 1991) and deer (Gallivan et al. 1998) are exposed to ticks carrying the bacterium (*Borrelia burgdorferi*) that causes Lyme disease in humans and so may be used as sentinels during disease surveillance. Certain taxa may be relatively more efficient at concentrating some pathogens, for instance predators at the top of food chains or scavengers that may be exposed to infectious carcasses (Smith 1994; Leighton et al. 1995). Plans for animal based surveillance of human infections have been considered in the field of public health, such as using the model of animal rabies surveillance (Childs et al. 2007), but as yet there are few practical projects that make use of animal sentinels for human health decision making (Rabinowitz et al. 2005). In addition, once a disease control campaign is underway, monitoring wildlife may be the only way to check whether pathogens are still circulating (Couacy-Hymann et al. 2005).

10.6.2 Risk Surveillance

As discussed earlier, surveillance can focus on critical transmission routes or on specific sites where the local ecology favours the probability of outbreaks. Risk-assessment methods can be used to inform the design of the surveillance approach such that it is optimised for the early detection and management of diseases (Stärk et al. 2006). This may involve targeting of wild animal populations that have a high probability of exposure to diseases or hazards. McKenzie et al. (2007) developed a methodology to prioritise pathogens for a wildlife disease surveillance strategy in New Zealand. The risk evaluation was based on the probability of importing pathogens using the framework recommended by the OIE (Murray 2004). The relative risks of different pathogens were represented by ranked scores for each of several taxonomic groups of hosts, allowing the priorities for surveillance to be clearly identified.

Observations of abnormal behaviour in terrestrial mammals have been used for decades in Europe to monitor rabies (WHO 2008b). This is an example of efficient risk-based surveillance. Surveillance of wild animals at rescue centres, or in sites at risk (as adopted for Highly Pathogenic Avian Influenza H5N1 in Europe: Pittman et al. 2007) are other examples of risk-based surveillance.

10.6.3 Syndromic Surveillance

Syndromic surveillance is designed to improve early detection of outbreaks by using existing data monitored in real time (Henning 2004). Efficient syndromic surveillance has to be based on clear definitions of cases, which could be recognised by computer programmes (medical informatics). As the data processing must be optimised to be efficient (standardisation of cases, data extraction and analysis), only well-established wildlife surveillance systems are likely to be able to operate such 'epidemio-surveillance'. Syndromic surveillance is an established approach in human epidemiology, but is still in its infancy as a tool for wildlife disease detection. Since clinical signs and lesions are difficult to observe in wildlife, syndromic surveillance may often be difficult to achieve in practice (Vourc'h et al. 2006). Also, the use of this approach to provide early identification of a potential risk to human or domestic animal health from diseases in wild hosts, assumes that the pathogen will have similar clinical effects in all these hosts. However, this of course could not be assessed until the causative agent was identified and described. Despite these potential limitations, syndromic surveillance holds much promise, especially where data from long term wildlife disease monitoring is already available (Zeng et al. 2005).

10.6.4 Future Challenges

The above approaches to disease surveillance are by no means mutually exclusive, and could be used in combination to improve detection and the prediction of future risks. New high throughput approaches such as microarray technology will enable more wildlife samples to be screened for more pathogens, and as this technology improves so more sophisticated surveillance systems may be developed. However, ultimately the reliable prediction of future outbreaks rests largely on our ability to understand the origins and drivers of disease emergence.

During the preparation of this chapter, the director general of OIE stated "*surveillance of wildlife diseases must be considered equally important as surveillance and control of diseases in domestic animals. Wildlife often acts as sentinels for animal diseases thus allowing an effective management and control of the diseases in domestic animals*" (Vallat 2008). The majority of human emerging infectious diseases (72%) originate in wildlife (e.g. SARS, Ebola), a trend which has increased significantly in recent times (Jones et al. 2008). In addition to these "*practical reasons*" it is the duty of humanity "*to maintain biological diversity, have better knowledge of animal sanitary statuses and prevent species at risk from disappearing while protecting the human and domestic animal populations from the introduction of diseases*" (Vallat 2008). Thus the road forward is laid out: scientists, veterinarians, game managers and wildlife conservationists, all have to build a new paradigm in the field of health and disease surveillance in wildlife.

With a few notable exceptions, wildlife disease monitoring or surveillance systems across the globe are largely in their infancy. New concepts are constantly being developed or adapted from experiences gained in public health. Novel technologies are emerging that can be applied to wildlife disease diagnostics, and these will require new approaches to collecting and processing data. Such developments will undoubtedly improve the efficiency of monitoring and surveillance of disease in wild mammals in the near future. Nevertheless, the costs of implementing such systems can be a major impediment, particularly in developing countries. The cooperation of the wealthier nations is therefore necessary to enlarge surveillance at a global scale. As most EIDs originated in the Southern hemisphere (Jones et al.

2008), the money spent there will represent a wise investment in the preservation of health in the more affluent nations of the North. If we are to successfully anticipate and manage the risks emanating from disease in wildlife in the future, then we must view the financial investment required to develop effective surveillance systems in relation to the potential costs of doing nothing.

Chapter 11
Disease Management in Endangered Mammals

Andrew C. Breed, Raina K. Plowright, David T. S. Hayman, Darryn L. Knobel, Fieke M. Molenaar, David Gardner–Roberts, Sarah Cleaveland, Dan T. Haydon, Richard A. Kock, Andrew A. Cunningham, Anthony W. Sainsbury, and Richard J. Delahay

11.1 Introduction

One quarter of all mammal species are considered threatened with extinction (IUCN 2007). The rate of loss of biodiversity is accelerating because increasing pressure from an expanding human population is shrinking natural habitat and over-exploiting wild animal populations. Although processes such as habitat loss and over-harvesting are usually identified as the major drivers of extinction, recent evidence suggests that disease can also be a significant threat to endangered species (Lyles and Dobson 1993; Daszak and Cunningham 1999; Daszak et al. 2000b; de Castro and Bolker

---

A.C. Breed
School of Veterinary Science, Australian Biosecurity Cooperative Research Centre, University of Queensland, Brisbane, Australia

R.K. Plowright
Consortium for Conservation Medicine, 460 West 34th Street, New York, NY 10001, USA

D.T.S. Hayman
Cambridge Infectious Diseases Consortium, University of Cambridge, Cambridge, CB3 0ES, UK

D.L. Knobel and S. Cleaveland
The Roslin Institute/Royal (Dick) School of Veterinary Studies, University of Edinburgh, Easter Bush, Roslin, Midlothian, EH25 9RG, UK

F.M. Molenaar, A.A. Cunningham and A.W. Sainsbury
Institute of Zoology, Zoological Society of London, Regent's Park, London, NW1 4RY, UK

D. Gardner-Roberts
The Mountain Gorilla Veterinary Project, BP 105 Musanze District, North Province, Rwanda

D.T. Haydon
Department of Environmental and Evolutionary Biology, Graham Kerr Bldg, University of Glasgow, Glasgow, Scotland, G12 8QQ, UK

R.A. Kock
Conservation Programmes, Zoological Society of London, Regent's Park, London, NW1 4RY, UK

R.J. Delahay
Central Science Laboratory, Sand Hutton, York, YO41 1LZ, UK

R.J. Delahay et al. (eds.), *Management of Disease in Wild Mammals*,
DOI:10.1007/978-4-431-77134-0_11, © Springer 2009

2004; Choisy and Rohani 2006; Lips et al. 2006; Smith et al. 2006). Disease has already been documented as a cause of extinction of a land snail (*Partula turgida*) (Cunningham and Dazsak 1998), and several amphibian species (Schloegel et al. 2006; Skerrat et al. 2007). Diseases are also known to cause significant population declines, as illustrated by the impact of canine distemper virus in black-footed ferrets (*Mustela nigripes*) (Williams et al. 1988) and lions (*Panthera leo*) (Roelke-Parker et al. 1996), rabies virus in African wild dogs (*Lycaon pictus*) (Woodroffe and Ginsberg 1999), Ebola virus in apes (Leroy et al. 2004), squirrelpox virus in red squirrels (*Sciurus vulgaris*) (Rushton et al. 2006) and transmissible facial tumour disease in Tasmanian devils (*Sarcophilus harrisii*) (Pearse and Swift 2006).

A recent study identified 54 species of mammal for which disease was considered a threatening process (Table 11.1). The majority of such species (88%) were from the orders Artiodactyla or Carnivora, with families containing the most familiar and widespread livestock and companion animals (i.e. cattle, sheep, goats, pigs, dogs and cats) most represented. This is probably because of the close taxonomic relationship of these wild animals with domestic species, increasing the likelihood of pathogen transfer, and due to the widespread distribution of large populations of domestic species, allowing exposure of wildlife to domestic animal pathogens. Viruses and bacteria with broad host ranges that include domestic animals have been identified as most likely to threaten wild mammal populations (Table 11.2). Close contact was the predominant mode of transmission (75%) amongst the listed

Table 11.1 Mammals for which parasites were identified as a major threat at either the species or subspecies level based on the IUCN Red List summary documentation (Pedersen et al. 2007)

Order	Family	Species	Common name
Artiodactyla	Bovidae	*Alcelaphus buselaphus*	Coke's Hartebeest
		Antidorcas marsupialis	Springbok
		Beatragus hunteri	Hirola
		Bos frontalis	Gaur
		Bos grunniens	Wild Yak
		Bos javanicus	Banteng
		Bos sauveli	Kouprey
		Bubalus bubalis	Asian Buffalo
		Bubalus depressicornis	Anoa
		Bubalus mindorensis	Tamaraw
		Bubalus quarlesi	Mountain Anoa
		Budorcas taxicolor	Takin
		Connochaetes taurinus	Blue Wildebeest
		Damaliscus lunatus	Tsessebe
		Hemitragus jayakari	Arabian Tahr
		Ovis canadensis	Bighorn Sheep
		Ovis orientalis	Punjab Urial
		Procapra gutturosa	Mongolian Gazelle
		Syncerus caffer	African Buffalo
		Tragelaphus imberbis	Lesser Kudu

(continued)

Table 11.1 (continued)

Order	Family	Species	Common name
Artiodactyla	Cervidae	*Dama dama*	Mesopotamian Fallow Deer
		Hippocamelus antisensis	North Andean Deer
		Hippocamelus bisulcus	Chilean Guemal
		Ozotoceros bezoarticus	Pampas Deer
Artiodactyla	Suidae	*Babyrousa babyrussa*	Babiroussa
		Phacochoerus aethiopicus	Cape Warthog
		Sus cebifrons	Visayan Warty Pig
		Sus philippensis	Philippine Warty Pig
Artiodactyla	Tayassuidae	*Catagonus wagneri*	Chacoan Peccary
Carnivora	Canidae	*Alopex lagopus*	Arctic Fox
		Atelocynus microtis	Short-eared Dog
		Canis lupus	Gray Wolf
		Canis simensis	Ethiopian Wolf
		Chrysocyon brachyurus	Maned Wolf
		Cuon alpinus	Dhole
		Lycaon pictus	African Wild Dog
		Nyctereutes procyonoides	Racoon Dog
		Otocyon megalotis	Bat-eared Fox
		Pseudalopex fulvipes	Darwin's Fox
		Urocyon littoralis	Channel Islands Fox
		Vulpes bengalensis	Bengal Fox
Carnivora	Felidae	*Felis silvestris*	Wild Cat
		Prionailurus bengalensis	Iriomote Cat
		Puma concolor	Florida Panther
Carnivora	Mustelidae	*Lontra felina*	Marine Otter
		Lontra provocax	Southern River Otter
Carnivora	Otariidae	*Eumetopias jubatus*	Steller Sea Lion
Carnivora	Phocidae	*Monachus monachus*	Mediterranean Monk Seal
Cetacea	Delphinidae	*Cephalorhynchus hectori*	Maui's Dolphin
Dasyuromorphia	Dasyuridae	*Dasyurus hallucatus*	Northern Quoll
		Parantechinus apicalis	Southern Dibbler
Peramelemorphia	Peramelidae	*Perameles gunnii*	Eastern Barred Bandicoot
Rodentia	Sciuridae	*Cynomys parvidens*	Utah Prairie Dog
Sirenia	Trichechidae	*Trichechus manatus*	West Indian Manatee

pathogens, with other routes such as indirect or arthropod-vectored transmission being considered less likely to cause extinction.

Disease may threaten an endangered mammal population by suppressing population growth rates, making them more vulnerable to extinction through stochastic factors. For example otodectic mange in the Mednyi arctic fox (*Alopex lagopus*) (Goltsman et al. 1996) and canine parvovirus in the gray wolf (*Canis lupus*) (Mech and Goyal 1995) reduced population growth by limiting recruitment. Synergistic interaction with other threatening processes, such as hunting, could increase the probability of population extinction (Choisy and Rohani 2006). Alternatively, disease may kill individuals more rapidly than they can reproduce, leading to deterministic extinction. This is most likely to occur in populations that are already small or fragmented.

Table 11.2 Parasites identified as causing population declines or reduced host fitness in mammals listed on the IUCN Red List as threatened by pathogens. The numbers in each column reflect the number of mammal species threatened by each pathogen. Names of diseases are in parentheses. Table from Pedersen et al. (2007)

Parasite name	Carnivores	Artiodactyls	Other
Viruses			
Morbillivirus, canine distemper virus	10	0	0
Parvovirus, canine parvovirus	4	0	0
Vesivirus, feline calicivirus	1	0	0
Coronavirus, feline infectious peritonitis virus	1	0	0
Parvovirus, feline panleukopenia virus	1	0	0
Gammaretrovirus, feline leukemia virus	0	0	0
Apthovirus, foot-and-mouth disease virus	0	7	0
Lentivirus, jembrana disease virus	0	1	0
Morbillivirus, monk seal morbillivirus	1	0	0
Rhadinovirus, ovine herpesvirus 2	0	1	0
Varicellovirus, pseudorabies virus	1	0	0
Lyssavirus, rabies virus	9	2	0
Morbillivirus, rinderpest virus	0	7	0
Bacteria			
Bacillus anthracis (anthrax)	0	5	0
Chlamydia sp. (infectious keratoconjuctivitis)	0	1	0
Fusobacterium necrophorum (hoof rot)	0	2	0
Mannheimia haemolytica (pasteurellosis)	0	1	0
Mycoplasma conjunctivae (infectious keratoconjuctivitis)	0	1	0
Mycobacterium bovis (bovine tuberculosis)	0	2	0
Pasteurella spp. (pasteurellosis)	0	2	0
Yersinia pestis (plague)	0	0	1
Helminths			
Angiocaulus gubernaculatus (nematode)	1	0	0
Dioctophyma renale, giant kidney worm	1	0	0
Dirofilaria immitis, heartworm	1	0	0
Protostrongylus spp., lungworm	0	1	0
Taenia hydatigena, thin-necked bladderworm	0	1	0
Arthropods			
Otodectes cynotis, ear canker mite	1	0	0
Psoroptes sp. (psoroptic mange)	0	1	0
Sarcoptes scabei (sarcoptic mange)	3	1	0
Protozoa			
Toxoplasma gondii (toxoplasmosis)	2	0	2
Fungi			
Encephalitozoon cuniculi (encephalitozoonosis)	1	0	0

Small and fragmented populations may themselves be more vulnerable to infection. Small populations might be below the critical threshold for pathogen maintenance, causing previously-endemic diseases to become locally extinct. When the population comes into contact with that pathogen again, the loss of herd immunity could result in heightened morbidity and mortality (Cunningham 1996). Also, small fragmented populations are likely to have reduced genetic variability, even if the population size subsequently increases. Susceptibility to infectious disease and neoplasia (tumours) in Californian sea lions (*Zalophus californianus*), for example, was positively correlated with inbreeding (Acevedo-Whitehouse et al. 2003). The mechanism responsible for this enhanced susceptibility is unknown, but heterozygosity of the major histocompatability complex (MHC) has been linked to effective immune response in other species (Penn 2002). For example, a reduction in the MHC region of the cheetah's (*Acinonyx jubatus*) genome after an historic population bottleneck may have contributed to the severity of an epidemic of feline infectious peritonitis in captive animals (Evermann et al. 1988). Tasmanian devils are another species in which low genetic diversity (Jones et al. 2004) has increased susceptibility to disease (Pyecroft et al. 2007; Woods et al. 2007). An invariably lethal transmissible tumour, not recognised as 'non-self' by the host, is spreading through Tasmanian devil populations, with current trends suggesting extinction could occur within 20 years (McCallum et al. 2007).

Dealing with disease in endangered mammals can be considered a special case within wildlife disease management for several reasons. First, the goal of management is principally the conservation of biodiversity (i.e. prevention of the extinction of populations and maintenance of genetic diversity) rather than disease control or eradication. Indeed, interaction between hosts and parasites is crucial for the healthy functioning of ecosystems and parasites are important components of biodiversity *per se*. Many parasites are host specific and, when treating endangered species in small populations, the inadvertent extinction of parasites is possible. Disease management actions, therefore, must be compatible with the over-arching aim of biodiversity conservation in its broadest sense; this may influence the choice of approach when working with endangered species. Second, in the case of endangered species, sufficient knowledge of the ecology and epidemiology of the host-disease system may be particularly difficult to acquire. Such information can be critical to the effective control of disease in any wildlife population, and consequently ill-informed *ad hoc* interventions to manage disease in endangered species have often done more harm than good. Therefore, the management of disease threats to endangered species needs to be considered as an integral component of the overall conservation plan, subjected to careful scrutiny and provided with adequate financial and logistical support.

Identifying when disease poses a real threat to endangered wildlife populations, and when management or intervention is appropriate, can be challenging for many reasons. The epidemiology of disease in species of conservation concern is often poorly understood because the basic ecology, behaviour and population dynamics of the hosts are usually not well described (Plowright et al. 2008); diseased or dead animals are frequently difficult to detect; and consequently substantial effort and expense is required to estimate disease impact and prevalence. Indeed the true impact of a disease on a population can only be determined through manipulation of the host-parasite relationship, for example by treating or vaccinating a portion of the population. Furthermore, diagnosis of disease is often limited by an absence of diagnostic tests or, where these are available, tests which have not been validated for the species concerned: most diagnostic tests used for wild mammals have been developed for use with domestic animals and may give poor or inaccurate results when used for wildlife

Despite the difficulties and expense, a thorough understanding of disease epidemiology, and the likely responses of host populations to management intervention, should ideally be gained prior to management intervention, to avoid wasted effort or even damaging interventions. Mathematical models can provide valuable insights into disease epidemiology and the potential impact of interventions, and as such is an important tool for those attempting to manage infectious disease threats in endangered mammal populations (see Chapter 4). In the context of endangered populations these outcomes are typically some measure of the likelihood of persistence of the population, in the face of varying levels of disease risk and over different time periods, or the quantitative demographic impact of disease on population abundance. Modelling infectious disease processes in these populations will be more uncertain. Traditional approaches using deterministic models predict the *average* progress of a disease through a population and often fail to capture key elements that influence the spread of infection in small populations. These elements depend on chance events in transmission. In models of small populations their inclusion will help to inform decision-makers of the range of possible outcomes associated with a disease outbreak. Such (stochastic) models have been successfully applied to the management of infectious disease risks to the world's most endangered canid species, the Ethiopian wolf (*Canis simensis*) (see Box 11.1).

Box 11.1 Modelling and the management of disease threats in endangered populations: the case of the Ethiopian wolf

Fewer than 600 Ethiopian wolves (*Canis simensis*) persist in seven populations confined to remnant fragments of Afroalpine habitat above 3,000m, in the Ethiopian highlands. Within these fragments, wolves live in discrete packs of 3–13 adults that communally share and defend an exclusive territory. The largest population of wolves (around 300 adults) is found in the Bale Mountains National Park in southeast Ethiopia. In the park, Ethiopian wolves occur in several subpopulations connected by narrow corridors of suitable habitat. The park and surrounding area are also occupied by pastoralists, their livestock and domestic dogs (*Canis lupus familiaris*). These dogs act as reservoirs for a number of infectious diseases, since their high numbers allow several generalist canid pathogens, including rabies and canine distemper, to persist within their populations. Sporadic spillover of these pathogens into the sympatric wolf population has been responsible for a number of large outbreaks – indeed, rabies is recognized as the most immediate threat to the short-term persistence of the Bale wolf population. Management decisions to mitigate this threat have recently taken into account results from mathematical models, which predict the consequences of rabies introduction into the population, and the effect of various intervention strategies on the outcome of such an introduction. This potentially powerful approach successfully combines elements of demographic monitoring, disease surveillance, contingency planning and reactive vaccination.

Models of disease dynamics in Ethiopian wolf populations have progressed from simple population viability analysis (Mace and Sillero-Zubiri 1997) to a sophisticated spatially-explicit demographically stochastic individual-based model (Haydon et al. 2006). The latter model incorporates the pack-based social structure of the wolves, an important advance as the composition and configuration of packs have been shown to play a critical role in the outcome of rabies introduction into the system. This model was able to quantify the threat posed by rabies to the persistence of wolf populations, an outcome that in itself was useful for galvanising support for a domestic dog rabies vaccination campaign in and around wolf habitats. The model has been used to make specific, practical recommendations to managers on the prevention of, and response to, future rabies outbreaks in the Bale Mountains wolf population. Traditional epidemiological theory is often used to predict the proportion of individuals that must be vaccinated in order to reduce the effective reproduction number (*R*; see Chapter 3) of the agent to less than one and eliminate infectious disease from a population – an approach which generally requires the vaccination of the majority of individuals (in domestic dogs, the coverage required for the elimination of rabies is estimated at 70%). However, the first priority of conservation biologists may be to ensure the long-term persistence of an endangered population. This objective may not require total elimination of all outbreaks, but perhaps only the largest that might compromise long-term population viability. Stochastic epidemiological and demographic models of the Bale wolf metapopulation, suggested that vaccination of as few as 20–40% of wolves against rabies might be sufficient to eliminate the largest outbreaks, and thus prevent populations from reaching low densities from which they would be unlikely to recover (Haydon et al. 2002a). These findings suggested that prophylactic vaccination of the wolf population against rabies could be a feasible and worthwhile undertaking.

The model has also informed contingency plans to deal with potential future outbreaks by showing that the impact of epidemics could be reduced by low-coverage reactive vaccination campaigns even after discovery of the outbreak. Model results have shown that vaccination within the infected zone could be effective and reduce mortality. Long-term persistence of wolf populations could be further improved by focusing reactive vaccination in the habitat corridors between sub-populations.

The Ethiopian wolf rabies model predicts that around 40% of spillover events will 'fade out', requiring no management action. If however more than four individuals become infected, the probability that an epidemic will occur increases. Hence, the model provides managers with a trigger threshold, above which action should be taken. Following the diagnosis of rabies in several wolves in the Bale Mountains in 2003, a vaccination programme was implemented which entailed the physical capture and injectable vaccination of wolves (Knobel et al. 2008). The virus did not progress beyond the initially infected subpopulation, and results of simulations based on the developed model demonstrated that the probability of the disease spreading into unaffected areas would have been much greater in the absence of intervention. Given the controversy surrounding the handling of endangered African canids (Woodroffe 2001), such evidence added valuable support to the benefits of this intervention.

(continued)

Box 11.1 (continued)

This case study clearly illustrates the potential utility of individual-based stochastic models in assisting managers of populations of endangered species in decision-making. The value of such models is dependent on the accuracy of the underlying data. A major strength of the model described here was the volume of detailed demographic and spatial data collected over a number of years by field staff of the Ethiopian Wolf Conservation Programme. This was enhanced by a major surveillance effort during the outbreak, which produced data on mortality patterns and the spatial distribution of cases. A detailed pre-existing database of genetic profiles of the animals within the outbreak area even allowed the pack membership of dead wolves to be ascertained. The ability of mathematical models to successfully inform management decisions for endangered populations thus depends on the synergistic interactions of field biologists and epidemiologists with modellers who have an understanding of the importance of underlying natural ecological processes to the outcome of pathogen introductions in small populations.

As described earlier in this book, approaches to disease management in wildlife can be categorised according to the proposed target of action. For those situations where the disease is better understood, interventions can be directed at the infectious agent through vaccination or medication (see Chapter 6 and Section 11.2); at the host population (see Chapter 7 and Section 11.3), or at the environment (see Chapter 8 and Section 11.4). Special cases arise when species are so valuable or endangered that animals are managed on an individual basis when they may require a combination of techniques (see Box 11.2), or when they are translocated as part of an integrated conservation plan (Section 11.5).

11.2 Targeting the Infectious Agent

Management actions targeting the infectious agent can involve (i) administration of anti-parasitic or antibiotic drugs or (ii) vaccination against the infectious agent. The use of anti-parasitic and antibiotic drugs in free ranging endangered mammals has been attempted on a few occasions, but with limited success. Treatment of ectoparasitic mites causing mange has been undertaken in cheetahs (Mwanzia et al. 1995), Mednyi arctic foxes (Goltsman et al. 1996), mountain gorillas (*Gorilla gorilla beringei*) (Graczyk et al. 2001) and southern hairy-nosed wombats (*Lasiorhinus latifrons*) (Ruykys et al. 2006). Individual cases showed a positive response to treatment but the long-term effects on populations are unknown. Intestinal and vascular nematodes have been treated with anthelmintics in red wolves (*Canis rufus*) (Phillips and Scheck 1991) and Florida panthers (*Puma concolor coryi*) (Roelke and Glass 1992).

Box 11.2 The Mountain Gorilla Veterinary project

Some endangered mammals are considered so valuable that individuals are monitored and treated if they become ill. The Mountain Gorilla Veterinary Project (MGVP), a non-profit group that provides veterinary care to mountain gorillas (*Gorilla gorilla beringei*) (Cranfield et al. 2001; Cranfield et al. 2005), is a prime example of this disease management strategy. MGVP considers the health of the gorillas not in isolation, but as part of an ecosystem that includes sympatric species such as local domestic livestock, wildlife and human populations (Nizeyi et al. 1999, 2002; Graczyk et al. 2002a; Graczyk et al. 2002b; The Mountain Gorilla Veterinary Project 2002 Employee Health Group 2004).

The mountain gorilla exists in two, geographically distinct, island populations: the Virunga Massif, a small body of forest at the intersection of the borders of the Democratic Republic of Congo (DRC), Rwanda and Uganda, and the Bwindi Impenetrable Forest in Uganda. The total estimated population of 700–750 individuals is divided equally between these two sites.

In Rwanda and Uganda, the development of protected areas, in the form of patrolled national parks, and a robust tourist industry has helped to reduce the threat from habitat degradation and poaching, leaving zoonotic disease as the major threat to the health of the gorilla population (Homsey 1999). However, in the DRC, political instability, militia forces and groups of internally displaced persons are currently more immediate threats to both gorilla health and habitat.

The majority of MGVP's routine work consists of health monitoring, preventative health procedures, education, research and the dissemination of information. To do this, MGVP works in partnership with the regional Protected Area Authorities and non-governmental organizations, to provide ongoing monitoring of the gorilla groups, disease monitoring and vaccination of domestic livestock and companion animals adjacent to the national parks, and health monitoring in the form of an annual Employee Health Programme (EHP) (Ali et al. 2004) for those people who work with the gorillas.

Emergency care in the field is provided to gorillas in the event of human induced conditions that are considered to be life-threatening. A 'decision tree' was developed to assist field vets in their choice of action in each case. Cases are usually identified during routine health monitoring visits or from feedback from partner organisations. Subsequent intensive, focal animal monitoring, to establish the nature, degree and progression of disease, is then undertaken. Data such as morbidity, current social status, demographic information (e.g. age, sex, relative 'genetic value' of an individual), geographic and meteorological information (e.g. proximity and interaction with other groups, altitude, season), any relevant history (recent, or likely, transfer to or from the group), and the perceived risks and benefits of intervention to the individual and to the group, all contribute to the decision on whether to monitor, or immobilize and treat. The final decision to immobilise an animal for treatment rests with the local

(continued)

Box 11.2 (continued)

protected area authority, and is based on factors such as the terrain, weather, time of day, and the availability of appropriate expertise.

The most significant causes of death in the mountain gorilla population are trauma and respiratory disease (Nutter et al. 2005). For example, the MGVP frequently treats injuries resulting from interactions with humans, such as snare removal and bacterial respiratory infections probably transmitted from humans. Intervention is more likely in 'high-risk' and 'high-value' cases such as infants.

Routine collection of blood, urine, faeces and tissue samples occurs during emergency interventions and at *post mortem* examination. All samples are stored in an in-country bio-bank of samples, with duplicates transported to the Biological Resource Center at the Maryland Zoo in Baltimore, USA. The international scientific community is encouraged to apply for access to samples for independent research projects. MGVP collaborates with various laboratories in Europe and the United States for rapid and accurate interpretation of pathological samples.

MGVP is also developing a contingency plan, to be enacted in the event of a catastrophe that threatens the survival of the gorillas. In collaboration with Mississippi State University, MGVP has developed the Internet-supported Management Program to Assist Conservation Technologies (IMPACT) database system, which integrates data collected on the gorillas (e.g. demographic, routine health monitoring, pathology: Cranfield and Minnis 2007). This database may be updated at any time via the Internet, to provide immediate, real-time information on health trends at the level of the individual, the group and the population. IMPACT has the potential to provide an early warning of disease trends associated with an outbreak.

Overall, the MGVP approach has proven to be effective, local political instability notwithstanding. During the period in which MGVP has been in operation there has been an overall increase in gorilla numbers (of 17% in the last 10 years), a significant milestone for the development of a sustainable population.

Both vaccination of the endangered host, and of domestic animal reservoir species, have been proposed as control strategies for minimising the transmission of pathogens to wildlife hosts. During the translocation of critically endangered hirola antelope (*Beatragus hunteri*) in Kenya in the mid 1990s there was an ongoing epizootic of rinderpest with virus circulating in the source and release areas (Kock, 2008). The translocated animals were vaccinated with the standard cattle vaccine without ill-effect and were monitored with no reports of dead animals with signs of the disease despite probable exposure (Butynski 2000). A second example was the vaccination of 65 mountain gorillas in the Virunga volcanoes region of Central Africa, during a measles-like respiratory disease outbreak in 1988 (Hastings et al. 1991). Signs of respiratory disease ceased after the vaccination program was initiated but because all non-pregnant animals had been treated, there was no control group so the role of measles vaccination in preventing the spread of this disease could not be rigorously evaluated. In this case it was considered impractical and unethical to withhold treatment from a control group of gorillas and other interventions in highly endangered populations are likely to be faced with similar dilemmas. Nevertheless, it is important that wherever possible attempts are made to assess the efficacy of disease management actions, as such information is crucial to the development of future management plans.

Vaccination of the host is also not without its risks. For example, vaccinating African wild dogs before conducting adequate vaccine trials may have led to the failure to control a rabies epidemic in the Serengeti. Subsequent trials demonstrated that multiple doses of rabies vaccine might be required for protection from disease (Woodroffe 1997; Hofmeyr et al. 2004). The use of live canine distemper vaccines in black-footed ferrets has induced fatal canine distemper in the past (Carpenter et al. 1976). In contrast, the use of a killed canine distemper vaccine in the same species failed to protect against fatal distemper infection (Williams et al. 1988).

The control program enacted against rinderpest in Africa highlights the enormous impact that vaccination of domestic animals *can* have on the prevalence of disease in wild mammal populations. Rinderpest caused catastrophic losses of wildlife and livestock after introduction of the virus into Africa in the late 1800s, however widespread vaccination of domestic cattle led to a rapid decline in the incidence of the disease in wild bovids and a marked increase in their abundance (Plowright 1982). By decreasing the number of susceptible domestic animals below the threshold required to sustain rinderpest virus, the cattle vaccination campaigns effectively reduced the distribution of the virus affecting both cattle and wildlife (Rossiter 2001). A similar approach was initiated in the Serengeti ecosystem, in Tanzania, where domestic dogs were vaccinated against pathogens that threaten endangered African canids (see Box 11.3). Domestic dogs were

Box 11.3 Managing disease threats from a domestic reservoir: rabies outbreaks in endangered African canids

Rabies has been responsible for a number of well-documented outbreaks in endangered African canids, including Ethiopian wolves (*Canis simensis*) and African wild dogs (*Lycaon pictus*). However, the virus appears incapable of persisting indefinitely within these populations, independent of other hosts. The high pathogenicity of the virus, coupled with small host population size, low connectivity between populations, and rapid transmission of the virus through packs facilitated by their social behaviour, ensure the rapid depletion of susceptible hosts and disappearance of the virus. Rabies epidemics in wild dogs and Ethiopian wolves are thus dependent on reintroduction of the virus from a population of one or more reservoir species. Prediction and prevention of these epidemics requires an understanding of the ecology of local reservoir hosts and the transmission dynamics of the virus within and between the reservoir and populations of endangered canids. Although rabies has a broad mammalian host range, within any given geographical area a single species is often principally responsible for its maintenance. Domestic dogs are the principal rabies hosts throughout most of the current distributions of African wild dogs and Ethiopian wolves.

(continued)

Box 11.3 (continued)

1 No intervention

In the face of a disease threat to an endangered population, a decision not to intervene may be valid. But this must be a proactive choice, based upon as full an understanding as possible of the consequences of inaction, rather than a decision made by default, through poor preparedness. In view of the shortage of detailed data on the epidemiology and dynamics of infectious disease outbreaks in wild populations, even if no direct intervention is contemplated managers should be prepared, in the event of pathogen spillover or an encroaching epidemic in the reservoir population, to collect as much information as possible on the spatial and temporal spread of disease, recent incidence in the reservoir population, clinical and pathological signs, morbidity and mortality rates, and molecular characteristics of the pathogen. Such data collection should be seen as the minimum adequate management response, and can be used to guide future disease management decisions. Utilising such information to develop mathematical models of possible outcomes of disease introduction into the target population (see Box 11.1) can be useful for future decision-making.

2 Reducing incidence in the reservoir population

Reducing the incidence of disease in the reservoir host population will decrease the force of infection acting on the population of interest. Three strategies can achieve this:

Reducing the density of susceptible animals

This can be achieved by reducing the survival rate of the population (e.g. culling of stray animals; limiting resource availability by for example burning, burying or otherwise reducing access to refuse), decreasing the fecundity of the population (through the sterilisation of females), or immunising susceptible animals through vaccination. Domestic dog populations can also be limited by reducing the need for people to keep dogs. The relative merit of each of these strategies will depend on local demographic and cultural circumstances, which will in turn affect their cost-effectiveness. In practice, lethal methods of controlling stray dog populations have met with limited success, and the resulting instability of dog populations and antagonism towards rabies control personnel within local communities may result in a net detrimental effect. The World Health Organization Expert Committee on Rabies (WHO 1992) concluded that removal of dogs should not be used in large-scale rabies control programmes unless ecological and socio-cultural studies show it to be feasible within the particular local context. Large-scale mass vaccination of dogs is now accepted as the control method of choice for rabies in most circumstances.

Eliminating infected animals from the population

Because of the danger they pose to human health, local authorities and the public attempt to kill clinically rabid dogs in rabies endemic areas. This probably results in some reduction in transmission to susceptible animals, but in isolation is unlikely to alter the course of an outbreak, since rabid dogs only exhibit clinical signs for a few days. Such actions may also compromise human safety and animal welfare standards, and will potentially miss infected animals that do not exhibit classical clinical signs.

Reducing contact between susceptible and infected animals

This can occur by encouraging owners to restrict the movement of their dogs by for example tying them up, confining them to a kennel or compound, and walking them on a leash. The adoption of these behaviours will however depend on the specific cultural reasons for dog ownership, implementation of education programmes, and possible enforcement by local legislation.

3 Reducing contact between reservoir and target populations

This may be achieved through the confinement methods described above or, in more extreme cases, by fencing off populations of endangered canids. Construction and maintenance of fences is costly and is usually implemented for management purposes other than disease control, for example to reduce human-carnivore conflicts, to prevent human encroachment, or to reduce disease transmission from wildlife to livestock. Reduced disease transmission to wild canids has seldom, if ever, been a primary reason for fencing (although in some small reserves in southern Africa income generated from eco-tourism centred around African wild dogs contributes to the upkeep cost of fences). In addition to the economic cost, the ecological implications of fence construction must obviously also be considered.

4 Vaccination of target populations

For both African wild dogs and Ethiopian wolves, effective vaccination regimens have been developed using commercially available inactivated domestic dog injectable vaccines. Hence, direct vaccination of endangered hosts against rabies is an option. Vaccination strategies may either be prophylactic (to prevent spillover) or reactive. As in all cases where an intervention is contemplated, the benefits of vaccination must be weighed up against the costs, both financial and in terms of risks to target and non-target species. Detailed contingency planning, ideally incorporating mathematical modelling of various outbreak and intervention scenarios, should be conducted in advance of spillover events. Improving delivery strategies for vaccines, particularly through the development of effective oral bait formulations, must be a priority for future research.

In all likelihood a combination of the above management options, dependent on the local context, will be necessary to ensure the persistence of all but the largest populations of Africa's endangered canids.

immunized against rabies, canine distemper, and parvovirus, with the goal of reducing disease outbreaks in lions, African wild dogs and bat-eared foxes (*Otocyon megalotis*) (Cleaveland et al. 2000; Cleaveland et al. 2003). Detailed plans have also been drawn up for the vaccination of Ethiopian wolves (*Canis simensis*) against rabies infection (see Box 11.1).

11.3 Targeting Hosts

Disease management action directed towards wild and domestic mammals has often included culling, under the assumption that a reduction in population density will reduce transmission rates. This is however almost always inappropriate when dealing with species of high conservation value. An exception is when infected individuals are culled to reduce the force of infection to susceptible individuals; as is the case during the management of facial tumour disease in Tasmanian devils (see Box 11.4).

Box 11.4 Facial tumour disease in Tasmanian devils

On the Australian island state of Tasmania, devil facial tumour disease (DFTD) threatens the survival of the Tasmanian devil (*Sarcophilus harrisi*), an endemic and endangered marsupial carnivore. Predictions for its future are gloomy, with all populations likely to be affected within five years, followed by extinction of the species in 20–30 years if no action is taken to mitigate the spread of disease (Jones et al. 2007).

DFTD is an emerging infectious disease found exclusively in wild devil populations that appears to be invariably fatal to affected individuals. It was first observed in the mid 1990s and its increasing prevalence and geographic distribution became rapidly apparent (McCallum and Jones 2006). It is a transmissible neoplasm (tumour) that appears to be an infectious allograft (the tumour cells are the infective agent), and is most likely spread by biting (Pearse and Swift 2006).

The disease management strategy currently in place is a multi-faceted approach based on information gathering and risk minimisation. All components address the three possible management options: maintaining insurance populations isolated from the disease for reintroduction in the event of extinction in the wild; in situ management (disease suppression; development of vaccines); and detecting and spreading devils that are resistant to the disease (Jones et al. 2007).

A disease suppression trial is currently underway, whereby any animals captured in the target area showing signs of the invariably fatal tumour are removed and euthanased (Jones et al. 2007). In the first trial, an intensive trapping programme is being implemented on the large isolated Tasman and

Forestier peninsulas (a combined area of 360 km^2), that are connected to mainland Tasmania by a single bridge. Site isolation, including the possibility of constructing a barrier to devil movement on the bridge that connects this peninsula to the mainland, reduces edge effects and will likely enhance the chance of disease eradication. It is too early to indicate whether disease suppression will be successful in either eradicating or controlling the disease, but initial reports indicate a reduction in the size of the tumours being detected (Jones et al. 2007).

Planning for the establishment of insurance populations of devils incorporates current knowledge of the epidemiology of this unique disease with the population biology of Tasmanian devils, in order to assess the risks and benefits of various translocation options. The genetic diversity of devils has already been reduced by about 50% (Jones et al. 2004), hence it is important to minimise any further reduction. There are currently separately managed captive populations of devils on the Australian mainland and Tasmania, and plans to translocate animals from disease free areas in western Tasmania to offshore islands. Close demographic and genetic monitoring of captive populations has the advantage of requiring a smaller effective and hence actual population size than translocated insurance populations as mating can be controlled to maximise genetic diversity. Captive management is more costly and labour intensive than management on offshore islands. Insurance populations on offshore islands also have the advantage of allowing a larger founder population size and the animals will retain their natural parasites and pathogens as well as behaviours that may be lost in captivity. The overall plan is for an insurance metapopulation comprised of multiple populations of captive and wild-living devils, with managed dispersal between populations with appropriate quarantine steps, to maintain a high level of genetic diversity for 50 years.

Where a disease that threatens an endangered mammal resides in a wild animal reservoir, it may be legitimate to cull the reservoir host in an effort to reduce the likelihood of transmission. For example, it has been suggested that culling introduced grey squirrels (*Sciurus carolinensis*) in the UK, could reduce transmission of squirrelpox virus (SQPV) (see Section 11.4) to the rare native red squirrel (Gurnell et al. 2006). However, culling can have complex effects on host behaviour that may influence transmission rates (see Chapter 7) so the wider ecological consequences of interventions should always be assessed before management action is implemented.

When the threat from disease is particularly severe, the establishment of 'insurance' populations, either in captivity, or free-living in isolation from the disease, may be necessary to prevent extinction (e.g. Williams et al. 1988). Caution must be exercised if disease vectors are involved, or if there is a long asymptomatic stage of infection, in which case thorough quarantine and testing is required before transferring individuals to the 'insurance' population.

11.4 Targeting the Environment

Close contact with domestic animals can risk disease transmission to endangered mammal hosts. Minimising such cross-species contact can be accomplished through the use of physical barriers, such as the buffer zones between agricultural areas and ranges of bighorn sheep (*Ovis canadensis*), which have been effective in reducing disease epidemics in susceptible wildlife (Jessup et al. 1991; Jessup et al. 1995). Conservation programmes may have policies that specifically seek to ensure that endangered species are not exposed to domestic animals. For example, domestic dogs are prevented from entering the remaining refuge areas of black-footed ferrets in Wyoming, to avoid transmission of canine distemper virus (CDV) (Williams et al. 1988); and the removal and subsequent banning of domestic dogs from Antarctica has been used to avoid possible transmission of CDV to pinnipeds (Anon 1994). Exposure to humans may pose particular disease threats to primates, and so tourists visiting habituated Mountain gorilla populations in the Virungas and Bwindi conservation areas in Central Africa are limited in group size (eight people), viewing time (one hour) and minimum distance (seven metres) to reduce direct and indirect contact (Ferris et al. 2005). Other measures to prevent transfer of disease from humans to gorillas include burying human faeces deeper than 30 cm and deterring gorillas from private land surrounding their habitat.

Another approach to reducing opportunities for inter-specific disease transmission is to limit temporal overlap in the use of shared water resources or grazing habitats. For example, separation of domestic cattle and bison (*Bison bison*) during the bison birthing season prevents cross-species transmission of brucellosis in the Greater Yellowstone ecosystem (Bienen and Tabor 2006). In Kruger National Park, South Africa, water holes provide focal points for the dissemination of anthrax (*Bacillus anthracis*) throughout ungulate populations. Control of this problem has been tackled by treating waterholes with antibiotics, which has successfully halted the spread of bacteria (Prins and Weyerhaeuser 1987).

The manipulation of habitat and landscape features has been used as an effective tool to make environments more attractive to species of conservation concern. Similarly, there may be opportunities to manage habitats to reduce disease transmission to endangered species, although such actions need to be consistent with the broader aims of natural habitat preservation. The presence of a squirrelpox virus (SQPV) reservoir in the grey squirrel population in England and Wales has been shown to accelerate the rate at which the rare native red squirrel has declined by 20-fold (Rushton et al. 2006). Minimising inter-specific contact is a crucial component of red squirrel conservation in Britain (Gurnell et al. 2006). Red and grey squirrels utilise forest habitats with differing efficiency. In particular the red squirrel, which is best adapted to mature boreal coniferous forests of Scots pine (*Pinus sylvestris*) and Norway spruce (*Picea abies*), is able to thrive in certain coniferous tree plantations, such as Sitka spruce (*Picea sitchensis*), which appear to be avoided by grey squirrels. The Kielder Forest is dominated by Sitka spruce and holds the largest remaining red squirrel population in England. This forest has been managed to maximise its suitability for red squirrels through tree species selection whilst minimising the likelihood of incursions by grey squirrels by trapping them around forest edges and habitat corridors in particular. Plantation management specifically includes minimising pine and large seeded broadleaves around Sitka plantations (Lurz et al. 2003).

Managing the movement of endangered species between fragmented subpopulations to limit disease transmission has recently been debated in the disease ecology literature. The development and use of 'corridors' of suitable habitat to facilitate movement between small and fragmented populations is increasingly advocated as a means of reducing the deleterious effects of isolation. However, while connectivity diminishes the loss of genetic diversity and allows recolonisation of local populations, it can also increase the risks of disease transmission (Hess 1996a). Nevertheless, recent modelling studies suggest that when a reservoir host (domestic or wild) occupies the matrix between patches, corridors may have relatively little effect on transmission of pathogens between populations of the endangered host, and that corridors should therefore provide a net conservation benefit (Woodroffe 1999; Gog et al. 2002). These investigations were extended to examine the situation where the endangered host and reservoir species occupied the same patches (McCallum and Dobson 2002). All studies concluded that too little connectivity always leads to extinction of the endangered host and the benefit of increased landscape connection far outweighs the risk of increased disease transmission.

11.5 Translocation and Reintroduction

Conservation programmes for endangered species usually aim to increase the genetic diversity of small populations, by enhancing the gene flow between fragmented populations and restoring a species historical range after local extinction. Translocation of individuals between different populations, reintroductions and re-stocking are important tools in many such programmes. However, these activities may themselves present high-risk opportunities for disease transmission, with potentially devastating implications for endangered populations. Consequently, it is essential that the disease risks of all translocations are effectively managed.

11.5.1 Why Do Translocations Represent a Disease Risk?

Animal translocations are thought to play a major role in the emergence of infectious diseases in wildlife (Daszak et al. 2000b; Williams et al. 2002a). Alien pathogens can be introduced with animals translocated into indigenous populations of the same or differing species where they may have a particularly severe impact if the recipients are naïve to infection (Cunningham 1996; De Leo and Dobson 2005). In the absence of effective immunity, the pathogen may cause disease and readily spread with potentially disastrous consequences. Both domestic and wild animal translocations present a disease risk to endangered species. The rinderpest pandemic in African ungulates described above represents a severe example of the consequences of alien parasite introduction with a translocated domestic mammal (cattle transported from Europe to Africa). There are several well-described examples where the translocation of wild mammals has resulted in anthropogenic spread of infectious diseases such as: the transmission of bovine tuberculosis to a local naïve population of wood bison (*Bison bison anthabascae*) after the introduction of plains bison (*Bison bison bison*) into a National Park in Canada (Carbyn and Watson 2001); the spread of the giant liver fluke (*Fascioloides magna*) to European ungulates when infected elk (*Cervus elaphus*) were introduced into Italy from the USA (Haigh 1988); the introduction of rabies into the raccoon (*Procyon lotor*) population in parts of the eastern United States following the translocation of raccoons to supplement hunted populations (Anthony et al. 1990). Programmes in which captive-bred animals, or animals held away from their geographic region of origin, are translocated are thought to represent a greater risk of alien parasite introduction, particularly where they have been in contact with exotic species, for example in zoological collections (Kirkwood and Sainsbury 1997).

The potential exposure of translocated animals to endemic pathogens in recipient populations, to which they have inadequate immunity, represents another disease hazard of translocations. Animals that have had no exposure to one or more parasites present in the destination environment are likely to be naïve and more susceptible to parasites they encounter after translocation. A classic example is mortality in reintroduced captive-reared black-footed ferrets caused by canine distemper, which was endemic in the wild (Williams et al. 1988). Other examples include: the development of neurological disease in eastern woodrats (*Neotoma floridana*) due to infection with *Baylisascaris procyonis* (a neurotropic roundworm of racoons) following their reintroduction to New York (Davidson and Nettles 1992); an outbreak of babesiosis in translocated sable antelope (*Hippotragus niger*) (McInnes et al. 1991); and mortality or disease due to cowdriosis (heartwater), trypanosomosis, babesiosis or theileriosis in African antelope, big horn sheep, mule deer (*Odocoileus hemionus*), white rhinoceros (*Ceratotherium simum*) and black rhinoceros (*Diceros bicornis*) (Kock et al. 2007; Kock et al. 1999b; Nijhof et al. 2003).

A further potential consequence of translocations is that pre-existing disease dynamics in the recipient ecosystem can be affected by the introduced species. By acting as new hosts, changing host-parasite dynamics through altering host density, or potentially forming new reservoirs of infection, translocated individuals could exacerbate disease caused by endemic pathogens. This scenario is most likely to occur among closely related wild and domestic species, since parasites are more likely to move between species of higher relatedness. For example, bacterial pneumonia caused by *Pasteurella* sp. resulted in high mortality rates in translocated bighorn sheep (Foreyt 1989), which were spatially correlated with high domestic sheep densities (Singer et al. 2001).

Translocation usually involves capture, transport, quarantine, introduction to a new environment, and subsequent competition for food, territory and mating opportunities. The associated stress experienced by individual animals can be considerable and may result in immunosuppression and greater susceptibility to infectious disease (Viggers et al. 1993; Kock et al. 2007).

11.5.2 Captivity and Exposure to Pathogens

The ex situ management of an endangered species may take place for the purpose of acquiring knowledge about the taxon, increasing public awareness of its plight, as a source for breeding and reintroduction, or any combination of these objectives. In critically endangered species, individuals from the few remaining populations are sometimes taken into captivity for captive breeding and reintroduction. Examples include the black-footed ferret in the USA (Thorne and Williams 1988; Williams et al. 1988; Dobson and Lyles 2000), Arabian oryx (*Oryx leucoryx*) in Oman (Spalding et al. 1999) and the golden lion tamarin (*Leontopithecus rosalia*) in Brazil (Gippoliti and Carpaneto 1997). Potential disease outbreaks in such small numbers of highly valuable individuals can have disastrous consequences for the success of conservation projects.

Unfortunately, disease-screening protocols are not always an inherent part of projects involving the captive management of endangered species. However, the time spent in captivity creates a situation of enhanced risk regarding the acquisition of novel diseases. The animals may be exposed to an array of pathogens that they would not normally encounter, such as those transmitted by commensal rodents that inhabit facilities and enclosures, or those carried by related host species in the direct vicinity, be they exotic animals in a zoo, domestic animals on neighbouring land or human caretakers. Pathogens with wide host ranges (which often include domestic animals) are the most likely to infect endangered species (Pedersen et al. 2007). For example, the grazing of domestic sheep along the perimeter fence of an endangered-species breeding centre resulted in transmission of capripox virus (pathogens of sheep, goats and cattle) to Arabian oryx reared for reintroduction purposes. Although only one case developed clinical signs, subsequent screening of the entire herd revealed a seroprevalence of 2% (Greth et al. 1992). However, contact with infected rodents in a captive breeding facility was identified as the source of outbreaks of toxoplasmosis and callitrichid hepatitis (caused by infection with lymphocytic choriomeningitis virus) in golden lion tamarins (Montali and Bush 1992). Captive animals also can be exposed to novel diseases via their food (e.g. an epidemic of toxoplasmosis decimated a captive colony of squirrel monkeys (*Saimiri sciureus*): Cunningham et al. 1992). A wide range of antelope and wild felid species died with neurodegenerative disease following exposure to the bovine spongiform encephalopathy (BSE) agent via commercially-available feed concentrate (bovids) or meat (felids) (Kirkwood & Cunningham 2006). As neither the degree of exposure of other zoo animals nor the biology of the disease (e.g. incubation period, transmissibility) in wild mammals is known, recommendations were made to minimise the risk of infected animals being translocated to disease-free regions or released to the wild (Kirkwood & Cunningham 1994, Kirkwood & Cunningham 2006). An example where infection acquired in captivity did reach the wild occurred when a pet orang-utan (*Pongo pygmaeus*) was released despite having previously tested positive for tuberculosis, which it was suspected to have acquired from its captors (Bonner 1995). These examples illustrate that where adequate information on risks and appropriate screening are absent, there may be significant opportunities for the transfer of pathogens into areas where they may pose a threat to both indigenous and translocated species.

11.5.3 Disease Risk Analysis for Translocations

Knowledge regarding the prevalence of pathogens in wild populations and susceptibility to clinical disease is often lacking for endangered species. Also, for most pathogens of wild mammals, reliable *ante mortem* diagnostic tests are unavailable (Kirkwood and Sainsbury 1997). Often infections are subclinical (the hosts may not necessarily develop clinical disease), which makes detection of the pathogen even more difficult. Consequently, for both translocated animals and recipient populations, enhanced exposure to novel pathogens is a realistic possibility in any translocation project. Although precautions should be taken when undertaking translocations, a 'zero risk' approach is simply not possible.

Although the IUCN provide guidelines that advocate disease monitoring during translocations, if there is no legal obligation to carry out a disease risk analysis, this requirement will often be ignored. However, governments may not be aware of the potentially serious risks of wildlife translocations and therefore often have no statutory regulations on such movements.

Standard disease control methods for any translocation project should include strict quarantine procedures, comprehensive health examinations (including *post mortem* examinations) with appropriate laboratory screening tests to detect a wide range of possible pathogens, vaccination protocols where appropriate, and clinical examinations, including haematological and plasma biochemical analyses where possible, prior to release to assess body condition and anticipate survival in the wild (Montali et al. 1995). To aid the identification of those pathogens that could be important during translocation projects, a risk assessment can be performed. This identifies the diseases that are prevalent in donor and recipient populations. After the major disease risks have been identified, screening for selected pathogens can be incorporated into the translocation project and suitable measures can be identified in the event of an outbreak (e.g. treatment, vaccination, euthanasia). This protocol does, however, rely on previous health studies on the donor and recipient ecosystems, which are often absent or incomplete. The incorporation of such studies should be considered as part of a translocation programme.

A disease risk analysis can be broken down as follows: (Macdairmid and Pharo 2003; Murray 2004).

(i) *Hazard identification*
All known pathogens that could potentially be imported with the species concerned are listed.

(ii) *Risk assessment*
An assessment of risk is carried out on each pathogen identified as a hazard. This evaluates the likelihood and possible consequences, both biological and economic, of entry, establishment or spread of the pathogen to the area of reintroduction.

(iii) *Risk management*
Based on the results of the risk assessments, decisions are made with regards to disease management protocols for the translocation procedure. Screening for the diseases of greatest risk can be planned for both the animals to be translocated and the recipient population. *Post mortem* examinations should be performed on animals found dead in captivity or post-release. Following disease screening, appropriate measures, such as (prophylactic) treatment for certain pathogens and vaccination protocols, should be implemented. In some circumstances it may be considered appropriate to expose the translocated animals to low levels of diseases they might encounter in their new habitat to build up herd immunity. This approach was employed for the reintroduction of black rhinoceros in southern Africa, where the animals were temporarily held in low-density tsetse fly (*Glossina* spp.) areas to permit low exposure rates to *Trypanosoma* spp. prior to release (Kock et al. 1999b). Treatment for certain diseases may also be considered before release. Post-release health surveillance is an important component of risk management because the results of surveillance can be used to refine risk management protocols. Close monitoring of animals' health and behaviour can be achieved through differing methods depending on the species, for example by radio-tracking or trapping.

(iv) *Risk communication*
At all stages during the risk analysis process all stakeholders should be involved in discussions on the potential disease risks and their consequences for the translocation project.

In some cases, the risk analysis may identify a risk that is of such significance that the intended translocation project should be abandoned. For instance, bovine tuberculosis has been identified in black rhinoceroses held in captivity in Western zoos, but as yet not in those in the wild. As there are currently no sufficiently reliable *ante-mortem* screening methods to detect infection, the risk of introducing this pathogen to the free-ranging population outweighs the potential conservation benefits of translocation (Osofsky et al. 2001).

Other approaches have been advocated to minimise or avoid the disease risks associated with animal translocations and ex situ breeding. For instance, translocation of germplasm rather than entire animals can be undertaken. Although disease transfer is still possible (Philpot 1993), it is considerably less likely. Also, where animals are captive-bred for local release within their natural range, they experience continuous exposure to the climate and endemic pathogens of the area.

Programmes for the reintroduction and translocation of endangered species are expensive and time consuming, and may require specialist facilities. The potential for negative outcomes in terms of transmission of novel pathogens to either the recipient population or translocated individuals is significant, and can have devastating consequences for the conservation project. Consequently, the extra effort and resources required to conduct sufficient research into the potential disease risks, to carry out the appropriate screening and to ensure adequate veterinary involvement throughout, constitute an essential investment in any translocation project.

11.6 Future Perspectives

In all likelihood more mammal species will become endangered throughout the world in the near future. At the same time the occurrence of new and emerging diseases are likely to increase. In fact, many of the same processes are likely to be driving both trends. Over-exploitation of natural resources, the disruption of ecosystems and continuing urban expansion bring humans and our livestock into increasingly closer contact with potential sources of disease from wildlife (see Chapter 1). Hence, we can expect to be more frequently challenged with the management of diseases in the small and fragmented populations of endangered mammals.

The imperative to act to safeguard the survival of many species, the heightened opportunities for disease transmission that these interventions incur and the impossibility of screening for all pathogens, mean that a ‘zero-tolerance’ approach to disease risk is unattainable. However, the thorough and systematic assessment of risk, based on current knowledge and the integration of disease management at all levels of conservation programmes provide the best available framework for continued action.

Most disease threats to endangered mammals are from well-known pathogens that also infect domestic mammals (Pedersen et al. 2007). However, recent experiences such as those in Tasmania (with devil facial tumour disease (DFTD)) and Central Africa (with Ebola and Marburg viruses: see Box 11.5) indicate that pathogens can arise from unexpected or unknown sources. This raises several points worthy of broader consideration in conservation biology and disease management. Firstly, early recognition of an infectious disease as the cause of population decline is crucial to development of a management plan, while identification of the causative agent is of lesser importance and hence syndromic surveillance can be effectively applied to detect emerging disease threats. Secondly, loss of genetic diversity can expose populations to unforeseen disease threats. With habitat loss and fragmentation increasingly leading to a reduction in genetic diversity of wild animal populations, more species may become susceptible to disease. Thirdly, both host-specific pathogens (e.g. DFTD), as well as the more familiar generalist pathogens (e.g. rabies and CDV) that reside in abundant reservoir species, are able to pose a significant extinction risk particularly when their transmission is frequency dependent (e.g. McCallum 2008).

Box 11.5 Emerging disease, human health and endangered species: Ebola in Central Africa

Marburg and Ebola virus are members of the filoviridae that cause acute viral haemorrhagic disease (Pourrut et al. 2005) and are a source of current concern for the health of humans and endangered primates. In Central Africa, Ebola Zaire virus (EBOV) has killed over 1,300 people, and populations of great apes (gorillas and chimpanzees) have declined by 80% in some of their last strongholds in Central Africa (Walsh et al. 2003; Leroy et al. 2004). Following a human EBOV outbreak along the Gabon-Congo border in late 2001-early 2002, the first gorilla (*Gorilla gorilla gorilla*) carcass was found in June 2002, and by October 130 animals out of 143 had disappeared. Of the 32 carcasses found, 10/12 gorilla and 3/3 common chimpanzee (*Pan troglodytes troglodytes*) carcasses tested were positive to EBOV by PCR, antigen capture or immunohistochemical staining (Bermejo et al. 2006). This has led to the presumption that the dramatic decline of gorilla and chimpanzee populations in the region was due to EBOV. It is estimated that these populations will take at least 75 years to recover to pre-EBOV outbreak densities (Walsh et al. 2005).

Identification of the reservoir hosts of filoviruses has proved challenging. Despite infection being identified in both primates and duikers (*Cephalophus* sp.), neither is thought to be the reservoir host due to their high disease-related mortality rates. Serological and antigen assays have provided further evidence of this. In an outbreak of EBOV-Reston subtype in a captive primate facility in the Philipines in 1996, 12.5% (131/1051) of the animals were antigen positive, but only 0.2% (3/1732) were seropositive (Miranda et al. 1999; Miranda et al. 2002). The index human filovirus cases had previously been linked to caves or buildings with resident bats, one such case reported a sting or bite from an arthropod, and there was some indication that the virus resembled certain plant viruses. Experimental studies with EBOV were therefore undertaken using 24 species of plants and 19 species of vertebrates and invertebrates (Swanepoel et al. 1996). Infection of Angola free-tailed bats (*Mops condylurus*), little free-tailed bats (*Chaerephon pumilus*), and Wahlberg's epauletted fruit bats (*Epomophorus wahlbergi*) resulted in virus replication without death (Swanepoel et al. 1996). Subsequently EBOV RNA was recovered from the liver and spleen tissues of wild forest dwelling fruit bats, following an outbreak of infection in humans (Leroy et al. 2005). This was followed by the discovery of specific immunoglobulin IgM antibodies in the same bat species (the hammer-headed fruit bat (*Hypsignathus monstrosus*), Franquet's epauletted bat (*Epomops franqueti*) and the little collared fruit bat (*Myonycteris torquata*)). The PCR and serological findings suggested acute infection followed by seroconversion, and together the evidence strongly implicates fruit bats as reservoir hosts for these viruses.

(continued)

Box 11.5 (continued)

It is unclear whether the fruit bat populations in Central Africa have a long standing association with EBOV and are endemically infected, or whether the virus may exhibit wave-like spread through the region (Walsh et al. 2005). Although the subtype EBOV-Zaire, probably diverged from EBOV-Ivory Coast subtype 700–1,300 years ago (Suzuki and Gojobori 1997), all Central African isolates identified subsequent to the Yambuku outbreak in 1976 are closely related descendents of a Yambuku-like virus suggesting a recent expansion in viral diversity (Walsh et al. 2005; Biek et al. 2006b). EBOV isolates from fruit bats show genetic variation, which suggests all strains in bats have descended from a common ancestor within the last 30 years (Biek et al. 2006b). Whether this is due to a genetic bottleneck, or there is another, as yet unidentified reservoir, is not known. There is, however, no evidence to date of an epidemic wave with associated spillovers occurring in other regions from where EBOV has been isolated. The death of chimpanzees (*Pan troglodytes verus*) in Tai National Park in Ivory Coast, suggest a single event (Le Guenno et al. 1999), with no local or regional epidemic causing further deaths in susceptible primate or duiker species.

Investigating EBOV in wild mammals has presented particular challenges in terms of diagnosing the cause of the decline in populations of endangered great apes, identifying the principal reservoir host and describing the pattern of infection. Another, perhaps even more challenging task, is to consider how to control or manage this virulent pathogen in an extensive and complex ecosystem such as the Central African forest. If the acute epidemic in Central Africa has been facilitated by the role of primates and duikers (Walsh et al. 2007), then it may quickly run out of susceptible hosts, particularly in those parks where over 80% of resident apes have already been lost (Walsh et al. 2003). This may also be true if the density of the putative fruit bat reservoir is low, the infectious period short and transmission is density dependent.

EBOV infection in African apes is an emerging zoonotic disease with potentially catastrophic consequences for endangered primates. There is compelling circumstantial evidence that EBOV has caused the decline of apes in the Congo-Gabon region, although the actual number of clinical cases diagnosed is small compared to the suggested level of mortality. There is little serological evidence of the disease in ape populations (Bermejo et al. 2006), however, this is unsurprising if apes are 'new' or spillover hosts with high mortality. When outbreaks occur in humans, survivors with detectable immunity are scarce (Busico et al. 1999; Jezek et al. 1999). It is therefore difficult to consider the potential merits of vaccination without a better understanding of the epidemiology and ecology of this disease. Even then, the practicality of vaccination, and impacts of intervention need to be carefully assessed. When the target population is small (e.g. a few tens of animals) vaccination may be a reasonable approach to consider, but where there are tens of thousands of animals in an extensive area, it may be impractical regardless of the conservation status of the species.

Wherever possible, efforts should be made to monitor and evaluate the efficacy of interventions to manage disease, to provide an evidence base for future work. This is true for the management of disease in any wildlife population, but for endangered species it takes on particular significance because of the potentially catastrophic effects of ineffective or counter-productive interventions. This may be particularly challenging for endangered species, as it is often not possible to collect scientifically robust data on the efficacy of interventions because there is no 'untreated' control group for comparison. The benefits of mathematical modelling are increasingly evident in these situations, particularly when epidemiological and host demographic data are available (see Box 11.1).

As recognition has increased for the role that parasites play in wildlife ecology and ecosystem health, including the value associated with their potential regulation of host numbers and contribution to biodiversity (Daszak and Kilpatrick 2008), so has the realisation that the health of humans, wild and domestic animals and ecosystems are inextricably connected (see Boxes 11.2 and 11.5). Hence, it is multidisciplinary teams that can best provide the necessarily broad range of knowledge and skills posed by the problems of the management of disease in wild mammals in the 21st century.

Glossary

This is a short glossary of the technical terms and abbreviations used in this book. For more comprehensive descriptions see Watt et al. 1995 and Thrusfield 2007.

Aetiology	The study of causative factors of disease
Age-structured model	A mathematical model in which the host population is partitioned into different age classes
Aggregation	Organisms cluster significantly more than would be expected at random.
AIDS	Acquired Immune Deficiency Syndrome, the disease caused by HIV
Antibody	A protein in the blood produced by the immune system in response to an antigen
Antigen	A substance, generally foreign, capable of inducing antibodies
Bacteraemia	The presence of live bacteria in the blood
Basic reproduction number (R_0)	The average number of secondary cases of infection resulting from one primary case introduced in to a population of susceptible individuals (for macroparasites, R_0 is the average number of female offspring produced by a mature female parasite)
bTB	Bovine tuberculosis, caused by *Mycobacterium bovis*
Carrying capacity	The number of individuals of a species that an area can support
CBA	Cost–benefit analysis
CDV	Canine Distemper virus
Commensal	A close association of two species in which one species benefits while the other is unaffected
Compartmental model	A mathematical model where the hosts are divided into different categories (e.g. susceptible, infected, recovered)
Contact rate	The average frequency per unit time during which infected individuals contact susceptible individuals

CWD	Chronic Wasting Disease; a TSE of deer
DALY	Disability-Adjusted Life Years
Definitive host	For macroparasites, the host in which the parasite sexually reproduces
Deterministic model	A mathematical model which assumes that all parameters and variables are not subject to variation
DFTD	(Tasmanian) Devil Facial Tumour Disease
EBLV	European Bat Lyssavirus
EBOV	Ebola virus
EIA	Environmental Impact Assessment
Endemic	A disease whose prevalence does not exhibit wide fluctuations through time in a defined location
Enzootic	'Endemic' with reference to animals
Epidemic	A rapid increase in the prevalence of a disease
Epizootic	'Epidemic' with reference to animals
Fecundity	The capacity to produce offspring
FIV	Feline Immunodeficiency Virus
FMD	Foot and Mouth Disease; a highly contagious viral disease
Force of infection	Lambda; the rate at which susceptible individuals become infected by disease
GIS	Geographical Information System
GMO	Genetically Modified Organism
GnRH	Gonadotropin-releasing hormone
GPS	Global Positioning System
Heterogeneity	Used in model terminology to demonstrate that factors (e.g. genetic, spatial) are not homogeneous across the whole population
HIV	Human Immunodeficiency Virus, the cause of AIDS
Incidence	Incidence is the number of new cases (as a proportion or percentage) that arise in a population per unit of time
Incubation period	The time between infection and the onset of disease
Infection	The presence of an organism within a host: it may or may not cause disease
Infectious period	The period during which an infected individual is able to transmit infection
IUD	Intra-Uterine Device
K_T	A density above which the disease is capable of establishing
Latent period	The period when an individual is infected but before it is capable of transmitting the infection

Latin Hypercube Sampling	A sampling method that involves splitting a variable distribution (e.g. normal) into equal areas and sampling the same number of times from each
Macroparasite	Parasites which in general do not multiply within their definitive hosts, but instead produce transmission stages (eggs and larvae) which pass into external environment or vectors
Map	*Mycobacterium avium* subspecies *paratuberculosis*; causative agent of paratuberculosis also known as Johne's Disease in cattle
MCMC	Markov chain Monte Carlo randomisation
Metapopulation	A group of spatially separated populations that are linked by dispersal
MGVP	Mountain Gorilla Veterinary Project
MHC	Major histocompatability complex; a portion of the chromosome liked to immune response
Microparasite	Parasites that undergo direct multiplication within their definitive host
Morbidity	The state of being diseased; from the Latin *morbidus*
Neophobia	Behavioural avoidance of novel objects or foods
Notifiable Diseases	Diseases that must, by law, be reported to an official authority
Pandemic	An epidemic occurring very widely
PCR	Polymerase Chain Reaction; the replication of DNA in the laboratory
Prevalence	The total number of cases (as a proportion or percentage) in an exposed population over a given sampling period
Prion	A protein capable of causing TSE
PZP	Porcine Zona Pellucida
R	Effective reproduction number. The actual number of secondary infections produced by an infectious individual. See also basic reproduction number
RBCT	Randomised Badger Culling Trial
RHD	Rabbit Haemorrhagic Disease
RNA	Ribonucleic Acid
RT-PCR	Reverse Transcription Polymerase Chain Reaction
SEIR	Susceptible, Exposed, Infected, Recovered: commonly used states within a compartmental model
Sensitivity	When an animal is known to be affected, the sensitivity of a test is its ability to give a positive response

SNA	Social Network Analysis
Specificity	When an animal is known to be unaffected, the specificity is the ability of the test to give a negative response
SQPV	Squirrelpox virus
STD	Sexually Transmitted Disease
Stochastic model	A mathematical model which takes into account random variation in one or more parameters or variables
Surveillance	From Chapter 10: collecting and analysing information on the health of wild animals to help manage disease
Syndrome	From Chapter 10: a collection of clinical signs, frequently observed in association and putatively linked with some aetiology or disease risk factors
TSE	Transmissible Spongiform Encephalopathy
Vector	In diseases with indirect lifecycles the intermediate host
VHF	Very High Frequency; 30–300 MHz
Viraemia	The presence of virus in the blood stream
Virulence	The case mortality rate of a parasite
VVIC	Virally Vectored Immunocontraceptive
WTP	Willingness to Pay
Zoonoses	A parasite naturally transmitted between humans and other vertebrate species
ZP	Zona Pellucida

References

Acevedo P, Escudero MA, Munoz R, Gortazar C (2006) Factors affecting wild boar abundance across an environmental gradient in Spain. Acta Theriol 51:327–336

Acevedo P, Ruiz-Fons F, Vicente J, Reyes-García AR, Alzaga V, Gortázar C (2008) Estimating red deer abundance in a wide range of management situations in Mediterranean habitats. J Zool 276:37–47

Acevedo-Whitehouse K, Gulland F, Greig D, Amos W (2003) Disease susceptibility in California sea lions. Nature 422:35

Aldwell FE, Keen DL, Stent VC, Thomson A, Yates GF, de Lisle GW, Buddle BM (1995a) Route of BCG administration in possums affects protection against bovine tuberculosis. NZ Vet J 43:356–359

Aldwell FE, Pfefer A, de Lisle GW, Jowett G, Heslop J, Keen D, Thomson A, Buddle BM (1995b) Effectiveness of BCG vaccination in protecting possums against bovine tuberculosis. Res Vet Sci 58:90–95

Aldwell FE, Tucker IG, de Lisle GW, Buddle BM (2003a) Oral delivery of *Mycobacterium bovis* BCG in a lipid formulation induces resistance to pulmonary tuberculosis in mice. Infect Immun 71:101–108

Aldwell FE, Keen DL, Parlane NA, Skinner MA, de Lisle GW, Buddle BM (2003b) Oral vaccination with *Mycobacterium bovis* BCG in a lipid formulation induces resistance to pulmonary tuberculosis in brushtail possums. Vaccine 22:70–76

Aldwell FE, Brandt L, Fitzpatrick C, Orme IM (2005a) Mice fed lipid-encapsulated *Mycobacterium bovis* BCG are protected against aerosol challenge with *Mycobacterium tuberculosis*. Infect Immun 73:1903–1905

Aldwell FE, Baird MA, Fitzpatrick CE, McLellan AD, Cross ML, Lambeth MR, Buchan GS (2005b) Oral vaccination of mice with lipid-encapsulated *Mycobacterium bovis* BCG: anatomical sites of bacterial replication and immune activity. Immunol Cell Biol 83:549–553

Aldwell FE, Cross ML, Fitzpatrick CE, Lambeth MR, de Lisle GW, Buddle BM (2006) Oral delivery of lipid-encapsulated *Mycobacterium bovis* BCG extends survival of the bacillus in vivo and induces a long-term protective immune response against tuberculosis. Vaccine 24:2071–2078

Alexander KA, Pleydell E, Williams MC, Lane EP, Nyange JFC, Michel AL (2002) *Mycobacterium tuberculosis*: an emerging disease of free-ranging wildlife. Emerg Infect Dis 8:598–601

Ali R, Cranfield M, Gaffikin L, Mudakikwa T, Ngeruka L, Whittier C (2004) Occupational health and gorilla conservation in Rwanda. Int J Occup Environ Health 10:319–325

Allan SA (2001) Ticks (class Arachnida: order Acarina). In: Sammuel WM, Pybus MJ, Kocan AA (eds) Parasitic diseases of wild mammals. 2nd edn. Iowa State University Press, Ames, USA, pp. 72–106

Anderson RM (1991) Discussion: the Kermack–McKendrick epidemic threshold theorem. Bull Math Biol 53:3–32

Anderson RM, May RM (1991) Infectious diseases of humans. Dynamics and control. Oxford University Press, Oxford, UK

Anderson RM, Trewhella W (1985) Population dynamics of the badger (*Meles meles*) and the epidemiology of bovine tuberculosis (*Mycobacterium bovis*). Phil Trans R Soc Lond B 310:327–381

Anderson RM, Jackson HC, May RM, Smith ADM (1981) Population dynamics of fox rabies in Europe. Nature 289:765–771

Angulo E, Cooke B (2002) First synthesize new viruses then regulate their release? The case of the wild rabbit. Mol Ecol 11:2703–2709

Anon. (1994) Last bark for Antarctic huskies. Science 263:606

Anon. (2002) SNOMED, HL7, LOINC the official informatics standards for veterinary medicine. J Am Vet Med Assoc News Archive. Available at: http://www.avma.org/onlnews/javma/jun02/s020601o.asp

Anthony JA, Childs JE, Glass GE, Korch GW, Ross L, Grigor JK (1990) Land use associations and changes in population indices of urban raccoons during a rabies epizootic. J Wildl Dis 26:170–180

Arneberg P (2002) Host population density and body mass as determinants of species richness in parasite communities: comparative analyses of directly transmitted nematodes of mammals. Ecography 25:88–94

Arneberg P, Skorping A, Grenfell B, Read AF (1998) Host densities as determinants of abundance in parasite communities. Proc R Soc B 265:1283–1289

Arnemo JM, Ahlqvist P, Andersen R, Berntsen F, Ericsson G, Odden J, Brunberg S, Segerström P, Swenson JE (2006) Risk of capture-related mortality in large free-ranging mammals: experiences from Scandinavia. Wildl Biol 12:109–114

Arthur AD, Pech RP, Singleton GR (2005) Predicting the effect of immunocontraceptive recombinant murine cytomegalovirus on population outbreaks of house mice (*Mus musculus domesticus*) in mallee wheatlands. Wildl Res 32:631–637

Artois M, Delahay R, Guberti V, Cheeseman C (2001) Control of infectious diseases of wildlife in Europe. Vet J 162:141–152

Artois M, Depner KR, Guberti V, Hars J, Rossi S, Rutili D (2002) Classical swine fever (hog cholera) in wild boar in Europe. Rev Sci Tech Off Int Epizoot 21:287–303

Aubert M (1994) Control of rabies in foxes: what are the appropriate measures? Vet Rec 134:55–59

Aubert M (2003) Du diagnostic de la rage vulpine à son élimination. Bull Acad Vét Fr 156:6–14

Aubert MFA (1999) Costs and benefits of rabies control in wildlife in France. Rev Sci Tech Off Int Epizoot 18:533–543

Augustine DJ (1998) Modelling *Chlamydia*-koala interactions: coexistence, population dynamics and conservation implications. J Appl Ecol 35:261–272

Austin EJ, Deary IJ, Edwards-Jones G, Arey D (2005) Attitudes to farm animal welfare: factor structure and personality correlates in farmers and agriculture students. J Individual Differences 26:107–120

Baer GM (ed) (1991) The natural history of rabies. 2nd edn. CRC Press, Inc., Boca Raton, FL, 620 p

Baer GM, Abelseth MK, Debbie JG (1971) Oral vaccination of foxes against rabies. Am J Epidemiol 93:487–490

Bailey NTJ (1957) The mathematical theory of epidemics. Griffin, London

Baker RHA, Black R, Copp GH, Hayson KA, Hulme PE, Thomas MB, Brown A, Brown M, Cannon RJC, Ellis J, Ellis M, Ferris R, Glaves P, Gozlan RE, Holt J, Howe L, Knight JD, MacLeod A, Moore NP, Mumford JD, Murphy ST, Parrott D, Sansford CE, Smith GC, St-Hilaire S, Ward NL (2008) The UK risk assessment scheme for all non-native species. Neobiota 7:46–57

Ballantyne EE, O'Donoghue JG (1954) Rabies control in Alberta. J Am Vet Med Assoc 125:316–326

Ballesteros C, Perez de la Lastra, M.J, la Fuente Jde (2007) Recent developments in oral bait vaccines for wildlife. Recent Patents on Drug Delivery & Formulation 1:230–235

Ballou JD, Traylor-Holzer K, Turner A, Malo AF, Powell D, Maldonado J, Eggert L (2008) Simulation model for contraception management of the Assateague Island feral horse population using individual-based data. Wildl Res 35:502–512

Barclay C (2001) Foot and mouth disease. House of Commons Library, Westminster, UK, 01/35

Barlow ND (1995) Crictical evaluation of wildlife disease models. In: Grenfell BT, Dobson AP (eds) Ecology of infectious diseases in natural populations. Cambridge University Press, Cambridge, UK, pp. 230–259

Barlow ND (1996) The ecology of wildlife disease control: simple models revisited. J Appl Ecol 33:303–314

Barlow ND (2000) Non-linear transmission and simple models for bovine tuberculosis. J Anim Ecol 69:703–713

Barr JJF, Lurz PWW, Shirley MDF, Rushton SP (2002) Evaluation of immunocontraception as a publicly acceptable form of vertebrate pest species control: the introduced grey squirrel in Britain as an example. Environ Manage 30:342–351

Bartelt G, Pardee J, Thiede K (2003) Environmental impact statement on rules to eradicate chronic wasting disease in Wisconsin's free-ranging white-tailed deer herd. Wisconsin Department of Natural Resources, Madison, USA

Bartlett MS (1957) Measles periodicity and community size. J R Stat Soc Ser A 120:48–71

Beaumont N, Townsend M, Mangi S, Austen MC (2006) Marine biodiversity. An economic valuation. Department for Environment and Rural Affairs, London. Available at: http://www.defra.gov.uk/wildlife-countryside/resprog/findings/mb-economic/index.htm

Becker NG (1989) Analysis of infectious disease data. Chapman and Hall, London

Begon M (1995) Beyond host–pathogen dynamics. In: Grenfell BT, Dobson AP (eds) Ecology of infectious diseases in natural populations. Cambridge University Press, Cambridge, pp. 478–509

Begon M, Hazel SM, Baxby D, Bown K, Cavanagh R, Chantrey J, Jones T, Bennett M (1999) Transmission dynamics of a zoonotic pathogen within and between wildlife host species. Proc R Soc B 266:1939–1945

Begon M, Bennett M, Bowers RG, French NP, Hazel SM, Turner J (2002) A clarification of transmission terms in host-microparasite models: numbers, densities and areas. Epidemiol Infect 129:147–153

Begon M, Hazel SM, Telfer S, Bown K, Carslake D, Cavanagh R, Chantrey J, Jones T, Bennett M (2003) Rodents, cowpox virus and islands: densities, numbers and thresholds. J Anim Ecol 72:343–355

Bengis RG, Thomson GR, Keet DF (1994) Foot and mouth disease in impala (*Aepyceros melampus*). Proceedings of the OIE Scientific Conference on the Control of Foot and Mouth Disease, African Horse Sickness and Contagious Bovine Pleuropneumonia, 20–23rd April, Gabarone (Botswana), pp. 13–14

Bengis RG, Kriek NPJ, Keet DF, Raath JP, De Vos V, Huchzermeyer HF (1996) An outbreak of bovine tuberculosis in a free-living African buffalo (*Syncerus caffer*) population in the Kruger National Park: a preliminary report. Onderstepoort J Vet Res 63:15–18

Bengis RG, Kock RA, Fischer J (2002) Infectious animal diseases: the wildlife/livestock interface. Rev Sci Tech Off Int Epizoot 21:53–65

Bennett R (2003) The 'direct costs' of livestock disease: the development of a system of models for the analysis of 30 endemic livestock diseases in Great Britain. J Agric Econ 54:55–71

Bennett R, Cooke R (2005) Control of bovine TB: preferences of farmers who have suffered a TB breakdown. Vet Rec 156:143–145

Bennett R, Cooke R, IJpelaar A (2004) Assessment of the economic impacts of TB and alternative control policies. Department for Environment, Food and Rural Affairs, London, UK

Bennett RM (1992) Case-study of a simple decision support system to aid livestock disease control decisions. Agric Syst 38:111–129

Bennett RM, IJpelaar J (2005) Updated estimates of the costs associated with 34 endemic livestock diseases in Great Britain. J Agric Econ 56:135–144

Berger KM, Gese E, Berger J (2008) Indirect effects and traditional trophic cascades: a test involving wolves, coyotes, and pronghorn. Ecology 89:818–828

Bermejo M, Rodriguez-Teijeiro JD, Illera G, Barroso A, Vila C, Walsh PD (2006) Ebola outbreak killed 5000 gorillas. Science 314:1564

Berry HH (1993) Surveillance and control of anthrax and rabies in wild herbivores and carnivores in Namibia. Rev Sci Tech Off Int Epizoot 12:137–146

Bertschinger HJ, Asa CS, Calle PP, Long JA, Bauman K, DeMatteo K, Jöchle W, Trigg TE, Human A (2001) Control of reproduction and sex related behaviour in exotic wild carnivores with the GnRH analogue deslorelin: preliminary observations. J Reprod Fertil Suppl 57:275–283

Bertschinger HJ, Jago M, Nöthling JO, Human A (2006) Repeated us of the GnRH analogue deslorelin to down-regulate reproduction in male cheetahs (*Acinonyx jubatus*). Theriogenology 66:1762–1767

Bicknell KB, Wilen JE, Howitt RE (1999) Public policy and private incentives for livestock disease control. Aust J Agric Resource Econom 43:501–521

Biek R, Drummond AJ, Poss M (2006a) A virus reveals population structure and recent demographic history of its carnivore host. Science 311:539–541

Biek R, Walsh PD, Leroy EM, Real LA (2006b) Recent common ancestry of Ebola Zaire virus found in a bat reservoir. PLoS Pathog 2:e90

Bienen L, Tabor G (2006) Applying an ecosystem approach to brucellosis control: can an old conflict between wildlife and agriculture be successfully managed? Front Ecol Environ 4:319–327

Black JG, Lawson KF (1980) The safety and efficacy of immunizing foxes (*Vulpes vulpes*) using bait containing attenuated rabies virus vaccine. Can J Comp Med 44:169–176

Blanchong JA, Scribner KT, Kravchenko AN, Winterstein SR (2007) TB-infected deer are more closely related than non-infected deer. Biol Lett 3:103–105

Blanchong JA, Samuel MD, Scribner KT, Weckworth BV, Langenberg JA, Filcek KB (2008) Landscape genetics and the spatial distribution of chronic wasting disease. Biol Lett 4:130–133

Blanco JC, Cortes Y, Virgos E (2005) Wolf responses to two kinds of barriers in an agricultural habitat in Spain. Can J Zool 83:312–323

Blancou J, Kieny MP, Lathe R, Lecocq JP, Pastoret PP, Soulebot JP, Desmettre P (1986) Oral vaccination of the fox against rabies using a live recombinant vaccinia virus. Nature 322:373–375

Blancou J, Pastoret PP, Brochier B, Thomas I, Bögel K (1988) Vaccinating wild animals against rabies. Rev Sci Tech Off Int Epizoot 7:1005–1013

Blancou J, Aubert MFA, Artois M (1991) Fox rabies. In: Baer GM (ed) The natural history of rabies. 2nd edn. CRC Press, Boston, pp. 257–290

Blancou J, Aubert M (1992) Vaccinia vectored rabies vaccine. In: Bögel K, Meslin FX, Kaplan M (eds) Wildlife rabies control. Wells Medical, Tunbridge Wells, UK, pp. 103–104

Blood DC, Studdert VP (1999) Saunders comprehensive veterinary dictionary. WB Saunders, London

Blower S, Roughgarden J (1989) Parasites detect host spatial pattern and density: a field experimental analysis. Oecologia 78:138–141

Blower SM, McLean AR (1991) Mixing ecology and epidemiology. Proc R Soc B 245:187–192

Boggs JF, McMurray ST, Leslie DM, Jr., Engle DM, Lochmillar RL (1991) Influence of habitat modification on the community of gastrointestinal helminths of cotton rats. J Wildl Dis 27:584–593

Bolker BM, Grenfell BT (1995) Space, persistence and dynamics of measles epidemics. Phil Trans R Soc Lond B 348:309–320

Bollo E, Ferroglio E, Dini V, Mignone W, Biolatti B, Rossi L (2000) Detection of *Mycobacterium tuberculosis* complex in lymph nodes of wild boar (*Sus scrofa*) by a target-amplified test system. J Vet Med B 47:337–342

Bomford M (1990) A role for fertility control in wildlife management? Bureau of Natural Resources Bulletin no. 7, Canberra

Bomford M, O'Brien P (1995) Eradication or control for vertebrate pests? Wildl Soc Bull 23:249–255

Bonner J (1995) Orang-utan release sparks off TB row. New Scient 148:9

Boots M, Sasaki A (1999) 'Small worlds' and the evolution of virulence: infection occurs locally and at a distance. Proc R Soc B 266:1933–1938

Boulding KE (1969) Economics as a moral science. Am Econ Rev 59:1–12

Bowen LH, Morgan DR, Eason CT (1995) Persistence of sodium monofluoroacetate (1080) in baits under simulated rainfall. NZ J Agric Res 38:529–531

Box GEP (1979) Robustness in the strategy of scientific model building. In: Launer RL, Wilkinson GN (eds) Robustness in statistics. Academic, New York

Brochier B, Blancou J, Thomas I, Languet B, Artois M, Kieny MP, Lecocq JP, Costy F, Desmettre P, Chappuis G, Pastoret PP (1989) Use of recombinant vaccinia-rabies glycoprotein virus for oral vaccination of wildlife against rabies: innocuity to several non-target bait consuming species. J Wildl Dis 25:540–547

Brochier B, Aubert MFA, Pastoret PP, Mason E, Schon J, Lombard M, Chappuis G, Languet B, Desmettre P (1996) Field use of a vaccinia-rabies recombinant vaccine for the control of sylvatic rabies in Europe and North America. Rev Sci Tech Off Int Epizoot 15:947–970

Brown KP, Sherley GH (2002) The eradication of possums from Kapiti Island, New Zealand. In: Veitch CR, Clout MN (eds) Turning the tide: the eradication of invasive species. IUCN, Gland, Switzerland and Cambridge, UK, pp. 46–52

Buddle BM, Aldwell FE, Keen DL, Parlane NA, Yates G, de Lisle GW (1997) Intraduodenal vaccination of brushtail possums with bacille Calmette-Guerin enhances immune responses and protection against *Mycobacterium bovis* infection. Int J Tuberc Lung Dis 1:377–383

Buddle BM, Aldwell FE, Skinner MA, de Lisle GW, Denis M, Vordermeier HM, Hewinson RG, Wedlock DN (2005) Effect of oral vaccination of cattle with lipid-formulated BCG on immune responses and protection against bovine tuberculosis. Vaccine 23:3581–3589

Buddle BM, Wedlock DN, Denis M (2006a) Progress in the development of tuberculosis vaccines for cattle and wildlife. Vet Microbiol 112:191–200

Buddle BM, Aldwell FE, Keen DL, Parlane NA, Hamel KL, de Lisle GW (2006b) Oral vaccination of brushtail possums with BCG: investigation into factors that may influence vaccine efficacy and determination of duration of protection. NZ Vet J 54:224–230

Burnham KP, Anderson DR (2002) Model selection and multi-model inference: a practical information-theoretic approach. Springer, New York

Burrows R (1992) Rabies in wild dogs. Nature 359:277

Burthe S, Telfer S, Lambin X, Bennett M, Carslake D, Smith A, Begon M (2006) Cowpox virus infection in natural field vole *Microtus agrestis* populations: delayed density dependence and individual risk. J Anim Ecol 75:1416–1425

Busico KM, Marshall KL, Ksiazek TG, Roels TH, Fleerackers Y, Feldmann H, Khan AS, Peters CJ (1999) Prevalence of IgG antibodies to Ebola virus in individuals during an Ebola outbreak, Democratic Republic of the Congo, 1995. J Infect Dis 179 (Suppl 1):S102–107

Butynski TM (2000) Independent evaluation of hirola antelope (Beatragus hunteri) conservation status and conservation action in Kenya. Nairobi: Kenya Wildlife Service and Hirola Management Committee

Cagnacci F, Massei G, Cowan DP, Walker N, Delahay RJ (2007) Effects of bait type and deployment strategy on uptake by free-living badgers. Wildl Res 34:454–460

Caley P (1997) Effects of eight years of possum maintenance control on levels of bovine tuberculosis in cattle and possums – the Hohotaka study. Landcare Research, Landcare Research Contract Report LC9697/141

Caley P (2006) Bovine tuberculosis in brushtail possums: models, dogma and data. NZ J Ecol 30:25–34

Caley P, Hone J (2002) Estimating the force of infection: *Mycobaterium bovis* infection in feral ferrets *Mustela furo* in New Zealand. J Appl Ecol 71:44–54

Caley P, Hone J (2005) Assessing the host disease status of wildlife and the implications for disease control: *Mycobacterium bovis* infection in feral ferrets. J Appl Ecol 42:708–719

Caley P, Ramsey D (2001) Estimating disease transmission in wildlife, with emphasis on leptospirosis and bovine tuberculosis in possums, and effects of fertility control. J Appl Ecol 38:1362–1370

Caley P, Spencer NJ, Cole RA, Efford MG (1998) The effect of manipulating population density on the probability of den-sharing among common brushtail possums, and the implications for transmission of bovine tuberculosis. Wildl Res 25:383–392

Caley P, Philp DJ, McCracken K (2008) Quantifying social distancing arising from pandemic influenza. J R Soc Interface 5:631–639

Calvete C, Estrada R, Osacar JJ, Lucientes J, Villafuerte R (2004a) Short-term negative effects of vaccination campaigns against myxomatosis and viral hemorrhagic disease (VHD) on the survival of European wild rabbits. J Wildl Manage 68:198–205

Calvete C, Estrada R, Lucientes J, Osacar JJ, Villafuerte R (2004b) Effects of vaccination against viral haemorrhagic disease and myxomatosis on long-term mortality rates of European wild rabbits. Vet Rec 155:388–392

Campbell K, Donlan CJ, Cruz F, Carrion V (2004) Eradication of feral goats *Capra hircus* from Pinta Island, Galapagos, Ecuador. Oryx 38:328–333

Capel-Edwards M (1970) Foot-and-mouth disease in the brown rat. J Comp Pathol 80:543–548

Carbyn LN, Watson D (2001) Translocation of plains bison to Wood Buffalo National Park: economic and conservation implications. In: Maehr DS, Noss RF, Larking JL (eds) Large mammal restoration: ecological and sociological challenges in the 21st century. Island Press, Covela, pp. 189–204

Carpenter JW, Appel MJG, Erickson RC, Novilla MN (1976) Fatal vaccine-induced canine distemper virus infection in black-footed ferrets. J Am Vet Med Assoc 169:961–964

Carpenter PJ, Pope LC, Greig C, Dawson DA, Rogers LM, Erven K, Wilson GJ, Delahay RJ, Cheeseman CL, Burke T (2005) Mating system of the Eurasian badger, *Meles meles*, in a high density population. Mol Ecol 14:273–284

Carter SP, Delahay RJ, Smith GC, Macdonald DW, Riordan P, Etherington TR, Pimley E, Walker NJ, Cheeseman CL (2007) Culling-induced social perturbation in Eurasian badgers *Meles meles* and the management of TB in cattle: an analysis of a critical problem in applied ecology. Proc R Soc B 274:2769–2777

Carvallo J, Gomes P (2003) Habitat suitability model for European wild rabbit (*Oryctolagus cuniculus*) with implications for restocking. Game Wildl Sci 20:287–301

Cassirer EE, Rudolph KM, Fowler P, Coggins VL, Hunter DL, Miller WM (2002) Evaluation of ewe vaccination as a tool for increasing bighorn lamb surviving following pasteurellosis epizootics. J Wildl Dis 37:49–57

Cattadori IM, Boag B, Bjornstad ON, Cornell SJ, Hudson PJ (2005) Peak shift and epidemiology in a seasonal host-nematode system. Proc R Soc B 272:1163–1169

Cauchemez S, Boelle P-Y, Donnelly CA, Ferguson NM, Thomas G, Leung GM, Hedley AJ, Anderson RM, Valleron A-J (2006) Real-time estimates in early detection of SARS. Emerg Infect Dis 12:110–113

Caughley G, Birch LC (1971) Rate of increase. J Wildl Manage 35:658–663

CCWHC (2008) http://wildlife.usask.ca

Chaddock HM (1998) Northeast Michigan surveillance activities for bovine tuberculosis in the livestock and free-ranging deer populations. Proceedings of 102nd Annual Meeting of the United States Animal Health Association, Minneapolis, USA, pp. 600–673

Change M, Glynn MK, Groseclose SL (2003) Endemic, notifiable bioterrorism-related diseases, United States, 1992–1999. Emerg Infect Dis 9:556–564

Chautan M, Pontier D, Artois M (2000) Role of rabies in recent demographic changes in red fox (*Vulpes vulpes*) populations in Europe. Mammalia 64:391–410

Cheeseman CL, Wilesmith JW, Ryan J, Mallinson PJ (1987) Badger population dynamics in a high-density area. Symp Zool Soc Lond 58:279–294

Cheeseman CL, Mallinson PJ, Ryan J, Wilesmith JW (1993) Recolonisation by badgers in Gloucestershire. In: Hayden TJ (ed) The badger. Royal Irish Academy, Dublin, pp. 78–93

Chenut G, Saintilan AF, Burger C, Rosenthal F, Crucière C, Picard M, Bruyère V, Albina E (1999) Oral immunisation of swine with a classical swine fever vaccine (Chinese strain) and transmission studies in rabbits and sheep. Vet Microbiol 64:265–276

Cheville NF, McCullough DR, Paulson LR (1998) Brucellosis in the Greater Yellowstone area. National Academy of Sciences, National Research Council, Washington, DC

Childs JE, Krebs JW, Real LA, Gordon ER (2007) Animal-based national surveillance for zoonotic disease: quality, limitations, and implications of a model system for monitoring rabies. Prev Vet Med 78:246–261

Chimbari MJ, Chirebvu E, Ndlela B (2004) Malaria and schistosomiasis risks associated with surface water and sprinkler irrigation systems in Zimbabwe. Acta Trop 89:205–213

Choisy M, Rohani P (2006) Harvesting can increase severity of wildlife disease epidemics. Proc R Soc B 273:2025–2034

Cimino L, Lovari S (2003) The effects of food or cover removal on spacing patterns and habitat use in Roe deer (*Capreolus capreolus*). J Zool 261:299–305

Citterio CV, Caslini C, Milani F, Sala M, Ferrari N, Lanfranchi P (2006) Abomasal nematode community in an alpine chamois (*Rupicapra r. rupicapra*) population before and after a die-off. J Parasitol 92:918–927

Clayton DH, Moore J (eds) (1997) Host-parasite evolution: general principles and avian models. Oxford University Press, Oxford, 473 p

Cleaveland S (2003) Emerging infectious diseases of wildlife. Microbiol Today 30:155–156

Cleaveland S, Appel MGJ, Chalmers WSK, Chillingworth C, Kaare M, Dye C (2000) Serological and demographic evidence for domestic dogs as a source of canine distemper virus infection for Serengeti wildlife. Vet Microbiol 72:217–227

Cleaveland S, Laurenson MK, Taylor LH (2001) Diseases of humans and their domestic mammals: pathogen characteristics, host range and the risk of emergence. Phil Trans R Soc Lond B 356:991–999

Cleaveland S, Hess GR, Dobson AP, Laurenson MK, McCallum HI, Robert MG, Woodroffe R (2002) The role of pathogens in biological conservation. In: Hudson PJ, Rizzoli A, Grenfell BT, Heesterbeek H, Dobson AP (eds) The ecology of wildlife diseases. Oxford University Press, Oxford, pp. 139–149

Cleaveland S, Kaare M, Tiringa P, Mlengeya T, Barrat J (2003) A dog rabies vaccination campaign in rural Africa: impact on the incidence of dog rabies and human dog-bite injuries. Vaccine 21:1965–1973

Clegg SB, Turnbull PCB, Foggin CM, Lindeque PM (2007) Massive outbreak of anthrax in wildlife in the Malilangwe Wildlife Reserve, Zimbabwe. Vet Rec 160:113–118

Cliquet F, Guiot AL, Munier M, Bailly J, Rupprecht CE, Barrat J (2006) Safety and efficacy of the oral rabies vaccine SAG2 in raccoon dogs. Vaccine 24:4386–4392

Clough P, Nixon C (2000) Coughing up for Bovine Tb control: economic review of the proposed National Pest Management Strategy for Bovine Tuberculosis. NZ Institute of Economic Research. Report to the Treasury, Wellington, NZ, 00/24 Treasury Working Paper

Clutton-Brock TH, Lonergan ME (1994) Culling regimes and sex-ratio biases in Highland red deer. J Appl Ecol 31:521–527

Clutton-Brock TH, Russell AF, Sharpe LL (2004) Behavioral tactics of breeders in cooperative meerkats. Anim Behav 68:1029–1040

Cohen S, Line S, Manuck SB, Rabin BS, Heise ER, Kaplan JR (1997) Chronic social stress, social status, and susceptibility to upper respiratory infections in nonhuman primates. Psychosomatic Med 59:213–221

Coleman J, Livingstone P (2000) Fewer possums: less bovine Tb. In: Montague TL (ed) The Brushtail Possum: Biology, Impact and Management of an Introduced Marsupial. Manaaki Whenua Press, Lincoln, New Zealand, pp. 220–231

Conover MR (1991) Reducing nuisance Canada geese problems through habitat manipulation. Proceedings of the Great Plains Wildlife Damage Conference, pp. 146–147

Cook AR, Otten W, Marion G, Gibson GJ, Gilligan CA (2007) Estimation of multiple transmission rates for epidemics in heterogeneous populations. Proc Natl Acad Sci USA 104:20392–20397

Cooke BD (2002) Rabbit haemorrhagic disease: field epidemiology and the management of wild rabbit populations. Rev Sci Tech Off Int Epizoot 21:347–358

Corner LA, Buddle BM (2005) Conjunctival vaccination of the brushtail possum (*Trichosurus vulpecula*) with bacille Calmette-Guerin. NZ Vet J 53:133–136

Corner LA, Barrett RH, Lepper AWD, Lewis V, Pearson CW (1981) A survey of mycobacteriosis of feral pigs in the Northern Territory. Aust Vet J 57:537–542

Corner LAL, Norton S, Buddle BM, Morris RS (2002) The efficacy of bacille Calmette-Guérin vaccine in wild brushtail possums (*Trichosurus vulpecula*). Res Vet Sci 73:145–152

Corner LAL, Pfeiffer DU, Morris RS (2003) Social-network analysis of *Mycobacterium bovis* transmission among captive brushtail possums (*Trichosurus vulpecula*). Prev Vet Med 59:147–167

Corner LAL, Costello E, Lesellier S, O'Meara D, Sleeman DP, Gormley E (2007) Experimental tuberculosis in the European badger (*Meles meles*) after endobronchial inoculation of *Mycobacterium bovis*: I. Pathology and bacteriology. Res Vet Sci 83:53–62

Côté IM, Poulin R (1995) Parasitism and group-size in social animals – a metaanalysis. Behav Ecol 6:159–165

Couacy-Hymann E, Bodjo C, Danho T, Libeau G, Diallo A (2005) Surveillance of wildlife as a tool for monitoring rinderpest and peste des petits ruminants in West Africa. Rev Sci Tech Off Int Epizoot 24:869–877

Coulon P, Lafay F, Leblois H, Tuffereau C, Artois M, Blancou J, Benmansour A, Flamand A (1992) The SAG: a new attenuated oral rabies vaccine. In: Bögel K, Meslin FX, Kaplan M (eds) Wildlife rabies control. Wells Medical, Tunbridge Wells, UK, pp. 105–111

Courchamp F, Say L, Pontier D (2000) Detection, identification, and correction of a bias in an epidemiological study. J Wildl Dis 36:71–78

Cousins DV, Roberts JL (2001) Australia's campaign to eradicate bovine tuberculosis: the battle for freedom and beyond. Tuberculosis 81:5–15

Cowan D, Dunsford G, Gill G, Jones E, Kerins G, MacNicoll A, Quy R (1995) The impact of resistance on the use of second-generation anticoagulants against rats on farms in southern England. Pesticide Sci 43:83–93

Cowan DP, Massei G, Mellows RJB (2006) A modeling approach to evaluating potential applications of emerging fertility control technologies in the UK. In: Timm RM, O'Brien JM (eds) Proceedings of the 22nd Vertebrate Pest Conference, 6–9th March, Berkely, California, pp. 55–62

Cowan PE (1993) Effects of intensive trapping on breeding and age structure of brushtail possums, *Trichosurus vulpecula*, on Kapiti Island, New Zealand. NZ J Zool 20:1–11

Cox DR (1972) Regression models and life-tables (with discussion). J R Stat Soc Ser B 34:187–220

Cranfield MJ, Minnis R (2007) An integrated health approach to the conservation of mountain gorillas *Gorilla beringei beringei*. Int Zoo Yearbook 41:110–121

Cranfield MJ, Gaffikin L, Cameron K (2001) Conservation medicine as it applies to the mountain gorilla. The Apes: challenges for the 21st Century; Brookfield, Brookfield Zoo, pp. 238–240

Cranfield MR, Gaffikin L, Whittier C, Nutter F, Minnis R, Lowenstine LJ (2005) The Mountain Gorilla Veterinary Project "Conservation Medicine". 56th Annual Meeting of the American College of Veterinary Pathologists and 40th Annual Meeting of the American Society for Veterinary Clinical Pathology. International Veterinary Information Service, Ithaca, NY (www.ivis.org)

Crawley MJ (2005) Statistics: an introduction using R. Wiley, Chichester

Creekmore TE, Rocke TE, Hurley J (2002) A baiting system for delivery of an oral plague vaccine to black-tailed prairie dogs. J Wildl Dis 38:32–39

Cross ML, Buddle BM, Aldwell FE (2007a) The potential of oral vaccines for disease control in wildlife species. Vet J 174:472–480

Cross PC (2005) Disease invasion and control in structured populations: bovine tuberculosis in the buffalo population of the Kruger National Park [PhD thesis]. University of California, Berkeley

Cross PC, Lloyd-Smith JO, Bowers J, Hay C, Hofmeyr M, Getz WM (2004) Integrating association data and disease dynamics in a social ungulate: bovine tuberculosis in African buffalo in the Kruger National Park. Ann Zool Fenn 41:879–892

Cross PC, Lloyd-Smith JO, Johnson PLF, Getz WM (2005) Duelling timescales of host movement and disease recovery determine invasion of disease in structured populations. Ecol Lett 8:587–595
Cross PC, Johnson PLF, Lloyd-Smith JO, Getz WM (2007b) Utility of R0 as a predictor of disease invasion in structured populations. J R Soc Interface 4:315–324
Cross PC, Edwards WH, Scurlock BM, Maichak EJ, Rogerson JD (2007c) Effects of management and climate on elk brucellosis in the Greater Yellowstone Ecosystem. Ecol Appl 17:957–964
Cunningham AA, Buxton D, Thomson KM (1992) An epidemic of toxoplasmosis in a captive colony of squirrel monkeys (*Saimiri sciureus*). J Comp Pathol 107:207–219
Cunningham AA (1996) Disease risks of wildlife translocations. Conserv Biol 10:349–353
Cunningham AA (2005) A walk on the wild side – emerging wildlife diseases. Br Med J 331:1214–1215
Cunningham AA, Dazsak P (1998) Extinction of a species of land snail due to infection with a microsporidian parasite. Conserv Biol 12:1139–1141
Curtis PD, Pooler RL, Richmond ME, Miller LA, Mattfeld GF, Quimby FW (2002) Comparative effects of GnRH and porcine zona pellucida (PZP) immunocontraceptive vaccines for controlling reproduction in white-tailed deer (*Odocoileus virginianus*). Reprod Suppl 60:131–141
Daels PF, Hughes JP (1995) Fertility control using intrauterine devices: an alternative for population control in wild horses. Theriogenology 44:629–639
Dalin AM, Andresen O, Malmgren L (2002) Immunization against GnRH in mature mares: antibody titres, ovarian function, hormonal levels and oestrus behaviour. J Vet Med A 49:125–131
Dalley D, Chambers MA, Cockle P, Pressling W, Gavier-Widen D, Hewinson RG (1999) A lymphocyte transformation assay for the detection of *Mycobacterium bovis* infection in the Eurasian Badger (*Meles meles*). Vet Immunol Immunopathol 70:85–94
Dalley D, Davé D, Lesellier S, Palmer S, Crawshaw T, Hewinson RG, Chambers M (2008) Development and evaluation of a gamma-interferon assay for tuberculosis in badgers (*Meles meles*). Tuberculosis 88:235–243
Daniels MJ, Hutchings MR, Greig A (2003a) The risk of disease transmission to livestock posed by contamination of farm stored feed by wildlife excreta. Epidemiol Infect 130:561–568
Daniels MJ, Hutchings MR, Beard PM, Henderson D, Greig A, Stevenson K, Sharp JM (2003b) Do non-ruminant wildlife pose a risk of paratuberculosis to domestic livestock and vice versa in Scotland? J Wildl Dis 39:10–15
Daszak P, Cunningham AA (1999) Extinction by infection. Trends Ecol Evol 14:279
Daszak P, Cunningham AA, Hyatt AD (2000a) Wildlife ecology – emerging infectious diseases of wildlife – threats to biodiversity and human health. Science 287:443–449
Daszak P, Cunningham AA, Hyatt AD (2000b) Emerging infectious diseases of wildlife – threats to biodiversity and human health. Science 287:443–449
Daszak P, Kilpatrick AM (2008) Pathogens, disease and ecosystem services. In: Kinzig A (ed) Princeton Guide to Ecology VI: Ecosystem Services. Princeton University Press, Princeton, New Jersey, USA. In press
Davidson WR, Nettles VF (1992) Relocation of wildlife: identifying and evaluating disease risks. Trans N Am Wildl and Nat Resour Conf, pp. 466–473
Davidson R, Marion G, White PCL, Hutchings MR (2008) Use of host population reduction to control wildlife disease: rabbits and paratuberculosis. Epidemiol Infect:doi:10.1017/S0950268808000642
Davies CR, Ayres JM, Dye C, Deane LM (1991) Malaria infection rate of Amazonian primates increases with body weight and group size. Funct Ecol 5:655–662
de Castro F, Bolker B (2004) Mechanisms of disease-induced extinction. Ecol Lett 8:117–126
De Jong MCM, Diekmann O, Heesterbeek H (1995) How does transmission of infection depend on population size? In: Mollison D (ed) Epidemic models: their structure and relation to data. Cambridge University Press, Cambridge, pp. 84–94
De Leo G, Dobson A (2005) Virulence management in wildlife populations. In: Dieckman U, Metz JAJ, Sabelis MW, Sigmund K (eds) Adaptive dynamics of infectious diseases: in pursuit of virulence anagement. Cambridge University Press, Cambridge, pp. 26–38

de Lisle GW, Mackintosh CG, Bengis RG (2001) *Mycobacterium bovis* in free-living and captive wildlife, including farmed deer. Rev Sci Tech Off Int Epizoot 20:86–111

de Lisle GW, Bengis RG, Schmitt SM, O'Brien DJ (2002) Tuberculosis in free-ranging wildlife: detection, diagnosis and management. Rev Sci Tech Off Int Epizoot 21:317–334

De Villiers MS, Meltzer DGA, Van Heerden J, Mills MGL, Richardson PRK, Van Jaarsveld AS (1995) Handling-induced stress and mortalities in African wild dogs (*Lycaon pictus*). Proc R Soc B 262:215–220

De Vos V, Bryden HB (1996) Anthrax in the Kruger National Park: temporal and spatial patterns of disease occurrence. Salisbury Med Bull 87 (Suppl):26–30

De Vos V, Rooyen G, Kloppers J (1973) Anthrax immunization of free-ranging roan antelope *Hippotragus equines* in the Kruger National park. Koede 16:11–25

De Vos V, Bengis RG, Kriek NPJ, Michel A, Keet DF, Raath JP, Huchzermeyer HF (2001) The epidemiology of tuberculosis in free ranging African buffalo (*Syncerus caffer*) in the Kruger National Park, South Africa. Onderstepoort J Vet Res 68:119–130

Debbie JG (1991) Rabies control of terrestrial wildlife by population reduction. In: Baer GM (ed) The natural history of rabies. 2nd edn. CRC Press, Boston, pp. 477–484

Delahay RJ, Langton S, Smith GC, Clifton-Hadley RS, Cheeseman CL (2000a) The spatio-temporal distribution of *Mycobacterium bovis* (bovine tuberculosis) infection in a high density badger population. J Anim Ecol 69:428–441

Delahay RJ, Brown JA, Mallinson PJ, Spyvee PD, Handoll D, Rogers LM, Cheeseman CL (2000b) The use of marked bait in studies of the territorial organisation of the European badger (*Meles meles*). Mammal Rev 30:73–87

Delibes M, Rodríguez A, Ferreras P (2000) Action plan for the conservation of the Iberian Lynx (*Lynx pardinus*) in Europe. Convention on the conservation of European wildlife and natural habitats. Council of Europe, Strasbourg

Delsink AK, Van Altena JJ, Grobler D, Bertschinger H, Kirkpatrick JF, Slotow R (2007) Implementing immunocontraception in free-ranging African elephants at Makalali Conservancy. S Afr Vet J 78:25–30

Delves PJ, Lund T, Roitt IV (2002) Antifertility vaccines. Trends Immunol 23:213–219

Demiris N, O'Neill PD (2005) Bayesian inference for stochastic multitype epidemics in structured populations via random graphs. J R Stat Soc Ser B 67:731–745

Dhabhar FS, McEwen BS (1999) Enhancing versus suppressive effects of stress hormones on skin immune function. Proc Natl Acad Sci USA 96:1059–1064

Diekmann O, Heesterbeek JAP (2000) Mathematical epidemiology of infectious diseases: model building, analysis, and interpretation. Wiley, Chichester

Dobson A (2004) Population dynamics of pathogens with multiple host species. Am Nat 164:S64–S78

Dobson A, Lyles A (2000) Black-footed ferret recovery. Science 288:985–988

Dobson A, Foufopoulos J (2001) Emerging infectious pathogens of wildlife. Phil Trans R Soc Lond B 356:1001–1012

Doherr MG, Audigé L (2001) Monitoring and surveillance for rare health-related events: a review from the veterinary perspective. Phil Trans R Soc Lond B 356:1097–1106

Dolan L (1993) Badgers and bovine tuberculosis in Ireland: a review. In: Hayden TJ (ed) The Badger. Royal Irish Academy, Dublin, pp. 108–116

Donnelly CA, Woodroffe R, Cox DR, Bourne J, Gettinby G, Le Fevre AM, McInerney JP, Morrison WI (2003) Impact of localized badger culling on tuberculosis incidence in British cattle. Nature 426:834–837

Donnelly CA, Wei G, Johnston WT, Cox DR, Woodroffe R, Bourne FJ, Cheeseman CL, Clifton-Hadley RS, Gettinby G, Gilks P, Jenkins HE, Le Fevre AM, McInerney JP, Morrison WI (2007) Impacts of widespread badger culling on cattle tuberculosis: concluding analyses from a large-scale field trial. Int J Infect Dis 11:300–308

Drewe JA, Dean GS, Michel AL, Pearce GP (2009) Accuracy of three diagnostic tests for determining *Mycobacterium bovis* infection status in live-sampled wild meerkats. J Vet Diagn Invest 21:in press

Duff JP, Whitwell K, Chasey D (1996) The emergence and epidemiology of European brown hare syndrome in the UK. In: Chasey D, Gaskell RM, Clarke IN (eds) Proceedings of a European Society for Veterinary Virology Meeting, Reading, United Kingdom, pp. 1

Dunbar BS, Kaul G, Prasad M, Skinner SM (2002) Molecular approaches for the evaluation of immune response to zona pellucida (ZP) and development of second generation ZP vaccines. Reprod Suppl 60:9–18

Dunnet GM, Jones DM, McInerney JP (1986) Badgers and bovine tuberculosis: review of policy. HMSO, London, UK

Durrett R (2007) Random graph dynamics. Cambridge University Press, Cambridge

Dutton A, Edwards-Jones G, Strachan R, Macdonald DW (2008) Ecological and social challenges to biodiversity conservation on farmland: reconnecting habitats on a landscape scale. Mammal Rev 38:205–219

Eason CT, Wickstorm M, Turck P, Wright GRG (1999) A review of recent regulatory and environmental toxicology studies on 1080: results and implications. NZ J Ecol 23:129–137

Eckert J, Gemmell MA, Meslin FX, Pawlowski ZS (eds) (2001) WHO/OIE Manual on Echinococcosis in Humans and Animals: a Public Health Problem of Global Concern. WHO/WOAH Ed, Paris (OIE), 265 p

EDEN (2008) http://www.eden-fp6project.net/

Edmonds MD, Samartino LE, Hoyt PG, Hagius SD, Walker JV, Enright FM, Schurig GG, Elzer P (2001) Oral vaccination of sexually mature pigs with Brucella abortus vaccine strain RB51. Am J Vet Res 62:1328–1331

EDQM (2008) www.edqm.eu

Edwards-Jones G (2006) Modelling farmer decision making: concepts, challenges and progress. Anim Sci 82:783–790

Edwards-Jones G, Deary I, Willock J (1998) Modelling farmer decision-making: what can psychology do for agricultural policy assessment models? Etud Res Syst Agraires Develop 31:153–173

Eisen RJ, Petersen JM, Higgins MS, Wong D, Levy CE, Mead PS, Schriefer ME, Griffith KS, Gage KL, Beard CB (2008) Persistence of *Yersinia pestis* in soil under natural conditions. Emerg Infect Dis 14:941–943

Eisinger D, Thulke HH (2008) Spatial pattern formation facilitates eradication of infectious diseases. J Appl Ecol 45:415–423

EMEA (2000) http://www.emea.europa.eu/pdfs/vet/vich/059598en.pdf

Epstein PR (2001) Climate change and emerging infectious diseases. Microb Infect 3:747–754

EU (2008) http://ec.europa.eu/food/animal/diseases/laboratories/index_en.htm.

European Commission (2002) The oral vaccination of foxes against rabies. Report of the Scientific Committee on Animal Health and Animal Welfare

Evermann JF, Heeney JL, Roelke ME, McKeirnan AJ, O'Brien SJ (1988) Biological and pathological consequences of feline infectious peritonitis virus infection in the cheetah. Arch Virol 102:155–171

EWDA (2008) http://www.ewda.org/

Exon JH, Talcott PA, Koller LD (1985) Effect of lead, polychlorinated biphenyls, and cyclophosphamide on rat natural killer cells, interleukin 2, and antibody synthesis. Fund Appl Toxicol 5:158–164

Ezenwa VO (2004) Selective defecation and selective foraging: antiparasite behavior in wild ungulates? Ethology 110:851–862

Fagerstone KA, Miller LA, Bynum KS, Eiseman JD, C. Y. (2006) When, where and for what wildlife species will contraception be a useful management approach? In: Timm RM, O'Brien JM (eds) Proceedings of the 22nd Vertebrate Pest Conference Berkley, California. Universtiy of California, Davis, pp. 45–54

Farrington CP, Kanaan MN, Gay NJ (2001) Estimation of the basic reproduction number for infectious diseases from age-stratified serological survey data. Appl Stat 50:251–292

Faustino CR, Jennelle CS, Connolly V, Davis AK, Swarthout EC, Dhondt AA, Cooch EG (2004) *Mycoplasma gallisepticum* infection dynamics in a House finch population: seasonal variation in survival, encounter and transmission rate. J Anim Ecol 73:651–669

Favia G, Cancrini G, Ferroglio E, Casiraghi M, Ricci I, Rossi L (2003) Molecular assay for the identification of *Setaria tundra*. Vet Parasitol 117:139–145

Fayrer-Hosken RA, Grobler D, Van Altena JJ, Kirkpatrick JF, Bertschinger H (2000) Immunocontraception of African elephants. Nature 407:149

Fearneyhough M, Wilson PJ, Clark KA, Smith DR, Johnston DH, Hicks BN, Moore GM (1998) Results of an oral vaccination program for coyotes. J Am Vet Med Assoc 212:498–502

Fehlner-Gardiner C, Nadin-Davis S, Armstrong J, Muldoon F, Bachmann P, Wandeler A (2008) ERA vaccine-derived cases of rabies in wildlife and domestic animals in Ontario, Canada, 1989–2004. J Wildl Dis 44:71–85

Fenner F (2002) Viruses, rabbits and wildlife. In: Saunders D, Spratt D, VanWensveen M (eds) Perspectives on wildlife research: celebrating 50 years of CSIRO wildlife and ecology. Surrey Beatty & Sons, NSW, Australia, pp. 1–7

Ferguson NM, Donnelly CA, Anderson RM (2001a) Transmission intensity and impact of control policies on the foot and mouth epidemic in Great Britain. Nature 413:542–548

Ferguson NM, Donnelly CA, Anderson RM (2001b) The Foot-and-Mouth epidemic in Great Britain: pattern of spread and impact of interventions. Science 292:1155–1160

Ferrari MJ, Bansal S, Meyers LA, Bjornstad ON (2006) Network frailty and the geometry of herd immunity. Proc R Soc B 273:2743–2748

Ferris S, Robbins MM, Williamson EA (2005) Eastern Gorilla (*Gorilla beringei*). In: Caldecott J, Miles L (eds) World atlas of great apes and their conservation. University of California Press, Berkeley, USA, pp. 129–152

Ferroglio E, Ragagli C (2008) *Physaloptera sibirica* in foxes from the North Western Alps (Italy). Parassitologia 50: 58

Ferroglio E, Rossi L (2001) Prevalence of *Neospora caninum* antibodies in wild ruminants from the Italian Alps. Vet Rec 148:754–755

Ferroglio E, Trisciuoglio A (2003) Antibodies to *Neospora caninum* in European brown hares. Vet Parasitol 115:75–78

Ferroglio E, Tolari F, Bollo E, Bassano B (1998) Isolation of *Brucella melitensis* from Alpine ibex. J Wildl Dis 34:400–402

Ferroglio E, Nebbia P, Robino P, Rossi L, Rosati S (2000) *Mycobacterium paratuberculosis* infection in two free-ranging Alpine ibex. Rev Sci Tech Off Int Epizoot 19:859–862

Ferroglio E, Bassano B, Trisciuoglio A, Rossi L (2001) Antibodies to *Neospora caninum* in Alpine ibex from the Italian Alps. Z Jagdwiss 47:226–228

Ferroglio E, Gennero S, Rossi L, Tolari E (2003) Monitoraggio di un focolaio di brucellosi nel camoscio Alpino. J Mt Ecol 7 (Suppl):229–232

Ferroglio E, Pasino M, Romano A, Grande D, Pregel P, Trisciuoglio A (2007) Evidence of *Neospora caninum* DNA in wild rodents. Vet Parasitol 148:346–349

Flanagan P (1993) In: Hayden TJ (ed) Badgers and bovine tuberculosis: epidemiology and case studies. Royal Irish Academy, Dublin

Foreyt WJ (1989) Fatal *Pasteurella haemolytica* pneumonia in bighorn sheep after direct contact with clinically normal domestic sheep. Am J Vet Res 50:341–344

Fornes A, Lord RD, Kuns ML, Larghi OP, Fuenzalida E, Lazara L (1974) Control of bovine rabies through vampire bat control. J Wildl Dis 10:310–316

Fraker MA, Brown RG, Gaunt GE, Kerr JA, Pohajdak B (2002) Long-lasting, single-dose immunocontraception of feral fallow deer in British Columbia. J Wildl Manage 66:1141–1147

Frank O, Straus D (1986) Markov graphs. J Am Stat Assoc 81:832–842

Fulford GR, Roberts MG, Heesterbeek JAP (2002) The metapopulation dynamics of an infectious disease: tuberculosis in possums. Theor Popul Biol 61:15–29

Fuller WA (2002) Canada and the "Buffalo", *Bison bison*: a tale of two herds. Can Field Nat 116:141–159

Gallagher J, Macadam I, Sayer J, Van Lavieren LP (1972) Pulmonary tuberculosis in free living Lechwe antelope in Zambia. Trop Anim Health Prod 4:204–213

Gallivan GJ, Barker IK, Artsob H, Magnarelli LA, Robinson JT, Voigt DR (1998) Serologic survey for antibodies to *Borrelia burgdorferi* in white-tailed deer in Ontario. J Wildl Dis 34:411–414

Garnett BT, Delahay RJ, Roper TJ (2002) Use of cattle farm resources by badgers (*Meles meles*) and risk of bovine tuberculosis (*Mycobacterium bovis*) transmission to cattle. Proc R Soc B 269:1487–1491

Genovesi P (2005) Eradications of invasive alien species in Europe: a review. Biol Invasions 7:127–133

Gionfriddo JP, Sullivan KJ, Healey RS, Miller LA, Fagerstone KA (2008) Field test of a single-injection GnRH immunocontraceptive vaccine on female white-tailed deer. Reprod Fert Dev:in press

Gippoliti S, Carpaneto GM (1997) Captive breeding, zoos and good sense. Conserv Biol 11:806–807

Godfroid J (2002) Brucellosis in wildlife. Rev Sci Tech Off Int Epizoot 21:277–286

Gog J, Woodroffe R, Swinton J (2002) Disease in endangered metapopulations: the importance of alternative hosts. Proc R Soc B 269:671–676

Gollier C (2001) The economics of risk and time. MIT Press, USA

Goltsman M, Kruchenkova EP, Macdonald DW (1996) The Mednyi arctic foxes: treating a population imperilled by disease. Oryx 30:251–258

Gomes-Solecki MJ, Brisson DR, Dattwyler RJ (2006) Oral vaccine that breaks the transmission cycle of the Lyme disease spirochete can be delivered via bait. Vaccine 24:4440–4449

Goodger J, Nolan A, Russell WP, Dalley DJ, Thomas CJ, Stuart FA, Croston P, Newall DG (1994) Serodiagnosis of *Mycobacterium bovis* infection in badgers – development of an indirect ELISA using a 25kD antigen. Vet Rec 135:82–85

Gordon ER, Krebs JW, Rupprecht CR, Real LA, Childs JE (2005) Persistence of elevated rabies prevention costs following post-epizootic declines in rates of rabies among raccoons (*Procyon lotor*). Prev Vet Med 68:195–222

Gortazar C, Acevedo P, Ruiz-Fons F, Vicente J (2006) Disease risks and overabundance of game species. Eur J Wildl Res 52:81–87

Gottstein B, Saucy F, Deplazes P, Reichen J, Demierre G, Busato A, Zuercher C, Pugin P (2001) Is high prevalence of *Echinococcus multilocularis* in wild and domestic animals assocaited with disease incidence in humans? Emerg Infect Dis 7:408–412

Graczyk TK, Mudakikwa AB, Cranfield MR, Eilenberger U (2001) Hyperkeratotic mange caused by *Sarcoptes scabiei* (Acariformes: Sarcoptidae) in juvenile human-habituated mountain gorillas (*Gorilla gorilla beringei*). Parasitol Res 87:1024–1028

Graczyk TK, Nizeyi JB, Ssebide B, Andrew Thompson RC, Read C, Cranfield MR (2002a) Anthropozoonotic *Giardia duodenalis* Genotype (Assemblage) A infections in habitats of free-ranging human-habituated gorillas, Uganda. J Parasitol 88:905–909

Graczyk TK, Nizeyi JB, DaSilva AJ, Moura INS, Pieniazek NJ, Cranfield MR, Lindquist HDA (2002b) A single genotype of *Encephalitozoon intestinalis* infects free-ranging gorillas and people sharing their habitats in Uganda. Parasitol Res 88:926–931

Grattarola C, Gennero MS, Ferroglio E, Chiavacci L, Bergagna S, Zoppi S, Dondo A (2006) Epidemiological study of *Brucella* infection in wild boar in la Mandria Park (Italy). VII European Wildlife Disease Association Congress, 27–30 September, Saint Vincent

Grear DA (2006) Chronic wasting disease infection patterns in female white-tailed deer related to demographics, genetic relationships, and spatial proximity of infected deer in southern Wisconsin. Masters Thesis, University of Wisconsin.

Grear DA, Samuel MD, Langenberg JA, Keane D (2006) Demographic patterns and harvest vulnerability of chronic wasting disease infected white-tailed deer in Wisconsin. J Wildl Manage 70:546–553

Green RE, Taggart MA, Das D, Pain DJ, Sashi KC, Cunningham AA, Cuthbert R (2006) Collapse of Asian vulture populations: risk of mortality from residues of the veterinary drug diclofenac in carcasses of treated cattle. J Appl Ecol 43:949–956

Green WQ, Coleman JD (1984) Response of a brush-tailed possum population to intensive trapping. NZ J Zool 11:319–328

Greenwald R, Esfandiari J, Lesellier S, Houghton R, Pollock J, Aagaard C, Andersen P, Hewinson RG, Chambers M, Lyashchenko K (2003) Improved serodetection of *Mycobacterium bovis* infection in badgers (*Meles meles*) using multiantigen test formats. Diagn Microbiol Infect Dis 46:197–203

Grenfell BT, Anderson RM (1985) The estimation of age related rates of infection from case notifications and serological data. J Hyg Camb 95:419–436

Greth A, Gourreau JM, Vassart M, Nguyen Ba V, Wyers M, LeFevre PC (1992) Capripoxvirus disease in an Arabian oryx (*Oryx leucoryx*) from Saudi Arabia. J Wildl Dis 28:295–300

Greth A, Flamand JR, Delhomme A (1994) An outbreak of tuberculosis in a captive herd of Arabian oryx (*Oryx leucoryx*): management. Vet Rec 134:165–167

Griffin JM, Williams DH, Kelly GE, Clegg TA, O'Boyle I, Collins JD, More SJ (2005) The impact of badger removal on the control of tuberculosis in cattle herds in Ireland. Prev Vet Med 67:237–266

Griffith B, Scott JM, Carpenter JW, Reed C (1993) Animal translocations and potential disease transmission. J Zoo Wildl Med 24:231–236

Griot C, Vandevelde M, Schobesberger M, Zurbriggen A (2003) Canine distemper, a re-emerging morbillivirus with complex neuropathogenic mechanisms. Anim Health Res Rev 4:1–10

Grosenbaugh DA, Maki JL, Rupprecht CE, Wall DK (2007) Rabies challenge of captive striped skunks (*Mephitis mephitis*) following oral administration of a live vaccinia-vectored rabies vaccine. J Wildl Dis 43:124–128

Gross JE, Miller MW (2001) Chronic wasting disease in mule deer: disease dynamics and control. J Wildl Manage 65:205–215

Guarner J, Johnson BJ, Paddock CD, Shieh WJ, Goldsmith CS, Reynolds MG, Damon IK, Regnery RL, Zaki SR, The Veterinary Monkeypox Virus Working Group (2004) Monkeypox transmission and pathogenesis in prairie dogs. Emerg Infect Dis 10:426–431

Guberti V, Rutili D, Ferrari G, Patta C, Oggiano A (1998) Estimate the threshold abundance for the persistence of the classical swine fever in the wild boar population of the eastern Sardinia. In: Rutili D (ed) Measures to control classical swine fever in European wild boar. Perugia, Italy, European Union, pp. 54–61

Guilbride PDL, Rollinson DHL, McAnulty EG, Alley JG, Wells EA (1963) Tuberculosis in the free living African (Cape) buffalo (*Syncerus caffer* Sparrman). J Comp Pathol 73:337–348

Gunson JR, Dorward WJ, Schowalter DB (1978) An evaluation of rabies control in Alberta. Can Vet J 19:214–220

Gurnell J, Rushton SP, Lurz PWW, Sainsbury AW, Nettleton P, Shirley MDF, Bruemmer C, Geddes N (2006) Squirrel poxvirus: landscape scale strategies for managing disease threat. Biol Conserv 131:287–295

Haigh JC (1988) Translocation and disease considerations. Proceedings of the American Association of Zoo Veterinarians, Ontario, Canada, pp. 119–120

Handcock MS, Jones JH (2004) Likelihood-based inference for stochastic models of sexual network formation. Theor Popul Biol 65:413–422

Hanlon CA, Niezgoda M, Morrill P, Rupprecht CE (2002) Oral efficacy of an attenuated rabies virus vaccine in skunks and raccoons. J Wildl Dis 38:420–427

Hansen F, Tackmann K, Jeltsch F, Wissel C, Thulke HH (2003) Controlling *Echinococcus multilocularis* – ecological implications of field trials. Prev Vet Med 60:91–105

Harris S (1981) An estimation of the number of foxes (*Vulpes vulpes*) in the city of Bristol, and some possible factors affecting their distribution. J Appl Ecol 18:455–465

Harris S, Morris P, Wray S, Yalden D (1995) A review of British mammals: population estimates and conservation status of British mammals other than cetaceans. Joint Nature Conservation Committee, Peterborough

Harris SL, Brookes SM, Jones G, Hutson AM, Racey P, Aegerter J, Smith GC, McElhinney LM, Fooks AR (2006) European bat lyssaviruses: distribution, prevalence and implications for conservation. Biol Conserv 131:193–210

Hars J, Gauthier D (1984) Suivi de l'évolution de la kératoconjonctivite sur le peuplement d'ongulés sauvages du Parc National de la Vanoise (département de la Savoie) en 1983. Travaux Sci Parc Natl Vanoise 15

Hashimoto S, Murakami Y, Taniguchi K, Nagai M (2000) Detection of epidemics in their early stage through infectious disease surveillance. Int J Epidemiol 29:905–910

Hastings BE, Kenny D, Lowenstine LJ, Foster JW (1991) Mountain gorillas and measles: ontogeny of a wildlife vaccination program. In: Junge RE (ed) Proceedings of the Annual Meeting of the American Association of Zoo Veterinarians, pp. 198–205

Haydon DT, Laurenson MK, Sillero-Zubiri C (2002a) Integrating epidemiology into population viability analysis: managing the risk posed by rabies and canine distemper to the Ethiopian wolf. Conserv Biol 16:1372–1385

Haydon DT, Cleaveland S, Taylor LH, Laurenson MK (2002b) Identifying reservoirs of infection: a conceptual and practical challenge. Emerg Infect Dis 8:1468–1473

Haydon DT, Randall DA, Matthews L, Knobel DL, Tallents LA, Gravenor MB, Williams SD, Pollinger JP, Cleaveland S, Woolhouse MEJ, Sillero-Zubiri C, Marino J, Macdonald DW, Laurenson MK (2006) Low-coverage vaccination strategies for the conservation of endangered species. Nature 443:692–695

Heesterbeek JAP, Roberts MG (1995) Mathematical models for microparasites of wildlife. In: Grenfell BT, Dobson AP (eds) Ecology of infectious diseases in natural populations. Cambridge University Press, Cambridge, UK, pp. 91–122

Hegglin D, Ward PI, Deplazes P (2003) Anthelmintic baiting of foxes against urban contamination with *Echinococcus multilocularis*. Emerg Infect Dis 9:1266–1272

Hegglin D, Bontadina F, Gloor S, Romer J, Müller U, Breitenmoser U, Deplazes P (2004) Baiting red foxes in an urban area: a camera trap study. J Wildl Manage 68:1010–1017

Heisey DM, Joly DO, Messier F (2006) The fitting of general force-of-infection models to wildlife disease prevalence data. Ecology 87:2356–2365

Henning KJ (2004) Overview of syndromic surveillance. What is syndromic surveillance? Morbidity Mortality Wkly Rep Surveillance Summaries 53 (Suppl):5–11

Hess G (1996a) Disease in metapopulation models: implications for conservation. Ecology 77:1617–1632

Hess GR (1996b) Linking extinction to connectivity and habitat destruction in metapopulation models. Am Nat 148:226–236

Hilliam RP, Ozkan N (1986) Comparison of local and systemic immunity after intratracheal, intraperitoneal, and intravenous immunization of mice exposed to aerosolised or ingested lead. Environ Res 39:265–277

Hobbs NT, Bowden DC, Baker DL (2000) Effects of fertility control on populations of ungulates: general, stage-structured models. J Wildl Manage 64:473–491

Hofmeyr M, Hofmeyr D, Nel L, Bingham J (2004) A second outbreak of rabies in African wild dogs (*Lycaon pictus*) in Madikwe Game Reserve, South Africa, demonstrating the efficacy of vaccination against natural rabies challenge. Anim Conserv 7:193–198

Holland EP, Aegerter J, Dytham C, Smith GC (2007) Landscape as a model: the importance of geometry. PLoS Computational Biol 3:e200, doi:210.1371/journal.pcbi.0030200

Holmala K, Kauhala K (2006) Ecology of wildlife rabies in Europe. Mammal Rev 36:17–36

Homsey J (1999) Ape tourism and human diseases: how close should we get? Report for the International Gorilla Conservation Programme (IGCP) Regional Meeting, Rwanda

Hone J (1992) Rate of increase and fertility control. J Appl Ecol 29:695–698

Hone J (1995) Analysis of vertebrate pest control. Cambridge University Press, Cambridge

Hone J, Pech R, Yip P (1992) Estimation of the dynamics and rate of transmission of classical swine fever (hog cholera) in wild pigs. Epidemiol Infect 108:377–386

Horan AD, Wolf CA (2005) The economics of managing infectious wildlife disease. Am J Agric Econ 87:537–551

Hounsome TD, Young RP, Davison J, Yarnell RW, Trewby ID, Garnett BT, Delahay RJ, Wilson GJ (2005) An evaluation of distance sampling to estimate badger (*Meles meles*) abundance. J Zool 266:81–87

Howald G, Donlan CJ, Galvan JP, Russell JC, Parkes J, Samaniego A, Wang Y, Veitch D, Genovesi P, Pascal M, Saunders A, Tershy B (2007) Invasive rodent eradication on islands. Conserv Biol 21:1258–1268

Hudson PJ, Greenman J (1998) Competition mediated by parasites: biological and theoretical progress. Trends Ecol Evol 13:387–390

Hugh-Jones ME, de Vos V (2002) Anthrax and wildlife. Rev Sci Tech Off Int Epizoot 21:359–383

Hunt RJ, Dall DJ, Lapidge SJ (2007) Effect of a synthetic lure on site visitation and bait uptake by foxes (*Vulpes vulpes*) and wild dogs (*Canis lupus dingo*, *Canis lupus familiaris*). Wildl Res 34:461–466

Hutchings MR, Harris S (1997) Effects of farm management practices on cattle grazing behaviour and the potential for transmission of bovine tuberculosis from badgers to cattle. Vet J 153:149–162
Hutchings MR, Athanasiadou S, Kyriazakis IG, I.J (2003) Can animals use foraging behaviour to combat parasites? Proc Nutrition Soc 62:361–370
Imboden I, Janett F, Burger D, Crowe MA, Hassig M, Thun R (2006) Influence of immunization against GnRH on reproductive cyclicity and estrous behavior in the mare. Theriogenology 66:1866–1875
Inayatullah C (1973) Wild boar in West Pakistan. Peshwar, Pakistan, Forest Institute Bulletin No. 1
Independent Scientific Group (2007) Bovine TB: the scientific evidence. Defra Publications, London, UK
Irsara A, Ruatti A, Gagliardi G, Orfei Z, Bellani L, Mantouani A (1982) Control of wildlife rabies in north-eastern Italy. Comp Immunol Microbiol Infect Dis 5:327–335
Isham V (1989) Estimation of the incidence of HIV infection. Phil Trans R Soc Lond B 325:113–121
IUCN (2007) http://www.iucnredlist.org
Jackson DB, Green RE (2000) The importance of the introduced hedgehog (*Erinaceus europaeus*) as a predator of the eggs of waders (*Charadrii*) on machair in South Uist, Scotland. Biol Conserv 93:333–348
Jackson M, White N, Giffard P, Timms P (1999) Epizootiology of *Chlamydia* infections in two free-range koala populations. Vet Microbiol 65:225–234
Jackson WB (1998) Ecology of pest rodents in the urban environment. In: Ambasht RS (ed) Modern trends in ecology and environment. Backhuys, Leiden, The Netherlands, pp. 13
Jenkins HE, Woodroffe R, Donnelly CA (2008) The effects of annual widespread badger culls on cattle tuberculosis following the cessation of culling. Int J Infect Dis: 12:457–465
Jenkins HE, Woodroffe R, Donnelly CA, Cox DR, Johnston WT, Bourne FJ, Cheeseman CL, Clifton-Hadley RS, Gettinby G, Gilks P, Hewinson RG, McInerney JP, Morrison WI (2007) Effects of culling on spatial associations of *Mycobacterium bovis* infections in badgers and cattle. J Appl Ecol 44:879–908
Jessup DA, Boyce WM, Clarke RK (1991) Diseases shared by wild, exotic and domestic sheep. In: Renecker LA, Hudson RJ (eds) Wildlife production: conservation and sustainable development. University of Alaska, Fairbanks, Alaska, pp. 438–445
Jessup DA, Boyce WM, Torres SG (1995) Bighorn sheep health management in California: a fifteen year retrospective. In: Junge RE (ed) Proceedings of the Joint Conference of the American Association of Zoo Veterinarians, the Wildlife Disease Association and the American Association of Wildlife Veterinarians, pp. 55–67
Jezek Z, Szczeniowski MY, Muyembe-Tamfum JJ, McCormick JB, Heymann DL (1999) Ebola between outbreaks: intensified Ebola hemorrhagic fever surveillance in the Democratic Republic of the Congo, 1981–1985. J Infect Dis 179 (Suppl 1):S60–64
Ji WH, Sarre SD, White PCL, Clout MN (2004) Population recovery of common brushtail possums after local depopulation. Wildl Res 31:543–550
Ji W, White PCL, Clout MN (2005) Contact rates between possums revealed by proximity data loggers. J Appl Ecol 42:595–604
Jolles AE, Cooper DV, Levin SA (2005) Hidden effects of chronic tuberculosis in African buffalo. Ecology 86:2258–2264
Joly DO, Messier F (2004) Factors affecting apparent prevalence of tuberculosis and brucellosis in wood bison. J Anim Ecol 73:623–631
Jones KE, Patel NG, Levy MA, Storeygard A, Balk D, Gittleman JL, Daszak P (2008) Global trends in emerging infectious diseases. Nature 451:990–993
Jones M, Jarman P, Lees C, Hesterman H, Hamede R, Mooney N, Mann D, Pukk C, Bergfeld J, McCallum H (2007) Conservation management of Tasmanian devils in the context of an emerging, extinction-threatening disease: devil facial tumor disease. EcoHealth 4:326–337
Jones ME, Paetkou D, Geffen E, Moritz C (2004) Genetic diversity and population structure of Tasmanian devils, the largest marsupial carnivore. Mol Ecol 13:2197–2209
Jones R, Saunders G, Balogh S (2006) An economic evaluation of a pest management control program: 'outfox the fox'. NSW Department of Primary Industries. Available at: www.agric.nsw.gov.au/reader/10550 – 21st May 2007

Judge J, Greig A, Kyriazakis I, Hutchings MR (2005a) Ingestion of faeces by grazing herbivores – risk of inter-species disease transmission. Agric Ecosys Environ 107:267–274

Judge J, Kyriazakis A, Greig A, Allcroft DJ, Hutchings MR (2005b) Clustering of *Mycobacterium avium* subsp. paratuberculosis in rabbits and the environment: how hot is a hotspot? Appl Environ Microbiol 71:6033–6038

Judge J, Kyriazakis I, Greig A, Davidson RS, Hutchings MR (2006) Routes of intraspecies transmission of *Mycobacterium avium* subsp paratuberculosis in rabbits (*Oryctolagus cuniculus*): a field study. Appl Environ Microbiol 72:398–403

Judge J, Davidson RS, Marion G, White PCL, Hutchings MR (2007) Persistence of *Mycobacterium avium* subspecies paratuberculosis in rabbits: the interplay between horizontal and vertical transmission. J Appl Ecol 44:302–311

Kaandorp J (2004) Transmissible diseases handbook. The Netherlands European Association of Zoo and Wildlife Veterinarians, Hilvarenbeek

Kaden V, Lange B (2001) Oral immunisation against classical swine fever (CSF): onset and duration of immunity. Vet Microbiol 82:301–310

Kaden V, Lange E (2008) Unpublished data.

Kaden V, Lange E, Fischer U, Stebelow G (2000) Oral immunisation of wild boar against classical swine fever: evaluation of the first field study in Germany. Vet Microbiol 73:239–252

Kaden V, Renner C, Rothe A, Lange E, Hanel A, Gossger K (2003) Evaluation of the oral immunisation of wild boar against classical swine fever in Baden-Wuttemberg. Berl Münch Tierärztl Wschr 116:362–367

Kaden V, Steyer H, Schnabel J, Bruer W (2005a) Classical swine fever (CSF) in wild boar: the role of the transplacental infection in the perpetuation of the CSF. J Vet Med B 52:161–164

Kaden V, Hänel A, Renner C, Gossger K (2005b) Oral immunisation of wild boar against classical swine fever in Baden-Württemberg: development of the seroprevalences based on the hunting bag. Eur J Wildl Res 51:101–107

Kaden V, Kramer M, Kern B, Hlinak A, Mewes L, Hänel A, Renner C, Dedek J, Bruer W (2006) Diagnostic procedures after completion of oral immunisation against classical swine fever in wild boar. Rev Sci Tech Off Int Epizoot 25:989–997

Kaden V, Lange E, Steyer H, Lange B, Klopfleisch R, Teifke JP, Bruer W (2008) Classical swine fever virus strain "C" protects the offspring by oral immunisation of pregnant sows. Vet Microbiol: 130:20–27

Kämpfer S, Dalley D, Hewinson RG, Chambers MA, Singh M (2003) Multi-antigen ELISA for enhanced diagnosis of tuberculosis in badgers. Vet Rec 153:403–404

Kao RR (2002) The role of mathematical modelling in the control of the 2001 FMD epidemic in the UK. Trends Microbiol 10:279–286

Kassilly FN (2002) The fence as a moderator of the wildlife menace in Kenya. Afr J Ecol 40:407–409

Keeling MJ (1999) The effects of local spatial structure on epidemiological invasions. Proc R Soc B 266:859–867

Keeling MJ, Eames KTD (2005) Networks and epidemic models. J R Soc Interface 2:295–307

Keeling MJ, Wilson HB, Pacala SW (2000) Reinterpreting space, time lags, and functional responses in ecological models. Science 290:1758–1761

Keet DF, Kriek NPJ, Penrith ML (1996) Tuberculosis in buffaloes (*Syncerus caffer*) in the Kruger National Park: spread of the disease to other species. Onderstepoort J Vet Res 63:239–244

Keet DF, Kriek NPJ, Bengis RG, Michel AL (2001a) Tuberculosis in kudus (*Tragelaphus strepsiceros*) in the Kruger National Park. Onderstepoort J Vet Res 68:225–230

Keet DF, Kriek NPJ, Dekker EH, Meltzer DGA (2001b) Tuberculosis in free ranging lions (*Panthera leo*) in the Kruger National Park. Proceedings of the Biennial South African Veterinary Association Congress, 21–23 September, Durban, pp. 232–241

Kemere P, Liddel MK, Evangelou P, Slate D, Osmek S (2000) Human conflicts with wildlife: economic considerations. USDA National Wildlife Research Center Symposia, University of Nebraska, Lincoln

Kennedy I, Aubert M, Bannister B, DeVile P, Dye C, Higgins A, Jarrett O, Sewell H, Smith J (1998) Quarantine & rabies: a reappraisal. MAFF Publications, London, UK

Kermack WO, McKendrick AG (1927) Contributions to the mathematical theory of epidemics. Proc R Soc Lond A 115:700–721

Kermack WO, McKendrick AG (1991) Contributions to the mathematical theory of epidemics – 1. Bull Math Biol 53:33–55

Killian G, Wagner D, Miller L (2005) Observations on the use of the GnRH vaccine GonaConTM in male white-tailed deer (*Odocoileus virginianus*). 11th Wildlife Damage Management Conference Traverse City, Michigan, pp. 256–263

Killian G, Miller LA, Diehl NK, Rhyan J, Thain D (2006) Long-term efficacy of three contraceptive approaches for population control of wild horses. In: Timm RM, O'Brien JM (eds) Proceedings of the 22nd Vertebrate Pest Conference, 6–9th March, Berkley, California, pp. 67–71

King AA, Fooks AR, Aubert M, Wandeler AI (eds) (2004) Historical perspectives of rabies in Europe and the Mediterranean basin. OIE, Paris, 359 p

King DR, Kirkpatrick WE, Wong DH, Kinnear JE (1994) Degradation of 1080 in Australian soils. In: Seawright AA, Eason CT (eds) Proceedings of the Science Workshop on 1080. The Royal Society of New Zealand Miscellaneous Series, pp. 1–9

Kirkpatrick JF, Liu IKM, Turner JW (1990) Remotely-delivered immunocontraception in feral horses. Wildl Soc Bull 18:326–330

Kirkwood JK, Cunningham AA (1994) Epidemiological observations on spongiform encephalopathies in captive wild animals in the British Isles. Vet Rec 135:296–303

Kirkwood JK, Sainsbury AW (1997) Diseases and other considerations in wildlife translocations and releases. Proceedings of the World Association of Wildlife Veterinarians Symposium on Veterinary Involvement with Wildlife Reintroduction and Rehabilitation, Ballygawley, UK, WAWV, pp. 12–16

Kirkwood JK, Cunningham AA (2006) Portrait of prion diseases in zoo animals. In: Hornlimann B, Riesner D, Kretzschmar H (eds) Prions in Humans and Animals. Walter de Gruyter, Berlin and New York. pp. 250–256

Klar N, Herrmann M, Kramer-Schadt S (2006) Effects of roads on a founder population of lynx in the biosphere reserve "Pfälzerwald – Vosges du Nord". A model as a planning tool. Naturschutz Landschaftsplanung 38:330–337

Klein G (1998) Sources of power: how people make decisions. MIT Press, Cambridge, Mass

Klein GO (2002) Standardization of health informatics. Results and challenges. Methods Inf Med 41:261–270

Knobel DL, Cleaveland S, Coleman PG, Fèvre EM, Meltzer MI, Miranda MEG, Shaw A, Zinsstag J, Meslin FM (2005) Re-evaluating the burden of rabies in Africa. Bull World Health Organ 83:360–368

Knobel DL, Fooks AR, Brookes SM, Randall DA, Williams SD, Argaw K, Shiferaw F, Tallents LA, Laurenson MK (2008) Trapping and vaccination of endangered Ethiopian wolves to control an outbreak of rabies. J Appl Ecol 45:109–116

Kock RA (2008) The Role of Wildlife in the Epidemiology of Rinderpest in East and Central Africa 1994–2004: A Study Based on Serological Surveillance and Disease Investigation. [DVM Thesis]. Cambridge, UK, University of Cambridge

Kock RA, Wambua JM, Mwanzia J, Wamwayi H, Ndungu EK, Barrett T, Kock ND, Rossiter PB (1999a) Rinderpest epidemic in wild ruminants in Kenya 1993–7. Vet Rec 145:275–283

Kock RA, Mihok SR, Wambua J, Mwanzia J, Saigawa K (1999b) Effects of translocation on hematologic parameters of free-ranging black rhinoceros (*Diceros bicornis michaeli*) in Kenya. J Zoo Wildl Med 30:389–396

Kock RA Wamwayi HM, Rossiter PB, Libeau G, Wambwa E, Okori J, Shiferaw FS, Mlengeya TD (2006) Rinderpest in East Africa: continuing re-infection of wildlife populations on the periphery of the Somali ecosystem. Prev Vet Med 75:63–80

Kock RA, Soorae PS, Mohammed OB (2007) Role of veterinarians in re-introductions. Int Zoo Yearbook 41:24–37

Kolomitsev A, Wishnyakov I, Zhesterev V, Khripunov E, Dubrovin V, Khukhorov I, Isakova N, Tcheryatnikov S, Fertickov V, Yegorov A, Semenikhine A, Pjsak Z (1998) Vaccination against classical swine fever in wild boar in Russia. In: Report on measures to control classical swine fever in European wild boar, 6–7 April Perugia, Italy, European Commission VI/196/98 AL, pp. 128–134

Kraabel BJ, Miller MW, Conlon JA, McNeil HJ (1998) Evaluation of a multivalent *Pasteurella haemolytica* vaccine in bighorn sheep: protection from experimental challenge. J Wildl Dis 34:325–333

Kramer-Schadt S, Fernández N, Thulke HH (2007) Potential ecological and epidemiological factors affecting the persistence of classical swine fever in wild boar *Sus scrofa* populations. Mammal Rev 37:1–20

Krause J, Ruxton GD (2002) Living in groups. Oxford University Press, Oxford

Krebs JR, Anderson RM, Clutton-Brock T, Morrison I, Young D, Donnelly C (1997) Bovine tuberculosis in cattle and badgers. MAFF Publications, London, UK, PB3423

Kroschewski K, Kramer M, Micklich A, Staubach C, Carmanns R, Conraths FJ (2006) Animal disease outbreak control: the use of crisis management tools. Rev Sci Tech Off Int Epizoot 25:211–221

Lacey LA, Lacey CM (1990) The medical importance of riceland mosquitoes and their control using alternatives to chemical insecticides. J Am Mosq Control Assoc 2:1–93

Laddomada A (2000) Incidence and control of CSF in wild boar in Europe. Vet Microbiol 73:121–130

Lambert MS, Quy RJ, Smith RH, Cowan DP (2008) The effect of habitat management on home-range size and survival of rural Norway rat populations. J Appl Ecol doi: 10.1111/j.1365-2664.2008.01543.x

Lane RS, Piesman J, Burgdorfer W (1991) Lyme borreliosis: relation of its causative agent to its vectors and hosts in North America and Europe. Annu Rev Entomol 36:587–609

Laurenson K, Cleaveland S, Artois M, Woodroffe R. (2004) Assessing and managing infectious disease threats to canids. In: Sillero-Zubiri C, Hoffmann M, Macdonald DW (eds) Canids: Foxes, wolves, jackals and dogs Status survey and conservation action plan. IUCN/SSC Canid Specialist Group, Gland, Switzerland and Cambridge, UK, pp. 246–256

Lawson AB, Kleinman K (2005) Spatial and syndromic surveillance for public health. Wiley, Chichester

Le Guenno B, Formenty P, Boesch C (1999) Ebola virus outbreaks in the Ivory Coast and Liberia, 1994–1995. Curr Top Microbiol Immunol 235:77–84

Leighton FA, Artois M, Capucci L, Gavierwiden D, Morisse JP (1995) Antibody-response to rabbit viral hemorrhagic disease virus in red foxes (*Vulpes vulpes*) consuming livers of infected rabbits (*Oryctolagus cuniculus*). J Wildl Dis 31:541–544

Leighton FA, Wobeser GA, Barker IK, Daoust PY, Martineau D (1997) The Canadian cooperative wildlife health centre and surveillance of wild animal diseases in Canada. Can Vet J 38:279–284

Lempert RJ, Popper SW, Bankes SC (2003) Shaping the next one hundred years – new methods for quantitative, long-term policy analysis. RAND Corporation, Santa Monica, CA

Leroy EM, Rouquet P, Formenty P, Souquiere S, Kilbourne A, Froment JM, Bermejo M, Smit S, Karesh W, Swanepoel R, Zaki SR, Rollin PE (2004) Multiple Ebola virus transmission events and rapid decline of central African wildlife. Science 303:387–390

Leroy EM, Kumulungui B, Pourrut X, Rouquet P, Hassanin A, Yaba P, Délicat A, Paweska JT, Gonzalez JP, Swanepoel R (2005) Fruit bats as reservoirs of Ebola virus. Nature 438:575–576

Lesellier S, Palmer S, Dalley DJ, Davé D, Johnson L, Hewinson RG, Chambers MA (2006a) The safety and immunogenicity of Bacillus Calmette-Guérin (BCG) vaccine in European badgers (*Meles meles*). Vet Immunol Immunopathol 112:24–37

Lesellier S, Corner LAL, Chambers MA, Aldwell FE, Costello E, Sleeman DP, Hewinson RG, Gormley E (2006b) Oral vaccination of badgers (*Meles meles*) against tuberculosis: immune responses and protection following BCG delivery via the oral route and pulmonary challenge with *Mycobacterium bovis*. 55th Annual Meeting of the Wildlife Disease Association;

University of Conneticut, pp. 64 [available online http://www.wildlifedisease.org/Documents/Proceedings/Storrs_06.pdf]

Leung LKP, Clark NM (2005) Bait avoidance and habitat use by the roof rat, *Rattus rattus*, in a piggery. Int Biodeterior Biodegrad 55:77–84

Lips KR, Brem F, Brenes R, Reeve JD, Alford RA, Voyles J, Carey C, Livo L, Pessier AP, Collins JP (2006) Emerging infectious disease and the loss of biodiversity in a Neotropical amphibian community. Proc Natl Acad Sci USA 103:3165–3170

Little TJ (2002) The evolutionary significance of parasitism: do parasite-driven genetic dynamics occur ex silico? J Evol Biol 15:1–9

Lively CM (1989) Adaptation by a parasitic trematode to local populations of its snail host. Evolution 43:1663–1671

Lloyd M (1967) Mean crowding. J Anim Ecol 36:1–30

Lloyd-Smith JO, Schreiber SJ, Kopp PE, Getz WM (2005a) Superspreading and the effect of individual variation on disease emergence. Nature 438:355–359

Lloyd-Smith JO, Cross PC, Briggs CJ, Daugherty M, Getz WM, Latto J, Sanchez MS, Smith AB, Swei A (2005b) Should we expect population thresholds for wildlife disease? Trends Ecol Evol 20:511–519

Lochmillar RL, Deerenberg C (2000) Trade-offs in evolutionary immunology: just what is the cost of immunity? Oikos 88:87–98

Long ES, Diefenbach DR, Rosenberry CS, Wallingford BD, Grund MRD (2005) Forest cover influences dispersal distance of white-tailed deer. J Mammal 86:623–629

Lord RD (1980) An ecological strategy for controlling bovine rabies through elimination of vampire bats. Proceedings of the 9th Vertebrate Pest Conference. University of Nebraska, Lincoln

Lurz PWW, Geddes N, Lloyd AJ, Shirley MDF, Rushton SP, Burlton B (2003) Planning a red squirrel conservation area: using a spatially explicit population dynamics model to predict the impact of felling and forest design plans. Forestry 76:95–108

Lyles AM, Dobson AP (1993) Infectious-disease and intensive management – population-dynamics, threatened hosts, and their parasites. J Zoo Wildl Med 24:315–326

Ma J, Earn DJD (2006) Generality of the final size formula for an epidemic of a newly invading infectious disease. Bull Math Biol 68:679–702

Macdairmid SC, Pharo HJ (2003) Risk analysis: assessment, management and communication. Rev Sci Tech Off Int Epizoot 22:397–408

Macdonald DW (1980) Rabies and wildlife. Oxford University Press, Oxford

Macdonald DW (1995) Wildlife rabies: the implications for Britain. Unresolved questions for the control of wildlife rabies: social perturbation and interspecific interactions. In: Beynon PH, Ednay ATB (eds) Rabies in a changing world. British Small Animal Veterinary Association, Cheltenham, pp. 33–48

Macdonald DW, Bunce RGH, Bacon PJ (1981) Fox populations, habitat characterization and rabies control. J Biogeogr 8:145–151

Macdonald DW, Buesching CD, Stopka P, Henderson J, Ellwood SA, Baker SE (2004) Encounters between two sympatric carnivores: red foxes (*Vulpes vulpes*) and European badgers (*Meles meles*). J Zool 263:385–392

Mace G, Sillero-Zubiri C (1997) A preliminary population viability analysis for the Ethiopian wolf. In: Sillero-Zubiri C, Macdonald DW (eds) The Ethiopian Wolf: Status Survey and Conservation Action Plan. World Conservation Union, Gland, Switzerland, pp. 51–60

Maillard D, Fournier P (1995) Effect of shooting with hounds on size of resting range of wild boar (*Sus scrofa* L.) groups in a mediterranean habitat. Ibex J Mt Ecol 3:102–107

Malcolm KD, Van Deelen TR (2007) Contraceptive efficacy of intrauterine devices for non-lethal management of overabundant deer. Proceedings of the Wildlife Society, 21–26 September, Tucson

Marano N, Argiun PN, Pappaioanou M (2007) Impact of globalization and animal trade on infectious disease ecology. Emerg Infect Dis 13:1807–1809

Marion G, Swain DL, Hutchings MR (2005) Understanding foraging behaviour in spatially heterogeneous environments. J Theor Biol 232:127–142

Marion G, Smith L, Swain DL, Davidson RS, Hutchings MR (2008) Understanding the impact of foraging behaviour on disease risks from faeces in spatially heterogeneous systems. J Agric Sci 146:507–520

Marks CA, Johnston MJ, Fisher PM, Pontin K, Shaw MJ (2006) Differential particle size ingestion: promoting target-specific baiting of feral cats. J Wildl Manage 70:1119–1124
Marshall A (1947) Principles of economics. Macmillan and Co. Ltd, London
Martin SW, Meek AH, Willeberg P (1987) Measurement of disease frequency and production. In: Martin SW, Meek AH, Willeberg P (eds) Veterinary epidemiology: principles and methods. Iowa State University, Ames, pp. 48–76
May RM, Anderson RM (1979) Population biology of infectious diseases: Part II. Nature 280:455–461
May RM, Anderson RM (1984) Spatial heterogeneity and the design of immunization programs. Math Biosci 72:83–111
McCallum H (1995) Modelling wildlife-parasite interactions to help plan and interpret field studies. Wildl Res 22:21–29
McCallum H (2000) In: Lawton JH, Likens GE (eds) Population parameters: estimation for ecological models. Blackwell Science, Oxford
McCallum H (2008) Tasmanian devil facial tumour disease: lessons for conservation biology. Trends Ecol Evol 23:631–637
McCallum H, Dobson A (2002) Disease, habitat fragmentation and conservation. Proc R Soc B 269:2041–2049
McCallum H, Jones M (2006) To lose both would look like carelessness: tasmanian devil facial tumour disease. PLoS Biol 4:e342
McCallum H, Barlow N, Hone J (2001) How should pathogen transmission be modelled? Trends Ecol Evol 16:295–300
McCallum H, Tompkins DM, Jones M, Lachish S, Marvanek S, Lazenby B, Hocking G, Wiersma J, Hawkins CE (2007) Distribution and impacts of tasmanian devil facial tumor disease. EcoHealth 4:318–325
McDonald RA, Larivière S (2001) Diseases and pathogens of *Mustela* spp., with special reference to the biological control of introduced stoat *Mustela erminea* populations in New Zealand. J R Soc NZ 31:721–744
McIlroy JC, Gifford EJ (1997) The 'Judas' pig technique: a method that could enhance control programmes against feral pigs, *Sus scrofa*. Wildl Res 24:483–491
McInerney J, Small KJ, Caley P (1995) Prevalence of *Mycobacterium bovis* infection in feral pigs in the Northern Territory. Aust Vet J 72:448–451
McInerney JP (1996) Old economics for new problems – livestock disease: Presidential address. J Agric Econ 47:295–314
McInnes EG, Steward CG, Penzhorn BL, Metlzer DG (1991) An outbreak of babesiosis in imported sable antelope (*Hippotragus niger*). J S Afr Vet Assoc 62:30–32
McKenzie J, Simpson H, Langstaff I (2007) Development of methodology to prioritise wildlife pathogens for surveillance. Prev Vet Med 81:194–210
Meagher M (1989) Evaluation of boundary control for bison of Yellowstone National Park. Wildl Soc Bull 17:15–19
Mech LD, Goyal SM (1995) Effects of canine parvovirus on gray wolves in Minnesota. J Wildl Manage 59:565–570
MedVetNet. (2008) http://www.medvetnet.org/cms/
Meerburg BG, Jacobs-Reitsma WF, Wagenaar JA, Kijlstra A (2006) Presence of *Salmonella* and *Campylobacter* spp. in wild small mammals on organic farms. Appl Environ Microbiol 72:960–962
Meltzer MI (1996) Assessing the costs and benefits of an oral vaccine for raccoon rabies: a possible model. Emerg Infect Dis 2:343–349
Meltzer MI, Rupprecht CE (1998) A review of the economics of the prevention and control of rabies part 2: Rabies in dogs, livestock and wildlife. Pharmacoeconomics 14:481–498
Mencher JS, Smith SR, Powell TD, Stinchcomb DT, Osorio JE, Rocke TE (2004) Protection of black-tailed prairie dogs (*Cynomys ludovicianus*) against plague after voluntary consumption of baits containing recombinant raccoon poxvirus vaccine. Infect Immun 72:5502–5505
Merrill JA, Cooch EG, P.D. C. (2003) Time to reduction: factors influencing management efficacy in sterilizing overabundant white-tailed deer. J Wildl Manage 67:267–279

Meyer DJ (ed) (2003) The economics of risk. Upjohn Institute, USA

Michel AL, Bengis RG, Keet DF, Hofmeyr M, de Klerk LM, Cross PC, Jolles AE, Cooper D, Whyte IJ, Buss P, Godfroid J (2006) Wildlife tuberculosis in South African conservation areas: Implications and challenges. Vet Microbiol 112:91–100

Mill JS (1848) Principles of political economy. Parker & Co., London

Miller LA, Fagerstone KA (2000) Induced infertility as a wildlife management tool. In: Salmon TP, Crabb AC (eds) Proceedings of the 19th Vertebrate Pest Conference, 6–9th March, San Diego, California, pp. 160–168

Miller LA, Johns BE, Elias DJ, Crane KA (1997) Comparative efficacy of two immunocontraceptive vaccines. Vaccine 15:1858–1862

Miller LA, Rhyan JC, Killian GJ (2003) Evaluation of GnRH contraceptive vaccine using domestic swine as a model for feral hogs. In: Timm RM, Fagerstone KA (eds) Wildlife Damage Management Conference. Hot Springs, Arkansas, pp. 120–127

Miller LA, Rhyan JC, Drew M (2004a) Contraception of bison by GnRH vaccine: a possible means of decreasing transmission of brucellosis in bison. J Wildl Dis 40:725–730

Miller LA, Rhyan J, Killian G (2004b) GonaCon(TM): a versatile GnRH contraceptive for a large variety of pest animal problems. In: Timm RM, Gorenzel WP (eds) Proceedings of the 21st Vertebrate Pest Conference. Visalia, California, pp. 269–273

Miller MW, Williams ES (2003) Horizontal prion transfer in mule deer. Nature 425:35–36

Miller MW, Corner MM (2005) Epidemiology of chronic wasting disease in free-ranging mule deer: spatial, temporal, and demographic influences on observed prevalence patterns. J Wildl Dis 41:275–290

Miller MW, Williams ES, McCarty CW, Spraker TR, Kreeger TJ, Larsen CT, Thorne ET (2000) Epizootiology of chronic wasting disease in free-ranging cervids in Colorado and Wyoming. J Wildl Dis 36:676–690

Miller MW, Williams ES, Hobbs NT, Wolfe LL (2004) Environmental sources of prion transmission in mule deer. Emerg Infect Dis 10:1003–1006

Miller MW, Hobbs NT, Tavener SJ (2006) Dynamics of prion disease transmission in mule deer. Ecol Appl 16:2208–2214

Miller R, Kaneene JB, Fitzgerald SD, Schmitt SM (2003) Evaluation of the influence of supplemental feeding of white-tailed deer (*Odocoileus virginianus*) on the prevalence of bovine tuberculosis in the Michigan wild deer population. J Wildl Dis 39:84–95

Miranda ME, Ksiazek TG, Retuya TJ, Khan AS, Sanchez A, Fulhorst CF, Rollin PE, Calaor AB, Manalo DL, Roces MC, Dayrit MM, Peters CJ (1999) Epidemiology of Ebola (subtype Reston) virus in the Philippines, 1996. J Infect Dis 179 (Suppl 1):S115–S119

Miranda ME, Yoshikawa Y, Manalo DL, Calaor AB, Miranda NL, Cho F, Ikegami T, Ksiazek TG (2002) Chronological and spatial analysis of the 1996 Ebola Reston virus outbreak in a monkey breeding facility in the Philippines. Exp Anim 51:173–179

Mishan EJ (1998) Cost–benefit analysis. George Allen and Unwin, London, UK

Mishan EJ, Quah E (2007) Cost–Benefit Analysis. Routledge, London

Moberly RL, White PCL, Webbon CC, Baker PJ, Harris S (2003) Factors associated with fox (*Vulpes vulpes*) predation of lambs in Britain. Wildl Res 30:219–227

Moennig V (2000) Introduction to classical swine fever: virus, disease and control policy. Vet Microbiol 73:93–102

Molyneux DH (1982) Trypanosomes, trypanosomiasis and tse tse control: impact on wildlife and its conservation. Symp Zool Soc Lond 50:29–55

Montague T, Warburton B (2000) Non-toxic techniques for possum control. In: Montague TL (ed) The brushtail possum: biology, impact and management of an introduced Marsupial. Manaaki Whenua Press, Lincoln, New Zealand, pp. 164–174

Montali RJ, Bush M (1992) Some diseases of Golden Lion tamarins acquired in captivity and their impact on reintroduction. Joint conference of the American Assocaition of Zoo Veterinarians and the American Association of Wildlife Veterinarians, pp. 14–16

Montali RJ, Bush M, Hess J, Ballou JD, Kleiman DG, Beck BB (1995) Ex situ diseases and their control for reintroduction of the endangered lion tamarin species (*Leontopithecus* spp.). Verhber Erkrg Zootiere 37:93–98

Moore SL, Wilson K (2002) Parasites as a viability cost of sexual selection in natural populations of mammals. Science 297:2015–2018

Moreno S, Villafuerte R, Cabezas S, Lombardi L (2004) Wild rabbit restocking for predator conservation in Spain. Biol Conserv 118:183–193

Morgan D, Hickling G (2000) Techniques used for poisoning possums. In: Montague TL (ed) The Brushtail Possum: Biology, Impact and Management of an Introduced Marsupial. Manaaki Whenua Press, Lincoln, New Zealand, pp. 143–153

Morgan ER, Shaikenov B, Torgerson PR, Medley GF, Milner-Gulland EJ (2005) Helminths of saiga antelope in Kazakhstan: implications for conservation and livestock production. J Wildl Dis 41:149–162

Morgan ER, Medley GF, Torgerson PR, Shaikenov BS, Milner-Gulland EJ (2006) Parasite transmission in a migratory multiple host system. Ecol Model 200:511–520

Morton M, Garmory HS, Perkins SD, O'Dowd AM, Griffin KF, Turner AK, Bennett AM, Titball RW (2004) A *Salmonella enterica* serovar Typhi vaccine expressing *Yersinia pestis* F1 antigen on its surface provides protection against plague in mice. Vaccine 22:2524–2532

Moutou F (1995) Les morbillivirus, virus d'actualité. Point Vét 168:133–140

Mörner T, Obendorf DL, Artois M, Woodford MH (2002) Surveillance and monitoring of wildlife diseases. Rev Sci Tech Off Int Epizoot 21:67–76

Müller J (1971) The effect of fox reduction on the occurrence of rabies. Observations from two outbreaks of rabies in Denmark. Bull Off Int Épizoot 75:7673–7776

Müller T, Selhorst T, Pötzsch C (2005) Fox rabies in Germany – an update. Eurosurveillance

Munson L (2006) Contraception in felids. Theriogenology 66:126–134

Murphy D, Corner LAL, Gormley E (2008) Adverse reactions to *Mycobacterium bovis* bacille Calmette-Guérin (BCG) vaccination against tuberculosis in humans, veterinary animals and wildlife species. Tuberculosis 88:344–357

Murray DL, Keith LB, Cary JR (1996) The efficacy of anthelmintic treatment on the parasite abundance of free-ranging snowshoe hares. Can J Zool 74:1604–1611

Murray JD, Stanley EA, Brown DL (1986) On the spatial spread of rabies among foxes. Proc R Soc Lond B 229:111–150

Murray N (ed) (2004) Handbook on import risk analysis for animals and animal products – 2 volumes. OIE, Paris 57 & 126 p

Mwanzia JM, Kock RA, Wambua JM, Kock ND, Jarrett O (1995) An outbreak of sarcoptic mange in free living cheetah (*Acinonyx jubatus*) in the Mara region of Kenya. Annual Conference – American Association of Zoo Veterinarians

Nash P, Furcolow CA, Bynum KS, Yoder CA, Miller LA, Johnston JJ (2007) Diazacholesterol as an oral contraceptive for blacktailed prairie dog population management. Hum-Wildlife Conflicts 1:60–67

Neal E, Cheeseman C (1996) Badgers. T & AD Poyser, London, UK

Nebbia P, Robino P, Ferroglio E, Rossi L, Meneguz G, Rosati S (2000) Paratuberculosis in red deer (*Cervus elaphus hippelaphus*) in the Western Alps. Vet Res Comm 24:435–443

Nijhof AM, Penzhorn BL, Lynen G, Mollel JO, Morker P, Bekker CP, Jongejan F (2003) *Babesia bicornis* sp. nov. and *Theileria bicornis* sp. nov.: tick-borne parasites associated with mortality in the black rhinoceros (*Diceros bicornis*). J Clin Microbiol 41:2249–2254

Nishi J, Shury T, Elkin BT (2006) Wildlife reservoirs for bovine tuberculosis (*Mycobacterium bovis*) in Canada: strategies for management and research. Vet Microbiol 112:325–338

Nizeyi JB, Mwebe R, Nanteza A, Cranfield MR, Kalema GRNN, Graczyk TK (1999) *Cryptosporidium* sp. and *Giardia* sp. infections in mountain gorillas (*Gorilla gorilla beringei*) of the Bwindi Impenetrable National Park, Uganda. J Parasitol 85:1084–1088

Nizeyi JB, Sebunya D, DaSilva AJ, Cranfield MR, Pieniazek NJ, Graczyk TK (2002) Cryptosporidiosis in people sharing habitats with free-ranging mountain gorillas (*Gorilla gorilla beringei*), Uganda. Am J Trop Med Hyg 66:442–444

Nogales M, Martín A, Tershy BR, Donlan CJ, Veitch D, Puerta N, Wood B, Alonso J (2004) A review of feral cat eradication on islands. Conserv Biol 18:310–319

Nugent G (2005) The role of wild deer in the epidemiology and management of bovine tuberculosis in New Zealand [PhD thesis]. Lincoln, New Zealand, Lincoln University

Nunn C, Altizer S (2004) Sexual selection, behaviour and sexually transmitted diseases. In: Kappler PM, van Schaik CP (eds) Sexual selection in primates: new and comparative perspectives. Cambridge University Press, Cambridge, pp. 117–130

Nunn CL, Heymann EW (2005) Malaria infection and host behaviour: a comparative study of Neotropical primates. Behav Ecol Sociobiol 59:30–37

Nutter FB, Whittier CA, Cranfield MR, Lownestine LJ (2005) Causes of death for mountain gorillas (*Gorilla beringei beringei* and *Gorilla beringei* undecided) from 1968–2004. In: Baerc K, Lawrence KS (eds) Wildlife Disease Association International Conference, Cairns, Queensland, Australia. Wildlife Disease Association, pp. 200–201

O'Brien DJ, Schmitt SM, Fierke JS, Hogle SA, Winterstein SR, Cooley TM, Moritz WE (2002) Epidemiology of *Mycobacterium bovis* in free ranging white-tailed deer, Michigan, USA, 1995–2000. Prev Vet Med 54:47–63

O'Brien DJ, Schmitt SM, Fitzgerald SD, Berry DE, Hickling GJ (2006) Managing the wildlife reservoir of *Mycobacterium bovis*: the Michigan, USA, experience. Vet Microbiol 112:313–323

OIE (2006) Terrestrial animal health code. 15th edn. Available at: http://www.oie.int/eng/normes/mcode/code2006/en_index.htm

OIE (2007) Health risk analysis in wild animal translocations guidelines. Available at: http://wildlife1.usask.ca/wildlife_health_topics/risk_analysis/rskguidintro.php

OIE (2008a) http://www.oie.int/wildlife/eng/en_reports.htm

OIE (2008b) http://www.oie.int/wildlife/eng/en_wildlife.htm

Olsen SC, Kreeger TJ, Schultz W (2002) Immune responses of bison to ballistic or hand vaccination with *Brucella abortus* strain RB51. J Wildl Dis 38:738–745

Olsen SC, Christie RJ, Grainger DW, Stoffregen WS (2006) Immunologic responses of bison to vaccination with *Brucella abortus* strain RB51: comparison of parenteral to ballistic delivery via compressed pellets or photopolymerized hydrogels. Vaccine 24:1346–1353

Oreskes N, Shrader-Frechette K, Belitz K (1994) Verification, validation, and confirmation of numerical models in the earth sciences. Science 263:641–646

Osofsky SA, Paglia DE, Radcliffe RW, Miller RE, Emslie RH, Foose TJ, Du Toit R, Atkinson MW (2001) First, do no harm: a precautionary recommendation regarding the movement of black rhinos from overseas zoos back to Africa. Pachyderm 30:17–23

OUP (2008) http://www.askoxford.com/

Palmer MV, Whipple DL, Waters WR (2001) Experimental deer-to-deer transmission of *Mycobacterium bovis*. Am J Vet Res 62:692–696

Palmer MV, Waters WR, Whipple DL (2004) Shared feed as a means of deer-to-deer transmission of *Mycobacterium bovis*. J Wildl Dis 40:87–91

Palmer MV, Thacker TC, Waters WR (2007) Vaccination of white-tailed deer (*Odocoileus virginianus*) with *Mycobacterium bovis* bacillus Calmette Guerín. Vaccine 25:6589–6597

Pastor-Satorras R, Vespignani A (2001) Epidemic dynamics and endemic states in complex networks. E 63, 066117. Phys Rev E 63:066117

Pastor-Satorras R, Vespignani A. (2002) Epidemic dynamics in finite-size scale free networks. Phys Rev E 65:035108(R)

Pastoret PP, Brochier B (1999) Epidemiology and control of fox rabies in Europe. Vaccine 17:1750–1754

Pastoret PP, Brochier B, Languet B, Thomas I, Paquot A, Bauduin B, Kieny MP, Lecocq JP, De Bruyn J, Costy F, Antoine H, Desmettre P (1988) First field trial of fox vaccination against rabies using a vaccinia-rabies recombinant virus. Vet Rec 123:481–483

Payton I (2000) Damage to native forest. In: Montague TL (ed) The brushtail possum: biology, impact and management of an introduced marsupial. Manaaki Whenua Press, Lincoln, New Zealand, pp. 111–125

Pearse AM, Swift K (2006) Allograft theory: transmission of devil facial-tumour disease. Nature 439:549

Pech RP, Hone J (1988) A model of the dynamics and control of an outbreak of foot and mouth disease in feral pigs in Australia. J Appl Ecol 25:63–77

Pedersen AB, Altizer S, Poss M, Cunningham AA, Nunn CL (2005) Patterns of host specificity and transmission among parasites of wild primates. Int J Parasitol 35:647–657

Pedersen AB, Jones KE, Nunn CL, Altizer S (2007) Infectious diseases and extinction risk in wild mammals. Conserv Biol 21:1269–1279

Pence DB, Ueckermann E (2002) Sarcoptic mange in wildlife. Rev Sci Tech Off Int Epizoot 21:385–398

Penn DJ (2002) The scent of genetic compatibility: sexual selection and the major histocompatibility complex. Ethology 108:1–22

Perry RD, Fetherston JD (1997) *Yersinia pestis* – etiologic agent of plague. Clin Microbiol Rev 10:35–66

Petavy AF (2008) personal communication

Phillips CJC (1993) Cattle Behaviour. Farming Press Books, Ipswich, UK

Phillips CJC, Foster CRW, Morris PA, Teverson R (2003) The transmission of *Mycobacterium bovis* infection to cattle. Res Vet Sci 74:1–15

Phillips MK, Scheck J (1991) Parasitism in captive and reintroduced red wolves. J Wildl Dis 27:498–501

Philpot M (1993) The dangers of disease transmission by artificial insemination and embryo transfer. Br Vet J 149:339–369

Pielke RA, Jr (2003) The role of models in prediction for decision. In: Canham C, Lauenroth W (eds) Understanding ecosystems: the role of quantitative models in observations, synthesis, and prediction. Princeton University Press, Princeton, NJ, pp. 113–137

Pielke RA, Jr., Sarewitz D, Byerly R, Jr., Jamieson D (1999) Prediction in the earth sciences and environmental policy making. Trans Am Geophys Soc 80:311–313

Pinzon J, Wilcon J, Tucker C, Arthur R, Jahrling P, Formenty P (2004) Trigger events: enviroclimatic coupling of Ebola haemorrhagic fever outbreaks. Am J Trop Med Hyg 71:664–674

Pittman M, Laddomada A, Freigofas R, Piazza V, Brouw A, Brown IH (2007) Surveillance, prevention, and disease management of avian influenza in the European Union. J Wildl Dis 43:S64–70

Pleydell DRJ, Raoul F, Tourneux F, Danson FM, Graham AJ, Craig PS, Giraudoux P (2004) Modelling the spatial distribution of *Echinococcus multilocularis* infection in foxes. Acta Trop 91:253–265

Plowright RK, Field HE, Smith C, Divljan A, Palmer C, Tabor G, Daszak P, Foley JE (2008) Reproduction and nutritional stress are risk factors for Hendra virus infection in little red flying foxes (*Pteropus scapulatus*). Proc R Soc B 275:861–869

Plowright W (1982) The effects of rinderpest and rinderpest control on wildlife in Africa. Symp Zool Soc Lond 50:1–28

Poole DW, Cowan DP, Smith GC (2003) Developing a census method based on sight counts to estimate rabbit (*Oryctolagus cuniculus*) numbers. Wildl Res 30:487–493

Post WM, DeAngelis DL, Travis CC (1983) Endemic disease in environments with spatially heterogeneous host populations. Math Biosci 63:289–302

Potter C, Gasson R (1988) Farmer participation in voluntary land division schemes: some predictions from a survey. J Rural Stud 4:365–375

Poulin R (1995) Phylogeny, ecology, and the richness of parasite communities in vertebrates. Ecol Monogr 65:283–302

Poulin R (1996) Helminth growth in vertebrate hosts: does host sex matter? Int J Parasitol 26:1311–1315

Pourrut X, Kumulungui B, Wittmann T, Moussavou G, Delicat A, Yaba P, Nkoghe D, Gonzalez JP, Leroy EM (2005) The natural history of Ebola virus in Africa. Microb Infect 7:1005–1014

Preedy KF, Schofield PG, Chaplain MAJ, Hubbard SF (2007) Disease induced dynamics in host-parasitoid systems: chaos and coexistence. J R Soc Interface 4:463–471

Prins HHT, Weyerhaeuser FJ (1987) Epidemics in populations of wild ruminants: anthrax and impala, rinderpest and buffalo in Lake Manyara National Park, Tanzania. Oryx 49:28–38

Prusiner SB (1991) Molecular biology of prion diseases. Science 252:1515–1522

Purse BV, McCormick BJJ, Mellor PS, Baylis M, Boorman JPT, Borras D, Burgu I, Capela R, Caracappa S, Collantes F, De Liberato C, Delgado JA, Denison E, Georgiev G, el Harak M, de la Rocque S, Lhor Y, Lucientes J, Mangana O, Miranda MA, Nedelchev N, Nomikou K, Ozkul A, Patakakis M, Pena I, Scaramozzino P, Torina A, Rogers DJ. (2007) Incriminating bluetongue virus vectors with climate envelope models. J Appl Ecol 44:1231–1242

Pyecroft SB, Pearse AM, Loh R, Swift K, Belov K, Fox N, Noonan E, Hayes D, Hyatt A, Wang L (2007) Towards a case definition for devil facial tumour disease: what is it? EcoHealth 4:346–351

Qureshi T, Drawe DL, Davis DS, Craig TM (1994) Use of bait containing Triclabenazole to treat *Fascioloides magna* infections in free-ranging white-tailed deer. J Wildl Dis 30:346–350

Qureshi T, Labes RE, Cross ML, Griffin JF, Mackintosh CG (1999) Partial protection against oral challenge with *Mycobacterium bovis* in ferrets (*Mustela furo*) following oral vaccination with BCG. Int J Tuberc Lung Dis 3:1025–1033

Rabinowitz PM, Gordon Z, Holmes R, Taylor B, Wilcox M, Chudnov D, Nadkarni P, Dein J (2005) Animals as sentinels of human environmental health hazards: an evidence-based analysis. EcoHealth 2:26–37

Rambozzi L, Curcio A, Ferroglio E, Otranto D, Sala L, Meloni D (2002) Prima segnalazione italiana di *Hypoderma diana* (Brauer, 1858) in un capriolo (*Capreolus capreolus*). Abstracts of the XXII SOIPA Congress. Parassitologia 44:149

Ramsey D (2007) Effects of fertility control on behaviour and disease transmission in brushtail possums. J Wildl Manage 71:109–116

Ramsey D, Spencer N, Caley P, Efford M, Hansen K, Lam M, Cooper D (2002) The effects of reducing population density on contact rates between brushtail possums: implications for transmission of bovine tuberculosis. J Appl Ecol 39:806–818

Ramsey DSL, Coleman JD, Coleman MC, Horton P (2006) The effect of fertility control on the transmission of bovine tuberculosis in wild brushtail possums. NZ Vet J 54:218–223

Randall DA, Williams SD, Kuzmin IV, Rupprecht CE, Tallents LA, Tefera Z, Argaw K, Shiferaw F, Knobel DL, Sillero-Zubiri C, Laurenson MK (2004) Rabies in endangered Ethiopian wolves. Emerg Infect Dis 10:2214–2217

Randolph SE (1973) A tracking technique for comparing individual home ranges of small mammals. J Zool 170:509–520

Real LA, Childs JE (2005) Spatial-temporal dynamics of rabies in ecological communities. In: Collinge SK, Ray C (eds) Disease ecology: community structure and pathogen dynamics. Oxford University Press, Oxford, UK, pp. 170–187

Real LA, Russell C, Waller L, Smith D, Childs J (2005) Spatial dynamics and molecular ecology of North American rabies. J Hered 96:253–260

Reynolds JC, Tapper SC (1995a) The ecology of the red fox *Vulpes vulpes* in relation to small game in rural southern England. Wildl Biol 1:105–119

Reynolds JC, Tapper SC (1995b) Predation by foxes *Vulpes vulpes* on brown hares *Lepus europaeus* in central southern England, and its potential impact on annual population growth. Wildl Biol 1:145–158

Ridpath MG, Waithman J (1988) Controlling feral Asian water buffalo in Australia. Wildl Soc Bull 16:385–390

Roberts MG, Heesterbeek JAP (1993) Bluff your way in epidemic models. Trends Microbiol 1:343–348

Roelke ME, Glass CM (1992) Strategies for the management of the endangered Florida panther (*Felis concolor coryi*) in an ever-shrinking habitat. In: Junge RE, (ed) Proceedings of the Joint Meeting of the American Association of Zoo Veterinarians and the American Association of Wildlife Veterinarians, pp. 38–43

Roelke-Parker ME, Munson L, Packer C, Kock R, Cleaveland S, Carpenter M, O'Brien SJ, Pospischil A, Hofmann-Lehmann R, Lutz H, Mwamengele GLM, Mgasa MN, Machange GA, Summers BA, Appel MJG (1996) A canine distemper virus epidemic in Serengeti lions (Panthero leo). Nature 379:441–445

Rogers LM, Cheeseman CL, Mallinson PJ, Clifton-Hadley R (1997) The demography of a high-density badger (*Meles meles*) population in the west of England. J Zool 242:705–728

Rogers LM, Delahay R, Cheeseman CL, Langton S, Smith GC, Clifton-Hadley RS (1998) Movement of badgers (*Meles meles*) in Woodchester Park, south west England: individual, population and disease effects. Proc R Soc B 265:1269–1276

Rogers LM, Delahay RJ, Cheeseman CL, Smith GC, Clifton-Hadley RS (1999) The increase in badger (*Meles meles*) density at Woodchester Park, south-west England: implications for disease (*Mycobacterium bovis*) prevalence. Mammalia 63:183–192

Roper TJ, Ostler JR, Conradt L (2003) The process of dispersal in badgers *Meles meles*. Mammal Rev 33:314–318

Rosatte RC, Power MJ, MacInnes CD, Campbell JB (1992) Trap-vaccinate-release and oral vaccination for rabies control in urban skunks, raccoons and foxes. J Wildl Dis 28:562–571

Rosatte RC, Donovan D, Allan M, Howes LA, Silver A, Bennett K, MacInnes C, Davies C, Wandeler A, Radford B (2001) Emergency response to raccoon rabies introduction into Ontario. J Wildl Dis 37:265–279

Rosenberry CS, Lancia RA, Conner MC (1999) Population effects of white tailed deer dispersal. Wildl Soc Bull 27:858–864

Rossi L, Ferroglio E (2001) *Camelostrongylus mentulatus* in a roe deer from the Italian Western Alps. Vet Rec 149:335

Rossi L, Ferroglio E, Frassetto D, Balbo T (2002) *Thelazia callipaeda* in foxes from North-West Italy. Abstracts of the XXII SOIPA Congress. Parassitologia 44:159

Rossi S, Artois M, Pontier D, Crucière C, Barrat J, Pacholek X, Fromont E (2005a) Long term monitoring of classical swine fever in wild boar (*Sus scrofa* sp.) using serological data. Vet Res 36:27–42

Rossi S, Fromont E, Pontier D, Cruciere C, Hars J, Barrat J, Pacholek X, Artois M (2005b) Incidence and persistence of classical swine fever in free-ranging wild boar (*Sus scrofa*). Epidemiol Infect 133:559–568

Rossi S, Hars J, Louguet Y, Masse-Provin N, Pol F, Le Potier MF (2006) Gestion d'un réservoir sauvage: la peste porcine du sanglier (*Sus scrofa*). Bull Acad Vét Fr 159:389–392

Rossiter PB (2001) Rinderpest. In: Williams ES, Barker IK (eds) Infectious diseases of wild mammals. 3rd edn. Iowa State University Press, Ames, USA, pp. 37–45

Roy SS, Macleod I, Moore NP (2006) The use of scent glands to improve the efficiency of mink (*Mustela vison*) captures in the Outer Hebrides. NZ J Zool 33:267–271

Rózsa L (2005) http://www.behav.org/qp/qp.htm

Rózsa L, Reiczigel J, Majoros G (2000) Quantifying parasites in samples of hosts. J Parasitol 86:228–232

Rupprecht CE, Smith JS, Fekadu M, Childs JE (1995) The ascension of wildlife rabies: a cause for public health concern or intervention? Emerg Infect Dis 1:107–114

Rupprecht CE, Blass L, Smith K, Orciari LA, Niezgoda M, Whitfield SG, Gibbons RV, Guerra M, Hanlon CA (2001) Brief report: human infection due to recombinant vaccinia – rabies glycoprotein virus. New Engl J Med 345:582–586

Rupprecht CE, Hanlon CA, Slate D (2006) Control and prevention of rabies in animals: paradigm shifts. Develop Biologicals 125:103–111

Rushton SP, Lurz PWW, Gurnell J, Nettleton P, Bruemmer C, Shirley MDF, Sainsbury AW (2006) Disease threats posed by alien species: the role of a poxvirus in the decline of the native red squirrel in Britain. Epidemiol Infect 134:521–533

Russell CA, Smith DL, Waller LA, Childs JE, Real LA (2004) A priori prediction of disease invasion dynamics in a novel environment. Proc R Soc B 271:21–25

Rutberg AT, Naugle RE, Thiele LA, Liu IKM (2004) Effects of immunocontraception on a suburban population of white-tailed deer *Odocoileus virginianus*. Biol Conserv 116:243–250

Rutili D, Tozzini F, Frescura T, Bandecci P (1987) Peste suina classica: Vaccinazione per via orale dei cinghiali. Atti Soc Ital Sci Vet 16:1078–1081

Ruykys L, Taggart DA, Breed WG, Schultz D (2006) Investigations of sarcoptic mange in southern hairy-nosed wombats (*Lasiorhinus latifrons*) in the Murraylands of South Australia. Australian Mammal Society Scientific Conference

Samuel MD, Takekawa JY, Baranyuk VV, Orthmeyer DL (1999) Effects of avian cholera on survival of lesser snow geese *Anser caerulescens*: an experimental approach. Bird Study 46:S239–247

Sapolsky RM (2005) The influence of social hierarchy on primate health. Science 308:648–652

Sawyer J, Mealing D, Dalley D, Davé D, Lesellier S, Palmer S, Bowen-Davies J, Crawshaw TR, Chambers MA (2007) Development and evaluation of a test for tuberculosis in live European badgers (*Meles meles*) based on measurement of gamma interferon mRNA by real-time PCR. J Clin Microbiol 45:2398–2403

Schalk G, Forbes MR (1997) Male biases in parasitism of mammals: effects of study type, host age, and parasite taxon. Oikos 78:67–74

Schauber EM, Woolf A (2003) Chronic wasting disease in deer and elk: a critique of current models and their application. Wildl Soc Bull 31:610–616

Schauber EM, Storm DJ, Neilson CK (2007) Effects of joint space use and group membership on contact rates among white-tailed deer. J Wildl Manage 71:155–163

Scheckelhoff MR, Telford SR, Hu LT (2006) Protective efficacy of an oral vaccine to reduce carriage of *Borrelia burgdorferi* (strain N40) in mouse and tick reservoirs. Vaccine 24:1949–1957

Schettler E, Steinbach F, Eschenbacher-Kaps I, Gerst K, Müssdörfer F, Risch K, Streich WJ, Frölich K (2006) Surveillance for prion disease in cervids, Germany. Emerg Infect Dis 12:319–322

Schloegel LM, Hero JM, Berger L, Speare R, McDonald K, Daszak P (2006) The decline of the sharp-snouted day frog (*Taudactylus acutirostris*): the first documented case of extinction by infection in a free-ranging wildlife species? EcoHealth 3:35–40

Schmidt RL, Hibler CP, Spraker TR, Rutherford WH (1979) An evaluation of drug treatment for lungworm in bighorn sheep. J Wildl Manage 43:461–467

Sen A (1987) On ethics and economics. Basil Blackwell Ltd., Oxford

Shang DQ, Xiao DL, Yin JM (2002) Epidemiology and control of brucellosis in China. Vet Microbiol 90:165–182

Shwiff SA, Sterner RT (2002) An economic framework for benefit-cost studies in wildlife damage studies. Proceedings of the 20th Vertebrate Pest Conference, pp. 240–344

Siegel S, Castellan NJ (1988) Nonparametric statistics for the behavioural sciences. McGraw Hill, New York

Sievers JD (2004) Factors influencing a declining pronghorn population in Wind Cave National Park, South Dakota [MSc thesis]. South Dakota State University

Simpson VR (2002) Wild animals as reservoirs of infectious diseases in the UK. Vet J 163:128–146

Singer FJ, Zeigenfuss LC, Spicer L (2001) Role of patch size, disease and movement in rapid extinction of bighorn sheep. Conserv Biol 15:1347–1354

Sivin I (1994) Contraception with Norplant implants. Hum Reprod 9:1818–1826

Skerratt LF, Berger L, Speare R, Cashins S, McDonald KR, Phillott AD, Hines HB, Kenyon L (2007) Spread of chytridiomycosis has caused the rapid global decline and extinction of frogs. EcoHealth 4:125–134

Skinner MA, Keen DL, Parlane NA, Hamel KL, Yates GF, Buddle BM (2005) Improving protective efficacy of BCG vaccination for wildlife against bovine tuberculosis. Res Vet Sci 78:231–236

Smith CR (1994) Wild carnivores as plague indicators in California. A cooperative interagency disease surveillance program. In: Halverson WS, Crabb AC (eds) Proceedings of 16th Vertebrates Pest Conference. University of California, Davies, pp. 192–199

Smith DL, Lucey B, Waller LA, Childs JE, Real LA (2002) Predicting the spatial dynamics of rabies epidemics on heterogeneous landscapes. Proc Natl Acad Sci USA 99:3668–3672

Smith GC (1995) Modelling rabies control in the U.K.: the inclusion of vaccination. Mammalia 59:629–637

Smith GC (2002) The role of the badger (*Meles meles*) in rabies epizootiology and the implications for Great Britain. Mammal Rev 32:13–26

Smith GC (2005) Modelling bovine tuberculosis in wildlife and cattle. In: Smithe LT (ed) Progress in tuberculosis research. Nova Science, New York, pp. 249–280

Smith GC (2006) Control of focal epizootics in wildlife. In: Feare CJ, Cowan DP (eds) Advances in vertebrate pest management IV. Filander Verlag, Fürth, Germany, pp. 111–124

Smith GC, Cheeseman CL (2007) Efficacy of trapping during the initial proactive culls in the Randomised Badger Culling Trial. Vet Rec 160:723–726

Smith GC, Fooks AR (2006) Wildlife rabies control policy in Great Britain. Develop Biologicals 125:113–118

Smith GC, Harris S (1989) The control of rabies in urban fox populations. In: Putman RJ (ed) Mammals as pests. Chapman & Hall, London, UK, pp. 209–224

Smith GC, Harris S (1991) Rabies in urban foxes (*Vulpes vulpes*) in Britain: the use of a spatial stochastic simulation model to examine the pattern of spread and evaluate the efficacy of different control regimes. Phil Trans R Soc Lond B 334:459–479

Smith GC, Wilkinson D (2002) Modelling disease spread in a novel host: rabies in the European badger *Meles meles*. J Appl Ecol 39:865–874

Smith GC, Wilkinson D (2003) Modeling control of rabies outbreaks in red fox populations to evaluate culling, vaccination, and vaccination combined with fertility control. J Wildl Dis 39:278–286

Smith GC, Woods JA (2007) Acceptance of baits, designed to carry oral rabies vaccines by foxes in Britain. Int J Pest Manage 53:323–328

Smith GC, Pugh B, Trout RC (1995) Age and sex bias in samples of wild rabbits, *Oryctolagus cuniculus*, from wild populations in southern England. NZ J Zool 22:115–121

Smith GC, Cheeseman CL, Clifton-Hadley RS (1997) Modelling the control of bovine tuberculosis in badgers in England: culling and the release of lactating females. J Appl Ecol 34:1375–1386

Smith GC, Cheeseman CL, Wilkinson D, Clifton-Hadley RS (2001) A model of bovine tuberculosis in the badger *Meles meles*: the inclusion of cattle and the use of a live test. J Appl Ecol 38:520–535

Smith GC, Prickett AJ, Cowan DP (2007a) Costs and benefits of rabbit control options at the local level. Int J Pest Manage 53:317–321

Smith GC, Bennet R, Wilkinson D, Cooke R (2007b) A cost–benefit analysis of culling badgers to control bovine tuberculosis. Vet J 173:302–310

Smith GC, Parrott D, Robertson PA (2008) Managing wildlife populations with uncertainty: cormorants *Phalacrocorax carbo*. J Appl Ecol, 45:1675–1682

Smith KF, Sax DF, Lafferty KD (2006) Evidence for the role of infectious disease in species extinction and endangerment. Conserv Biol 20:1349–1357

Southey A, Costello E, Gormley E (2002) Detection of *Mycobacterium bovis* infection and production of interleukin-2 by in vitro stimulation of badger lymphocytes. Vet Immunol Immunopathol 87:73–78

Southey AK, Sleeman DPS, Lloyd K, Dalley D, Chambers MA, Hewinson RG, Gormley E (2001) Immunological responses of Eurasian badgers (*Meles meles*) vaccinated with *Mycobacterium bovis* BCG (bacillus calmette guerin). Vet Immunol Immunopathol 79:197–207

Spalding JA, Brend SA, Lawrance W (1999) Arabian oryx reintroductions in Oman: successes and setbacks. Oryx 33:168–175

Spiegelhalter DJ, Best NG, Carlin BR, van der Linde A (2002) Bayesian measures of model complexity and fit. J R Stat Soc Ser B 64:583–639

Stärk KDC, Regula G, Hernandez J, Knopf L, Fuchs K, Morris RS, Davies P (2006) Concepts for risk-based surveillance in the field of veterinary medicine and veterinary public health: review of current approaches. BMC Health Serv Res 6:20

Steck F (1982) Rabies in wildlife. Symp Zool Soc Lond 50:57–75

Stephens PA, Russell AF, Young AJ, Sutherland WJ, Clutton-Brock TH (2005) Dispersal, eviction, and conflict in meerkats (*Suricata suricatta*): an evolutionarily stable strategy model. Am Nat 165:120–135

Sterner RT, Smith GC (2006) Modelling wildlife rabies: transmission, economics, and conservation. Biol Conserv 131:163–179

Streftaris G, Gibson GJ (2004) Bayesian analysis of experimental epidemics of foot-and-mouth disease. Proc R Soc B 271:1111–1117

Suppo C, Naulin JM, Langlais M, Artois M (2000) A modelling approach to vaccination and contraception programmes for rabies control in fox populations. Proc R Soc B 267:1575–1582

Sutherland WJ, Pullin AS, Dolman PD, Knight TM (2004) The need for evidence-based conservation. Trends Ecol Evol 19:305–308

Suzuki Y, Gojobori T (1997) The origin and evolution of Ebola and Marburg viruses. Mol Biol Evol 14:800–806

Swanepoel R (1994) Rabies. In: Coetzer JAW, Thomson GR, Tustin RC (eds) Infectious diseases of livestock with special reference to Southern Africa. Oxford University Press, Cape Town, Oxford & New York, pp. 493–552

Swanepoel R, Leman PA, Burt FJ, Zachariades NA, Braack LE, Ksiazek TG, Rollin PE, Zaki SR, Peters CJ (1996) Experimental inoculation of plants and animals with Ebola virus. Emerg Infect Dis 2:321–325

Swinton J, Harwood J, Grenfell BT, Gilligan CA (1998) Persistence thresholds for phocine distemper virus infection in harbour seals *Phoca vitulina* metapopulations. J Anim Ecol 67:54–68

Taylor LH, Latham SM, Woolhouse MEJ (2001) Risk factors for human disease emergence. Phil Trans R Soc Lond B 356:983–989

Taylor RD, Martin RB (1987) Effects of veterinary fences on wildlife conservation in Zimbabwe. Environ Manage 11:327–334

Terrier ME, Picard E, Barrat J, Guibe A, Cliquet F (2006) Surveillance sanitaire de la faune sauvage en France: réseau SAGIR et épidémio-surveillance de la rage des chiroptères. Bull Acad Vét Fr 159:383–387

Tessaro SV (1986) The existing and potential importance of Brucellosis and tuberculosis in Canadian wildlife: a review. Can Vet J 27:119–124

The Mountain Gorilla Veterinary Project 2002 Employee Health Group (2004) Risk of Disease Transmission between Conservation Personnel and the Mountain Gorillas: Results from an Employee Health Program in Rwanda. EcoHealth 1:351–361

Theis JH, Schwab RG (1992) Seasonal prevalence of *Taenia taeniaeformis*: Relationship to age, sex, reproduction and abundance of an intermediate host (*Peromyscus maniculatus*). J Wildl Dis 28:42–50

Thomas I, Brochier B, Languet B, Blancou J, Peharpre D, Kieny MP, Desmettre P, Chappuis G, Pastoret PP (1990) Primary multiplication site of the vaccinia-rabies glycoprotein recombinant virus administered to foxes by the oral route. J Gen Virol 71:37–42

Thompson HV (1994) The rabbit in Britain. In: Thompson HV, King CM (eds) The European rabbit: the history and biology of a successful colonizer. Oxford University Press, Oxford, pp. 64–107

Thompson WL, White GC, Gowan C (1998) Monitoring vertebrate populations. Academic Press, San Diego, USA

Thomson GR, Vosloo W, Esterhuysen JJ, Bengis RG (1992) Maintenance of foot and mouth disease virus in buffalo (*Syncerus caffer* Sparrman) in Southern Africa. Rev Sci Tech Off Int Epizoot 11:1097–1107

Thorburn JA, Thomas AD (1940) Tuberculosis in the Cape kudu. J S Afr Vet Med Assoc 11:3–10

Thorne ET, Williams ES (1988) Disease and endangered species: the black-footed ferret as a recent example. Conserv Biol 2:66–74

Thrall PH, Antonovics J, Dobson AP (2000) Sexually transmitted diseases in polygynous mating systems: prevalence and impact on reproductive success. Proc R Soc B 267:1555–1563

Thrusfield M (1995) Veterinary epidemiology. Blackwell Science, Oxford

Thrusfield M (2007) Veterinary epidemiology. Blackwell Science, Oxford

Thulke HH, Tischendorf L, Staubach C, Selhorst T, Jeltsch F, Muller T, Schluter H, Wissel C (2000) The spatio-temporal dynamics of a post-vaccination resurgence of rabies in foxes and emergency vaccination planning. Prev Vet Med 47:1–21

Tisdel C, Wilson C (2006) Information, wildlife valuation, conservation: experiments and policy. Contemp Econ Policy 24:144–159

Toma B, Dufour B, Sanaa M, Benet JJ, Shaw A, Moutou F, Louza A (2001) Epidémiologie appliquée à la lutte collective contre les maladies animales transmissibles majeures. AEEMA, Maisons-Alfort

Train K (2003) Discrete choice methods with simulation. Cambridge University Press, Cambridge

Trewby ID, Wilson GJ, Delahay RJ, Walker NJ, Young RP, Davison J, Cheeseman CL, Robertson PA, Gorman ML, McDonald RA (2008) Experimental evidence of competitive release in sympatric carnivores. Biol Lett 4:170–172

Trout RC, Ross J, Tittensor AM, Fox AP (1992) The effect on a British wild rabbit population (*Oryctolagus cuniculus*) of manipulating myxomatosis. J Appl Ecol 29:679–686

Tsao JI, Wootton JT, Bunikis J, Luna MG, Fish D, Barbour AG (2004) An ecological approach to preventing human infection: vaccinating wild mouse reservoirs intervenes in the Lyme disease cycle. Proc Natl Acad Sci USA 101:18159–18164

Tsukada H, Hamazaki K, Ganzorig S, Iwaki T, Konno K, Lagapa JT, Matsuo K, Ono A, Shimizu M, Sakai H, Morishima Y, Nonaka N, Oku Y, Kamiya M (2002) Potential remedy against *Echinococcus multilocularis* in wild red foxes using baits with anthelmintic distributed around fox breeding dens in Hokkaido, Japan. Parasitol 125:119–129

Turnbull PC, Tindall BW, Coetzee JD, Conradie CM, Bull RL, Lindeque PM, Huebschle OJ (2004) Related vaccine-induced protection against anthrax in cheetah (*Acinonyx jubatus*) and black rhinoceros (*Diceros bicornis*). Vaccine 22:3340–3347

Turner A (2003) Tuberculosis freedom assurance program. Animal Health Australia, Final Report 1998–2002

Turner JW, Liu IKM, Flanagan DR, Bynum KS, Rutberg AT (2002) Porcine zona pellucida (PZP) immunocontraception of wild horses (*Equus cabalus*) in Nevada: a 10 year study. Reprod Suppl 60:177–186

Turner JW, Jr., Liu IKM, Flanagan DR, Rutberg AT, Kirkpatrick JF (2007) Immunocontraception in wild horses: one inoculation provides two years of infertility. J Wildl Manage 71:662–667

Tuyttens FAM, Macdonald DW (1998) Sterilisation as an alternative strategy to control wildlife diseases: bovine tuberculosis in European badgers as a case study. Biodiversity Conserv 7:705–723

Tweddle NE, Livingstone P (1994) Bovine tuberculosis and eradication programs in Australia and New Zealand. Vet Microbiol 40:23–39

Uhaa IJ, Data VM, Sorhage FE, Beckley JW, Roscoe DE, Gorsky RD, Fishbein DB (1992) Benefits and costs of using an orally absorbed vaccine to control rabies in raccoons. J Am Vet Med Assoc 201:1873–1882

Vallat B (2008) Improving wildlife surveillance for its protection while protecting us from the diseases it transmits. On line Editorial of OIE web site, OIE: 1 p

Van Dyke F, Darragh JA (2007) Response of elk to changes in plant production and nutrition following prescribed burning. J Wildl Manage 71:23–29

Vannie P, Capua I, Le Potier MF, Mackay DK, Muylkens B, Parida S, Paton DJ, Thiry E (2007) Marker vaccines and the impact of their use on diagnosis and prophylactic measures. Rev Sci Tech Off Int Epizoot 26:351–372

Vassant J, Brandt S, Jullien JM (1993) Influence du passage de l'autoroute A5 sur les populations cerf et sanglier du Massif d'Arc-en-Barrois. Bull Mensuel Off Nat Chase 183:15–25

Vercauteren KC, Hygnstrom SE, Timm RM, Corrigan RM, Beller J, Bitney LL, Brumm MC, Meyer D, Virchow DR, Wills RW (2002) Development of a model to assess rodent control in swine facilities. In: Clark L, Hone J, Shivik JA, Watkins RA, Vercauteren KC et al. (eds) Human conflicts with wildlife: economic considerations. Proceedings of the Third NWRC Special Symposium. National Wildlife Research Center, Fort Collins, Colorado, USA. pp. 59–64. Available at: www.aphis.usda.gov/ws/nwrc/symposia/economics/vercauteren.pdf

Vercauteren KC, Shivik JA, Lavalle MJ (2005) Efficacy of an animal-activated frightening device on urban elk and mule deer. Wildl Soc Bull 31:1282–1287

Vercauteren KC, Lavalle MJ, Hygnstrom SE (2006) Fences and deer-damage management: a review of designs and efficacy. J Wildl Manage 34:191–200

Vere DT, Jones RE, Saunders G (2004) The economic benefits of rabbit control in Australian temperate pastures by the introduction of rabbit haemorrhagic disease. Agric Econ 30:143–155

Vicente J, Höfle U, Garrido JM, Fernandez-de-Mera IG, Juste R, Barral M, Gortazar C (2006) Wild boar and red deer display high prevalences of tuberculosis-like lesions in Spain. Vet Res 37:107–119

Vicente J, Delahay RJ, Walker N, Cheeseman CL (2007a) Social organization and movement influence the incidence of bovine tuberculosis in an undisturbed high-density badger *Meles meles* population. J Anim Ecol 76:348–360

Vicente J, Hofle U, Garrido JM, Fernandez-De-Mera IG, Acevedo P, Juste R, Barral M, Gortazar C (2007b) Risk factors associated with the prevalence of tuberculosis-like lesions in fenced wild boar and red deer in south central Spain. Vet Res 38:451–464

Viel JF, Giradoux P, Abrial V, Bresson-Hadni S (1999) Water vole (*Arvicola terrestris* Scherman) density as risk factor for human alveolar echinococcosis. Am J Trop Med Hyg 61:559–565

Viggers KL, Lindenmayer DB, Spratt DM (1993) The importance of disease in re-introduction programmes. Wildl Res 20:687–698

Vignon V, Joyeux A, Dumas JL (2002) Problématique du besoin de pénétration des sangliers dans les emprises autoroutières. Le cas de l'autoroute A57. Rev Gen Routes Aerodromes 806:56–60

Villafuerte R, Calvete C, Gortazar C, Moreno S (1994) First epizootic of rabbit hemorrhagic disease in free-living populations of *Oryctolagus cuniculus* at Donana National Park, Spain. J Wildl Dis 30:176–179

Voigt DR, Macdonald DW (1984) Variation in the spatial and social behaviour of the red fox, *Vulpes vulpes*. Acta Zool Fenn 171:261–265

Voigt DR, Tinline RR, Broekhoven LH (1985) A spatial simulation model for rabies control. In: Bacon PJ (ed) Population dynamics of rabies in wildlife. Academic, London, UK, pp. 311–349

Von Rüden S, Staubach C, Kaden V, Hess RG, Blicke J, Kühne S, Sonneburg J, Fröhlich A, Teuffert J, Moennig V (2008) Retrospective analysis of the oral immunisation of wild boar populations against classical swine fever virus (CSFV) in region Eifel of Rhineland-Palatinate. Vet Microbiol 132:29–38

Vos A, Neubert A, Aylan O, Schuster P, Pommerening E, Müller T, Chai Chivatsi D (1999) An update on safety studies of SAD B19 rabies virus vaccine in target and non-target species. Epidemiol Infect 123:165–175

Vose D (2001) Risk analysis: a quantitative guide. Wiley, Chichester

Vourc'h G, Bridges VE, Gibbens J, de Groot DB, McIntyre L, Poland R, Barnouin J (2006) Detecting emerging diseases in farm animals through clinical observations. Emerg Infect Dis 12:204–210

Vuillaume P, Aubert M, Demerson JM, Cliquet F, Barrat J, Breitenmoser U (1997) Vaccination des renards contre la rage par dépôt d'appats vaccinaux à l'entrée des terriers. Annls Méd Vét 141:55–62

Wallinga J, Teunis P (2004) Different epidemic curves for severe acute respiratory syndrome reveal similar impacts of control measures. Am J Epidemiol 160:509–516

Walsh PD, Abernethy KA, Bermejo M, Beyers R, De Wachter P, Akou ME, Huijbregts B, Mambounga DI, Toham AK, Kilbourn AM, Lahm SA, Latour S, Maisels F, Mbina C, Mihindou Y, Obiang SN, Effa EN, Starkey MP, Telfer P, Thibault M, Tutin CE, White LJ, Wilkie DS (2003) Catastrophic ape decline in western equatorial Africa. Nature 422:611–614

Walsh PD, Biek R, Real LA (2005) Wave-like spread of Ebola Zaire. PLoS Biol 3:e371

Walsh PD, Breuer T, Sanz C, Morgan D, Doran-Sheehy D (2007) Potential for Ebola transmission between gorilla and chimpanzee social groups. Am Nat 169:684–689

Walter WD, Bryant RL, Leslie DM, Jr (2005) Unusual documentation of elk behaviour using automated cameras. Proc Oklahoma Acad Sci 85:81–83

Ward AI, Tolhurst BA, Roper TJ, Delahay RJ (2008a) A survey of badger access to farm buildings and facilities in relation to contact with cattle. Vet Rec 163:107–111

Ward A, Minja E, Blackie M, Edwards-Jones G (2008b) Beyond participation – building farmer confidence: experience from Sub-Saharan Africa? Outlooks Agric 36:259–266

Ward AI, Smith GC, Etherington TR, Delahay RJ (submitted) Cattle exposure to TB from wild deer in England and Wales: a quantitative risk assessment. J Wildl Dis (submitted)

Wasserman S, Faust K (1994) Social network analysis: methods and applications. Cambridge University Press, Cambridge

Waters WR, Palmer MV, Whipple DL, Slaughter RE, Jones SL (2004) Immune responses of white-tailed deer (*Odocoileus virginianus*) to *Mycobacterium bovis* BCG vaccination. J Wildl Dis 40:66–78

Watt C, Dobson AP, Grenfell BT (1995) Glossary. In: Grenfell BT, Dobson AP (eds) Ecology of infectious diseases in natural populations. Cambridge University Press, Cambridge, pp. 521

Watts DJ (1999) Small worlds: the dynamics of networks between order and randomness. Princeton University Press, Princeton

Webster JP, Macdonald DW (1995) Parasites of wild brown rats (*Rattus norvegicus*) on UK farms. Parasitol 111:247–255

Wedlock DN, Aldwell FE, Keen D, Skinner MA, Buddle BM (2005) Oral vaccination of brushtail possums (*Trichosurus vulpecula*) with BCG: immune responses, persistence of BCG in lymphoid organs and excretion in faeces. NZ Vet J 53:301–306

Westergaard JM (1982) Measures applied in Denmark to control the rabies epizootic in 1977–1980. Comp Immunol Microbiol Infect Dis 5:383–387

White J, Horskins K, Wilson J (1998) The control of rodent damage in Australian macadamia orchards by manipulation of adjacent non-crop habitats. Crop Prot 17:353–357

White PCL, Lewis AJG, Harris S (1997) Fertility control as a means of controlling bovine tuberculosis in badger (*Meles meles*) populations in south-west England: predictions from a spatial stochastic simulation model. Proc R Soc B 264:1737–1747

White PJ, Trout RC, Moss SR, Desai A, Armesto M, Forrester NL, Gould EA, Hudson PJ (2004) Epidemiology of rabbit haemorrhagic disease virus in the United Kingdom: evidence for seasonal transmission by both virulent and avirulent modes of infection. Epidemiol Infect 132:555–567

Whitehouse AM, Kerley GIH (2002) Retrospective assessment of long-term conservation management of elephants in Addo Elephant National Park, South Africa. Oryx 36:243–248

WHO (1992) Expert committee on rabies, 8th Report. WHO, Geneva

WHO (2008a) www.who.int/immunization_safety

WHO (2008b) http://www.who-rabies.bulletin.org

Whyte IJ, Joubert SCJ (1988) Blue wildebeest population trends in the Kruger National Park and the effects of fencing. S Afr J Wildl Res 18:78–87

Wilkinson D, Smith GC, Delahay RJ, Cheeseman CL (2004) A model of bovine tuberculosis in the badger *Meles meles*: an evaluation of different vaccination strategies. J Appl Ecol 41:492–501

Williams E, Yuill T, Artois M, Fischer G, Haigh SA (2002a) Emerging infectious diseases in wildlife. Rev Sci Tech Off Int Epizoot 21:139–157

Williams ES, Miller MW, Kreeger TJ, Kahn RH, Thorne ET (2002b) Chronic wasting disease of deer and elk: a review with recommendations for management. J Wildl Manage 66:551–563

Williams ES, Thorne ET, Appel MJG, Belitsky DW (1988) Canine distemper in black-footed ferrets (*Mustela nigripes*) from Wyoming. J Wildl Dis 24:385–398

Willock J, Deary I, Edwards-Jones G, Gibson GJ, McGregor MJ, Sutherland A, Dent JB, Morgan O, Grieve R (1999) The role of attitudes and objectives in farmer decision making: business and environmentally-orientated behaviour in Scotland. J Agric Econ 50:286–303

Wilson G, Delahay R (2001) A review of methods to estimate the abundance of terrestrial carnivores using field sign and observation. Wildl Res 28:151–164

Wilson GJ, Frantz AC, Pope LC, Roper TJ, Burke TA, Cheeseman CL, Delahay RJ (2003) Estimation of badger abundance using faecal DNA typing. J Appl Ecol 40:658–666

Wilson JC, Fuller SJ, Mather PB (2002) Formation and maintenance of discrete wild rabbit (*Oryctolagus cuniculus*) population systems in arid Australia: habitat heterrogeneity and management implications. Austral Ecol 27:183–191

Wittenberg R, Cock MJW (eds) (2001) Invasive alien species: a toolkit of best prevention and management practises. CAB International, Wallingford, Oxford, 228 p

Wobeser G (2002) Disease management strategies for wildlife. Rev Sci Tech Off Int Epizoot 21:159–178

Wobeser G (2006) Essentials of disease in wild animals. Blackwell Publishing, Iowa, USA

Wobeser G (2007) Disease in wild animals: investigation and management. Springer-Verlag, Berlin

Wolfe LL, Miller MW, Williams ES (2004) Feasibility of "test-and-cull" for managing chronic wasting disease in urban mule deer. Wildl Soc Bull 32:500–505

Wood CL, Byers JE, Cottingham KL, Altman I, Donahue MJ, Blakeslee AMH (2007) Parasites alter community structure. Proc Natl Acad Sci USA 104:9335–9339

Wood SN, Thomas MB (1996) Space, time and persistence of virulent pathogens. Proc R Soc B 263:673–680

Woodford M, Kock RA (1991) Veterinary considerations in re-introduction and translocation projects. In: Gipps JWH (ed) Beyond captive breeding re-introducing captive mammals to the wild. Clarendon, Oxford, pp. 101–110

Woodford M, Keet DF, Bengis RG (2000) Post-mortem procedures for wildlife veterinarians and field biologists. IUCN/SSC Veterinary Specialist Group, Office International des Epizooties (OIE), Care for the Wild International Paris (France) and West Sussex (United Kingdom)

Woodford MH (1982) Tuberculosis in wildlife in the Ruwenzori National Park, Uganda (Part II). Trop Anim Health Prod 14:155–160

Woodford MH, Rossiter PB (1993) Disease risks associated with wildlife translocation projects. Rev Sci Tech Off Int Epizoot 12:115–135

Woodroffe R (1997) The conservation implications of immobilizing, radio-collaring and vaccinating free-ranging wild dogs. In: Woodroffe R, Ginsberg JR, Macdonald DW (eds) The African wild dog: status survey and conservation action plan. IUCN, Gland, Switzerland

Woodroffe R (1999) Managing disease threats to wild mammals. Anim Conserv 2:185–193

Woodroffe R (2001) Assessing the risks of intervention: immobilization, radio- collaring and vaccination of African wild dogs. Oryx 35:234–244

Woodroffe R, Ginsberg JR (1999) Conserving the African wild dog *Lycaon pictus*. I. Diagnosing and treating causes of decline. Oryx 33:132–142

Woodroffe R, Donnelly CA, Cox DR, Bourne FJ, Cheeseman CL, Delahay RJ, Gettinby G, McInerney JP, Morrison WI (2006a) Effects of culling on badger *Meles meles* spatial organization: implications for the control of bovine tuberculosis. J Appl Ecol 43:1–10

Woodroffe R, Donnelly CA, Jenkins HE, Johnston WT, Cox DR, Bourne FJ, Cheeseman CL, Delahay RJ, Clifton-Hadley RC, Gettinby G, Gilks P, Hewinson RG, McInerney JP, Morrison WI (2006b) Culling and cattle controls influence tuberculosis risk for badgers. Proc Natl Acad Sci USA 103:14713–14717

Woodroffe R, Gilks P, Johnston WT, Le Fevre AM, Cox DR, Donnelly CA, Bourne FJ, Cheeseman CL, Gettinby G, McInerney JP, Morrison WI (2008) Effects of culling on badger abundance: implications for tuberculosis control. J Zool 274: 28–37

Woods GM, Kreiss A, Belov K, Siddle HV, Obendorf DL, Muller HK (2007) The immune response of the tasmanian devil (*Sarcophilus harrisii*) and devil facial tumour disease. EcoHealth 4:338–345

Woolhouse MEJ, Haydon DT, Antia R (2005) Emerging pathogens: the epidemiology and evolution of species jumps. Trends Ecol Evol 20:238–244

Wright AN, Gompper ME (2005) Altered parasite assemblages in raccoons in response to manipulated resource availability. Oecologia 144:148–156

Yip P (1989) Estimating the initial relative infection rate for a stochastic epidemic model. Theor Popul Biol 36:202–213

Young AJ, Carlson AA, Clutton-Brock T (2005) Trade-offs between extraterritorial prospecting and helping in a cooperative mammal. Anim Behav 70:829–837

Zanardi G, Macchi C, Sacchi C, Rutili D (2003) Classical swine fever in wild boar in the Lombardy region of Italy from 1997 to 2002. Vet Rec 152:461–465

Zeng D, Chen H, Lynch C, Eidson M, Gotham I (2005) Infectious disease informatics and outbreak detection. In: Chen H (ed) Medical informatics: knowledge management and data mining in biomedicine. Springer, New York, pp. 359–395

Zou L, Miller SN, Schmidtmann ET (2006) Mosquito larval habitat mapping using remote sensing and GIS: implications of coalbed methane development and West Nile virus. J Med Entomol 43:1034–1041

Zou L, Miller SN, Schmidtmann ET (2007) A GIS tool to estimate West Nile virus risk based on a degree-day model. Environ Monitoring Assess 129:413–420

Zuk M, McKean KA (1996) Sex differences in parasite infections: patterns and processes. Int J Parasitol 26:1009–1023

Index

GPSR Compliance

The European Union's (EU) General Product Safety Regulation (GPSR) is a set of rules that requires consumer products to be safe and our obligations to ensure this.

If you have any concerns about our products, you can contact us on ProductSafety@springernature.com

In case Publisher is established outside the EU, the EU authorized representative is:

Springer Nature Customer Service Center GmbH
Europaplatz 3
69115 Heidelberg, Germany

Batch number: 10844669

Printed by Printforce, the Netherlands